CALIFORNIA MATHEMATICS REVIEW

Supports Content Standards 6-Algebra I

Revised to
CAHSEE July 2003
Mathematics Blueprints

COLLEEN PINTOZZI

AMERICAN BOOK COMPANY
P O BOX 2638
WOODSTOCK, GEORGIA 30188-1383
TOLL FREE: 1 (888) 264-5877   PHONE: 770-928-2834   FAX: 770-928-7483
Web site: www.americanbookcompany.com

ACKNOWLEDGEMENTS

In preparing this book, I would like to acknowledge the editorial assistance of the following individuals: Dr. Frank Pintozzi, Devin Pintozzi, Maria Struder, and Marsha Klosowski. I am also grateful to Kelly Berg for her contributions in editing, developing graphics, and formatting this book. I would also like to express my appreciation to Dr. Tim Teitloff of Kennesaw State University for writing the following sections of Chapter 19 (Patterns and Problem Solving): Inductive Reasoning and Patterns, Finding a Rule for Patterns, and Proportional Reasoning. Finally, I would like to thank my many students whose needs and questions inspired me to write this text.

Copyright © 2003
by American Book Company
P.O. Box 2638
Woodstock, GA 30188-1383

ALL RIGHTS RESERVED

The text of this publication, or any part thereof, may not be reproduced or transmitted in any form or by any means, electronic or mechanical, including photocopying, recording, storage in an information retrieval system, or otherwise, without the prior permission of the publisher.

Printed in the United States of America

w0801 10/03 04/04 07/04

TABLE OF CONTENTS

CALIFORNIA MATHEMATICS REVIEW

PREFACE	vi
INTRODUCTION	vii
DIAGNOSTIC TEST	1
EVALUATION CHART	14

CHAPTER 1 — 15
Fractions
Simplifying Improper Fractions	15
Changing Mixed Numbers to Improper Fractions	16
Greatest Common Factor	17
Reducing Proper Fractions	18
Multiplying Fractions	19
Dividing Fractions	20
Finding Numerators	21
Least Common Multiple	22
Adding Fractions	23
Subtracting Mixed Numbers from Whole Numbers	24
Subtracting Mixed Numbers with Borrowing	25
Deduction–Fraction Off	26
Fraction Word Problems	27
Chapter 1 Review	28

CHAPTER 2 — 30
Decimals
Adding Decimals	30
Subtracting Decimals	31
Multiplication of Decimals	32
Division of Decimals by Whole Numbers	34
Changing Fractions to Decimals	35
Changing Mixed Numbers to Decimals	36
Changing Decimals to Fractions	37
Changing Decimals with Whole Numbers to Mixed Numbers	37
Division of Decimals by Decimals	38
Estimating Division of Decimals	39
Decimal Word Problems	40
Finding a Profit	41
Chapter 2 Review	42

CHAPTER 3 — 43
Percents
Changing Percents to Decimals and Decimals to Percents	43
Changing Percents to Fractions and Fractions to Percents	44
Changing Percents to Mixed Numbers and Mixed Numbers to Percents	45
Representing Rational Numbers Graphically	46
Finding the Percent of the Total	47
Tips and Commissions	48
Finding the Amount of Discount	49
Finding the Discounted Sale Price	50
Sales Tax	51
Finding the Percent	52
Finding the Percent Increase and Decrease	53
Understanding Simple Interest	54
Compound Interest	55
Chapter 3 Review	56

CHAPTER 4 — 58
Problem Solving and Critical Thinking
Missing Information	58
Exact Information	60
Extra Information	62
Estimated Solutions	63
Two-Step Problems	64
Time of Travel	65
Rate	66
Product Measurements	68
Distance	69
Miles Per Gallon	70
Chapter 4 Review	71

CHAPTER 5 — 73
Integers and Order of Operations
Integers	73
Absolute Value	73
Adding Integers	74
Rules for Adding Integers with the Same Signs	75
Rules for Adding Integers with Opposite Signs	76
Rules for Subtracting Integers	77
Multiplying and Dividing Integers	78
Mixed Integer Practice	79
Properties of Addition and Multiplication	79
Understanding Exponents	80
Square Root	81
Order of Operations	82
Chapter 5 Review	84

CHAPTER 6 — 85
Exponents and Roots
Multiplying Exponents with the Same Base	85
Multiplying Exponents Raised to an Exponent	85
Fractions Raised to a Power	86
More Multiplying Exponents	86
Negative Exponents	87
Multiplying with Negative Exponents	87

Dividing with Exponents	88	**CHAPTER 11**	**152**
Simplifying Square Roots	89	**Introduction to Graphing**	
Estimating Square Roots	89	Cartesian Coordinates	152
Adding and Subtracting Roots	90	Identifying Ordered Pairs	153
Multiplying Roots	91	Drawing Geometric Figures on a	
Dividing Roots	92	Cartesian Coordinate Plane	155
Scientific Notation	93	Chapter 11 Review	158

CHAPTER 7 — Introduction to Algebra (97)

Algebra Vocabulary	97
Substituting Numbers for Variables	98
Understanding Algebra Word Problems	99
Setting Up Algebra Word Problems	103
Matching Algebraic Expressions	104
Changing Algebra Word Problems to Algebraic Equations	105
Chapter 7 Review	106

CHAPTER 8 — Solving One-Step Equations and Inequalities (108)

One-Step Algebra Problems with Addition and Subtraction	108
One-Step Algebra Problems with Multiplication and Division	110
Multiplying and Dividing with Negative Numbers	112
Variables with a Coefficient of Negative One	113
Graphing Inequalities	114
Solving Inequalities by Addition and Subtraction	115
Solving Inequalities by Multiplication and Division	116
Chapter 8 Review	117

CHAPTER 9 — Solving Multi-Step Equations and Inequalities (119)

Two-Step Algebra Problems	119
Combining Like Terms	122
Solving Equations with Like Terms	122
Removing Parentheses	125
Multi-Step Algebra Problems	127
Multi-Step Inequalities	129
Solving Equations and Inequalities with Absolute Values	131
Chapter 9 Review	134

CHAPTER 10 — Algebra Word Problems (135)

Geometry Word Problems	136
Age Problems	137
Mixture Word Problems	139
Coin and Stamp Problems	141
Uniform Motion Problems	142
Return Trip Motion Problems	143
Working Together Problems	145
Consecutive Integer Problems	147
Inequality Word Problems	148
Chapter 10 Review	150

CHAPTER 11 — Introduction to Graphing (152)

(listed above)

CHAPTER 12 — Graphing and Writing Equations (159)

Graphing Linear Equations	159
Graphing Horizontal and Vertical Lines	161
Finding the Intercepts of a Line	163
Understanding Slope	165
Slope-Intercept Form of a Line	168
Verify That a Point Lies on a Line	169
Graphing a Line Knowing a Point and Slope	170
Finding the Equation of a Line Using Two Points or a Point and Slope	171
Writing an Equation From Data	172
Graphing Linear Data	173
Identifying Graphs of Linear Equations	175
Graphing Non-Linear Equations	177
Chapter 12 Review	178

CHAPTER 13 — Graphing Inequalities (182)

Chapter 13 Review	185

CHAPTER 14 — Systems of Equations and Systems of Inequalities (186)

Systems of Equations	186
Finding Common Solutions for Intersecting Lines	188
Solving Systems of Equations by Substitution	189
Graphing Systems of Inequalities	191
Chapter 14 Review	192

CHAPTER 15 — Polynomials (193)

Adding and Subtracting Monomials	193
Adding Polynomials	194
Subtracting Polynomials	196
Adding and Subtracting Polynomials Review	200
Multiplying Monomials	201
Multiplying Monomials with Different Variables	202
Dividing Monomials	203
Extracting Monomial Roots	204
Monomial Roots with Remainders	205
Multiplying Monomials by Polynomials	206
Dividing Polynomials by Monomials	207
Removing Parentheses and Simplifying	208
Multiplying Two Binomials	209
Simplifying Expressions with Exponents	211
Chapter 15 Review	212

Also in Chapter 6:
- Using Scientific Notation for Large Numbers — 93
- Using Scientific Notation for Small Numbers — 94
- Chapter 6 Review — 95

CHAPTER 16	**214**	**CHAPTER 21**	**275**
Statistics		**Ratios, Proportions, and Scale Drawings**	
Mean	214	Ratio Problems	275
Finding Data Missing from the Mean	215	Solving Proportions	276
Median	216	Ratio and Proportion Word Problems	277
Mode	217	Maps and Scale Drawings	278
Stem-and-Leaf Plots	218	Using a Scale to Find Distances	279
Quartiles and Extremes	220	Using a Scale on a Blueprint	280
Box-And-Whisker Plots	221	Chapter 21 Review	281
Scatter Plots	222		
Misleading Statistics	224	**CHAPTER 22**	**282**
Chapter 16 Review	226	**Plane Geometry**	
		Perimeter	282
CHAPTER 17	**230**	Area of Squares and Rectangles	283
Data Interpretation		Area of Triangles	284
Reading Tables	230	Area of Trapezoids and Parallelograms	285
Bar Graphs	231	Circumference	286
Line Graphs	232	Area of a Circle	287
Circle Graphs	234	Two-Step Area Problems	288
Chapter 17 Review	235	Perimeter and Area With Algebraic Expressions	290
		Estimating Area	292
CHAPTER 18	**237**	Geometric Relationships of Plane Figures	293
Probability		Congruent Figures	295
Probability	237	Similar Triangles	297
Independent and Dependent Events	239	Pythagorean Theorem	299
More Probability	241	Finding the Missing Leg of a Right Triangle	300
Tree Diagrams	242	Chapter 22 Review	301
Chapter 18 Review	244		
		CHAPTER 23	**303**
CHAPTER 19	**246**	**Solid Geometry**	
Patterns and Problem Solving		Understanding Volume	303
Number Patterns	246	Volume of Rectangular Prisms	304
Using Diagrams to Solve Problems	247	Volume of Cubes	305
Trial and Error Problems	248	Volume of Spheres, Cones, Cylinders,	
Making Predictions	249	and Pyramids	306
Inductive Reasoning and Patterns	250	Two-Step Volume Problems	308
Finding a Rule for Patterns	254	Estimating Volume	309
Proportional Reasoning	258	Geometric Relationships of Solids	310
Mathematical Reasoning/Logic	259	Surface Area	312
Chapter 19 Review	262	Solid Geometry Word Problems	316
		Chapter 23 Review	317
CHAPTER 20	**266**		
Measurement		**CHAPTER 24**	**320**
Customary Measure	266	**Reflections, Translations, and Plotted Shapes**	
Approximate English Measure	267	Reflections	320
Converting Units Using Dimensional Analysis	267	Translations	323
The Metric System	269	Finding Lengths Of Plotted Shapes On	
Understanding Meters	270	Cartesian Planes	325
Understanding Liters	270	Chapter 24 Review	327
Understanding Grams	270		
Estimating Metric Measurements	271	**PROGRESS TEST 1**	**329**
Converting Units in the Metric System	272	**PROGRESS TEST 2**	**343**
Chapter 20 Review	274		
		CHART OF STANDARDS	**356**

Preface

THE CALIFORNIA MATHEMATICS REVIEW (Content Standards 6 –Algebra I) will help you review and learn important concepts and skills related to middle and high school mathematics. Some of this material will be a review of skills you have already learned, while other sections will present you with new applications in Arithmetic, Data Analysis/Statistics, Pre-Algebra, Algebra, and Geometry. To help identify which areas are of greater challenge for you, begin by taking the Diagnostic Test at the beginning of this book. Once you have taken the test, complete the evaluation chart with your instructor in order to help you identify the chapters which require your careful attention. When you have finished your review of all of the material your teacher assigns, take the progress tests to evaluate your understanding of the California Mathematics Content Standards 6 - Algebra I. **The materials in this book are based on the standards and content descriptions for mathematics published by the California Department of Education.**

This book contains several sections. These sections are as follows: 1) A Diagnostic Test; 2) Chapters that teach the concepts and skills for Content Standards Grades 6 - Algebra I; 3) Two Progress Tests. Answers to the tests and exercises are in a separate manual.

The diagnostic and progress tests are divided into the following topics:

Topic and Grade Level	Number of Questions
Statistics, Data Analysis, and Probability - 6th grade level	6 questions
Number Sense - 7th grade level	14 questions
Algebra and Functions - 7th grade level	17 questions
Measurement and Geometry - 7th grade level	17 questions
Statistics, Data Analysis, and Probability - 7th grade level	6 questions
Mathematical Reasoning - 7th grade level	8 questions
Algebra I - 8th - 9th grade level.	12 questions
Total	**80 questions**

We welcome comments and suggestions about the book. Please contact the author at

American Book Company
PO Box 2638
Woodstock, GA 30188-1383

Toll Free: 1 (888) 264-5877
Phone: (770) 928-2834
Fax: (770) 928-7483
Web site: www.americanbookcompany.com

ABOUT THE AUTHOR

Colleen Pintozzi has taught mathematics at the middle school, junior high, senior high, and adult level for 22 years. She holds a B.S. degree from Wright State University in Dayton, Ohio and has done graduate work at Wright State University, Duke University, and the University of North Carolina at Chapel Hill. She is the author of eight mathematics books including such best-sellers as ***Basics Made Easy: Mathematics Review, Passing the New Alabama Graduation Exam in Mathematics, Passing the Georgia High School Graduation Test in Mathematics, Writing, and English Language Arts, Passing the TCAP Competency Test in Mathematics, Passing the Louisiana LEAP 21 Graduation Exit Exam, Passing the Indiana ISTEP+ Graduation Qualifying Exam in Mathematics, Passing the Minnesota Basic Standards Test in Mathematics,*** and ***Passing the Nevada High School Proficiency Exam in Mathematics.***

California Math Review Diagnostic Test

1. Find the <u>mean</u> of **36, 54, 66, 45, 36, 36, and 63.**

   A. 36
   B. 45
   C. 48
   D. 63

2.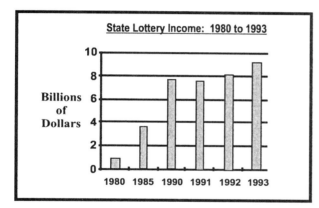

   "People spent more money in 1986 than any other year." According to the graph above, this statement is

   A. true based on the data.
   B. not supported by the data.
   C. false based on the data.
   D. None of the above

3. One light year is equal to about 9,461,000,000,000 kilometers. How would you express this in scientific notation?

   A. 9461×10^9
   B. 946.1×10^{10}
   C. 9.461×10^9
   D. 9.461×10^{12}

4. Sarah deposits 50¢ into Miss Clucky, a machine that makes chicken squawks and gives Sarah one plastic egg with a toy surprise. In the machine, 30 eggs contain a rubber frog, 43 eggs contain a plastic ring, 23 eggs contain a necklace, and 18 eggs contain a plastic car. What is the probability that Miss Clucky will give Sarah a necklace in her egg?

   A. $\frac{1}{114}$
   B. $\frac{23}{114}$
   C. $\frac{23}{91}$
   D. $\frac{1}{23}$

5. The probability of flipping a coin 8 times and coming up with heads each time is 1/256. What is the probability against flipping heads 8 times in a row?

   A. $\frac{256}{1}$
   B. $\frac{255}{256}$
   C. $\frac{8}{256}$
   D. $\frac{1}{8}$

6. In problem 5, each time you flip the coin is a(n)

   A. independent event.
   B. dependent event.
   C. pattern.
   D. probability.

7. Each lap around the lake is $\frac{4}{5}$ of a mile. How many miles did Teri run if she ran $4\frac{1}{2}$ laps?

   A. $3\frac{3}{5}$
   B. $3\frac{1}{5}$
   C. $4\frac{2}{5}$
   D. $5\frac{5}{8}$

8. Right now, the temperature is 5° above zero. Tonight the temperature is expected to be 12° cooler. What is the temperature expected to be?

   A. $-17°$
   B. $-7°$
   C. $-12°$
   D. $-5°$

9. $5^3 =$

   A. 15
   B. 25
   C. 125
   D. 555

10. $3\frac{2}{5}$ is the same as

    A. 3.4
    B. 3.25
    C. 0.34
    D. 0.034

11. $\frac{5}{8}$ written as a percent is

    A. 0.58%
    B. 0.625%
    C. 6.25%
    D. 62.5%

12. A table and chairs set that normally sells for $450.00 is on sale this week for 30% off the regular price. How much money would Trina save if she bought the set this week?

    A. $30.00
    B. $31.50
    C. $135.00
    D. $315.00

13. Last year, there were 96 students in marching band. This year, the band's size has increased by 25%. How many students are in marching band this year?

    A. 121
    B. 120
    C. 24
    D. 125

14. Cheryl borrowed $7,500.00 to buy a used car at 12% simple interest per year. Cheryl made no payments during the first year. How much interest did she owe at the end of the year?

    A. $120.00
    B. $750.00
    C. $850.00
    D. $900.00

15. Li-kim places two red and two white chips in a bag. He draws two chips from the bag at random. If the first chip he draws is red, what is the probability the next chip he draws will also be red?

    A. $\frac{1}{4}$
    B. $\frac{1}{2}$
    C. $\frac{1}{3}$
    D. $\frac{2}{3}$

16. $5^{15} \times 5^{-12} =$

    A. 25
    B. 125
    C. 243
    D. 0.125

17. Which of the following can be used to compute $\frac{5}{6} + \frac{2}{9}$?

    A. $\frac{5}{6 \times 9} + \frac{2}{6 \times 9}$
    B. $\frac{5+2}{6+9}$
    C. $\frac{5}{6 \times 3} + \frac{2}{9 \times 2}$
    D. $\frac{5 \times 3}{6 \times 3} + \frac{2 \times 2}{9 \times 2}$

18. Between what two integers does the square root of 7 lie?

    A. 1 and 2
    B. 2 and 3
    C. 3 and 4
    D. 4 and 9

19. The absolute value of a number is equal to 2. Which of the following sets of numbers represents all of the possible values of that number?

    A. {2}
    B. {−2}
    C. {−2, 2}
    D. {−2, −1, 0, 1, 2}

20. Which of the following equations describes "two *x* minus three *y* equals five *z* squared"?

    A. $2x - 3y = 5z^2$
    B. $2^x - 3^y = 5z^2$
    C. $2(x - 3y) = 5z^2$
    D. $2x - 3z = 5y^2$

21. Andrea has 10 more jellybeans than her friend Chelsea, but she has half as many as Rebecca. Which expression below best describes Rebecca's jelly beans?

    A. $R = C + 10$
    B. $R = 2C + 20$
    C. $R = A + \frac{1}{2}C$
    D. $R = 2A + 10$

22. To properly evaluate an algebraic expression like $4(3x + 5)^2$, the first step is to

    A. multiply 4 times $(3x + 5)$.
    B. square $(3x + 5)$.
    C. add.
    D. divide.

23. Simplify the expression shown below:

    $$3x^{-2}$$

    A. $(3x)^{-1}(3x)^{-1}$
    B. $\frac{9}{x^2}$
    C. $\frac{1}{3x^2}$
    D. $\frac{3}{x^2}$

24.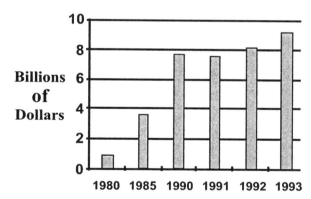

    How much was the increase in income from 1985 to 1990?

    A. 2 Billion
    B. 3 Billion
    C. 4 Billion
    D. 5 Billion

25. According to the chart below, how many more feet does it take to stop a car traveling at 70 miles per hour than at 55 miles per hour?

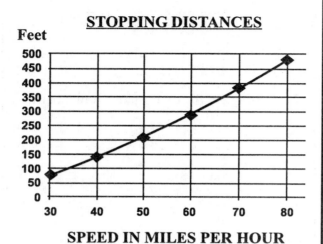

SPEED IN MILES PER HOUR

A. 100 feet
B. 125 feet
C. 150 feet
D. 175 feet

26. Simplify the expression shown below:

$$\frac{8x^4}{2x^2}$$

A. $2x^4$

B. $4x^2$

C. $\frac{1}{4x^2}$

D. $\frac{4x^2}{x}$

27. Simplify the expression shown below:

$$\frac{3^{-3}}{2}$$

A. $-\frac{9}{2}$

B. $-\frac{27}{2}$

C. $\frac{1}{54}$

D. $\frac{1}{216}$

28. Todd graphed his income according to the number of hours he worked.

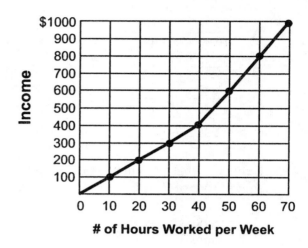

Which statement can be supported by the graph above?

A. Todd always earns $20 per hour.
B. Todd earns double pay after 40 hours per week.
C. Todd's pay is the same per hour no matter how many hours he works.
D. Todd never works more than 40 hours per week.

29. Which of the following graphs represents $y = 2x^2$?

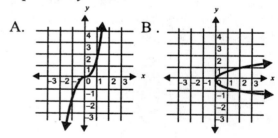

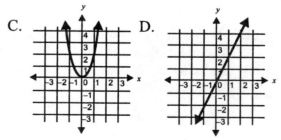

30. Which of the following graphs shows a line with a slope of $-\frac{1}{3}$ that passes through the point $(0, -1)$?

A. 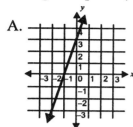 B.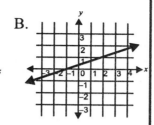

C. 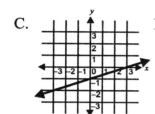 D.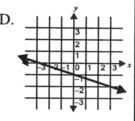

31. Which of the following is the graph of the equation $y = x - 3$?

A. 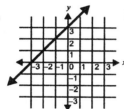 B.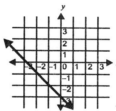

C. 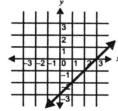 D.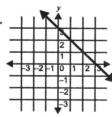

32. Based on the graph below, how much income would 150 compact discs sold generate?

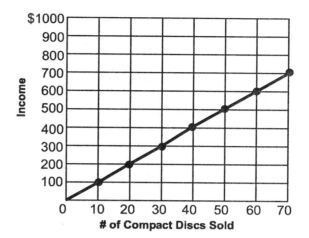

A. $15
B. $150
C. $1,500
D. $15,000

33. Nihar jogged at 7 miles per hour for forty-five minutes and 3 miles per hour for thirty minutes. How far did he jog?

A. $6\frac{3}{4}$ miles
B. $5\frac{3}{4}$ miles
C. $6\frac{1}{4}$ miles
D. $7\frac{1}{2}$ miles

34. The T-Shirt Factory sells tie-dyed shirts for $8.00 a shirt. White shirts cost $200 per pack of 100 and the dye costs $2.50 per shirt. Based on the cost of the dye and the shirts, what is the profit per shirt?

A. $5.00 per shirt
B. $3.75 per shirt
C. $4.50 per shirt
D. $3.50 per shirt

35. Chantal is going to visit a friend who lives 392 miles away. If she averages 56 miles per hour and leaves home at 9 AM, when will she arrive at her friend's house?

   A. 4:00 PM
   B. 5:00 PM
   C. 6:00 PM
   D. 7:00 PM

36. A bricklayer uses 7 bricks per square foot of surface covered. If he covers an area 28 feet wide and 10 feet high, about how many bricks will he use?

   A. 40
   B. 1,960
   C. 2,800
   D. 3,000

37. Linda is starting a pet grooming business to pay back the $250.75 she borrowed from her mom. She figured out that the supplies are going to cost $81.95. She plans to groom 8 dogs per week. How much will Linda have to charge for each dog in order to cover her expenses and pay her mom back in 3 weeks?

   A. $34.10
   B. $10.44
   C. $15.63
   D. $13.86

38. 2 kiloliters is equal to:

   A. 2000 liters
   B. 200 liters
   C. 20 liters
   D. .002 liters

39. A piece of wire is 4 feet long. 1 inch = 2.54 centimeters. How many centimeters long is the wire?

   A. 60 cm
   B. 91.42 cm
   C. 121.92 cm
   D. 10.16 cm

40. Below is a drawing of a farm plot:

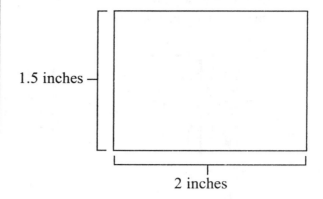

   Scale: 1 inch = 0.75 miles

   What is the perimeter of this farm plot?

   A. $2\frac{1}{4}$ miles
   B. 3 miles
   C. $5\frac{1}{4}$ miles
   D. 7 miles

41. It takes 48 worker·weeks to complete a book in a publishing company. If three writers are involved in a book project, how many weeks should it take them to complete the book?

   A. 12 weeks
   B. 48 weeks
   C. 20 weeks
   D. 16 weeks

42. Lupe has an accident while cooking. She wants to add a dash of pepper to two liters of boiling water for a soup she is making. Instead, the cap falls off and three hundred grams, the entire bottle of pepper, falls into her soup. What is the density of the pepper/boiling water mixture?

   A. 15 grams/liter
   B. 30 grams/liter
   C. 150 grams/liter
   D. 300 grams/liter

43. Find the area of the trapezoid below.

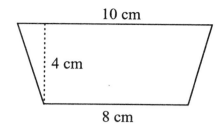

Use the formula $A = \dfrac{(b_1 + b_2) \times h}{2}$

A. 22 square centimeters
B. 36 square centimeters
C. 72 square centimeters
D. 320 square centimeters

44. What is the <u>area</u> of the following circle? Use $A = \pi r^2$, and $\pi = 3.14$.

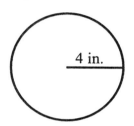

A. 12.56 in^2
B. 16 in^2
C. 25.12 in^2
D. 50.24 in^2

45. What is the area of the shaded region below?

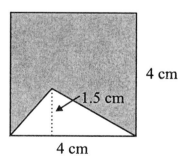

A. 16 cm^2
B. 13 cm^2
C. 10 cm^2
D. 6 cm^2

46. Find the volume of the pyramid. The formula for the volume of a pyramid is:

$V = \tfrac{1}{3}LWH$

where L = length
       W = width
       H = height

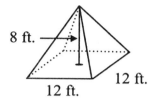

A. 384 ft^3
B. 36 ft^3
C. 32 ft^3
D. 1 ft^3

47. How many square feet is the area of the great room below?

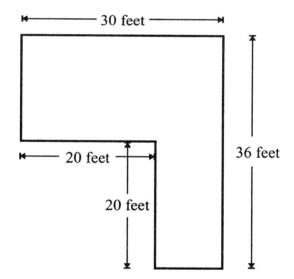

A. 480 feet^2
B. 680 feet^2
C. 720 feet^2
D. 1080 feet^2

48. In the figures below, an edge of the larger cube is twice as big as an edge of the smaller cube. What is the ratio of the volume of the smaller cube to that of the larger cube?

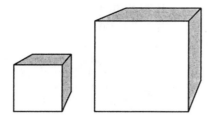

A. 1:2
B. 1:4
C. 1:8
D. 1:16

49. The ceramic tile below has an area of 25 square inches. How many square centimeters is the tile?

1 square inch = 6.45 square centimeters.

A. 3.876 cm^2
B. 10.85 cm^2
C. 15.525 cm^2
D. 161.25 cm^2

50. What is the mean of the following set of data?

11, 13, 17, 15, 11, 14, 17

A. 14
B. 15
C. 16
D. 17

51. If the figure below were reflected across the y-axis, what would be the coordinates of point A?

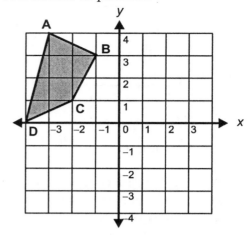

A. (2, 4)
B. (3, 4)
C. (4, 3)
D. (4, 2)

52. What is the area of the triangle below in graph units?

Use the formula $A = \frac{1}{2}bh$.

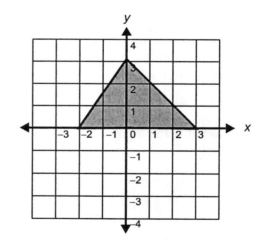

A. 2.5 units
B. 4.5 units
C. 7.5 units
D. 9.5 units

53. Find the length of the hypotenuse, *h*, of the following triangle.

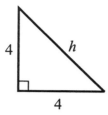

A. 16
B. $4\sqrt{2}$
C. 6
D. 32

54. Find the <u>perimeter</u> of the triangle below.

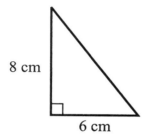

A. 14 cm
B. 24 cm
C. 32 cm
D. 48 cm

55.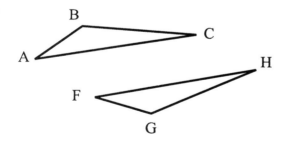

△ABC is congruent to △FGH. ∠B is congruent to

A. ∠C.
B. ∠F.
C. ∠G.
D. ∠H.

56. Examine the following two data sets:

Set #1: 49, 55, 68, 72, 98
Set #2: 20, 36, 47, 68, 75, 82, 89

Which of the following statements is true?

A. They have the same mode.
B. They have the same median.
C. They have the same mean.
D. None of the above.

57. Which of the following accurately describes the relationship between age and value of cars in the graph below?

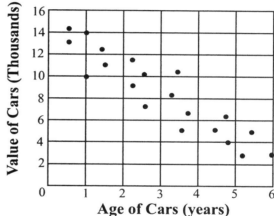

A. Positive
B. Negative
C. No relationship
D. Cannot be determined

58. If you were going to make a stem-and-leaf plot for the following set of data, how many stems would be needed?

21	51	64	29	33
42	57	72	15	39
51	28	36	30	52

A. 6
B. 7
C. 8
D. 9

59. Based on the information below, what is the relationship between rainfall and water usage?

Daily Rainfall (in.)	Daily Household Water Usage (gallons)
.5	175
3	100
5	90
2	120
4	110
1	160

A. No relationship
B. Negative
C. Positive
D. Cannot be determined

60. What is the median of the following set of data?

33, 31, 35, 24, 38, 30

A. 32
B. 31
C. 30
D. 29

61. K.B. Toys sells skateboards in three colors: black, green and blue. It sold 12 skateboards over the weekend, 5 of them green. You could find out how many blue skateboards were sold if you knew

A. the price of the blue skateboard.
B. the average number of blue skateboards sold.
C. the number of black skateboards sold.
D. the total number of blue and black skateboards sold.

62. Amin has a part-time job earning $10.00 per hour. He made a chart of his hours, earnings, and federal taxes taken out of his paycheck.

Hours	Earned	Taxes
25	$250	$23
26	$260	$25
27	$270	$26
28	$280	$28
29	$290	$29

If the pattern continues, how much will be taken from his check for federal taxes if he works 32 hours?

A. $32
B. $33
C. $34
D. $35

63. Heather ate 2 cookies with 90 calories each, 1 glass of skim milk with 110 calories, and a sandwich with 212 calories. She thinks she consumed 502 calories. Which estimation would verify her calculations?

A. 90 + 100 + 200
B. 100 + 100 + 100 + 200
C. 90 + 90 + 110 + 200
D. 180 + 110 + 200

64.

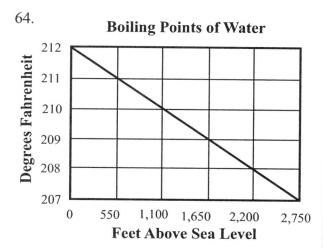

A certain mountain is 6,050 feet high. Based on the information given by the line graph above, what would be the boiling point of water to the nearest degree on top of the mountain?

A. 199°
B. 200°
C. 201°
D. 202°

65.

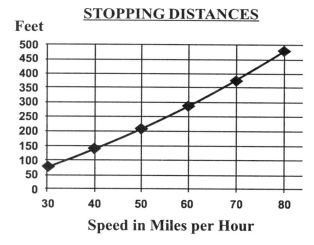

If Karen were traveling 65 mph, about how many feet would it take her to stop?

A. 310 feet
B. 315 feet
C. 325 feet
D. 345 feet

66. According to Enrique, the following statements are true:

If I don't do well in driving school, I won't be able to pass the driving test.

If I cannot pass the driving test, I won't receive my driver's license.

If I receive my driver's license, I can get a job this summer.

Based on Enrique's statements, which of the following is <u>not</u> a logical conclusion?

A. If Enrique does well in driving school, he will get a job this summer.
B. If Enrique does not receive his driver's license, he did not do well in driving school.
C. If Enrique gets a job this summer, he did well in driving school.
D. If Enrique does not get a job this summer, he did pass the driving test.

67.

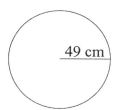

Caritina calculated the area of the circle above and got an answer of 307.72 cm^2. She knew she was wrong because the correct answer should be about

A. $50 \times 50 \times 3 = 7{,}500$ cm^2
B. $25 \times 25 \times 3 = 1{,}875$ cm^2
C. $50 \times 100 \times 3 = 15{,}000$ cm^2
D. $100 \times 100 \times 3 = 30{,}000$ cm^2

68. If the pattern below continues, how many blocks will Figure 6 contain?

Figure 1

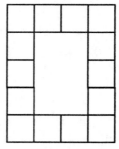
Figure 2

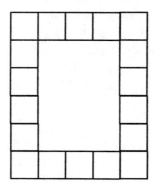
Figure 3

A. 22
B. 28
C. 30
D. 36

69. Simplify: $\sqrt{45} \times \sqrt{27}$

A. $3\sqrt{15}$
B. $\sqrt{72}$
C. $\sqrt{1215}$
D. $9\sqrt{15}$

70. Assume v is an integer and solve for v.

$$12 - 3|v| \geq 6$$

A. $\{-3, -2, -1, 0, 1, 2, 3\}$
B. $\{-2, -1, 0, 1, 2\}$
C. $\{-2, -1, 0, 1\}$
D. $\{-1, 0, 1\}$

71. Which of the following is equivalent to $5(x - 5) > 4x - 20$?

A. $5x + x - 5 < 4x - 20$
B. $5x + 5 < 4x + 20$
C. $5x - 5 > 4x - 20$
D. $5x - 25 > 4x - 20$

72. Solve the following inequality:
$-3(4x + 5) > 2(5x + 6) + 13$

A. $x < -\frac{20}{11}$
B. $x > 20$
C. $x > \frac{20}{20}$
D. $x < \frac{20}{20}$

73. A builder is constructing a fence 85 feet long. Each section of fence contains 6 beams of wood and takes up $2\frac{1}{2}$ feet. How many beams of wood will the builder need?

A. 102 beams
B. 204 beams
C. 308 beams
D. 420 beams

74. What is the x-intercept of the following linear equation?

$$3x + 4y = 12$$

A. (0, 3)
B. (3, 0)
C. (0, 4)
D. (4, 0)

75. Compute the x-intercept and y-intercept for the equation $x + 2y = 6$.

   A. x-intercept = (0, 6)
      y-intercept = (3, 0)
   B. x-intercept = (4, 1)
      y-intercept = (2, 2)
   C. x-intercept = (0, 6)
      y-intercept = (0, 3)
   D. x-intercept = (6, 0)
      y-intercept = (0, 3)

76. What is the equation of the line that includes the point (4, −3) and has a slope of −2?

   A. $y = -2x - 5$
   B. $y = -2x - 2$
   C. $y = -2x + 5$
   D. $y = 2x - 5$

77. Which of the following lines is parallel to $y = -4x + 6$?

   A. $y = -2x + 6$
   B. $y = -4x + 2$
   C. $y = 4x + 6$
   D. $y = 2x + 6$

78. What is the intercept of the following linear equations?

   $y = 3x - 1$
   $y = 4x + 2$

   A. (−3, 10)
   B. (−3, −10)
   C. (10, −3)
   D. (3, −10)

79. The triangle and rectangle have dimensions as shown. Which of the following expressions represents the area of the shaded region?

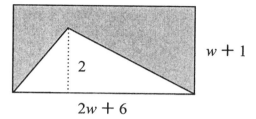

   A. $2w + 6$
   B. $3w + 6$
   C. $2w^2 + 6w$
   D. $3w^2 + 6w$

80. A farmer has some cream which is 20% butterfat and some which is 15% butterfat. How many gallons of each must be mixed to produce 50 gallons of cream which is 18% butterfat?

   A. 15 gallons of 15% butterfat and 35 gallons of 20% butterfat
   B. 15 gallons of 20 % butterfat and 35 gallons of 15% butterfat
   C. 20 gallons of 15% butterfat and 30 gallons of 20% butterfat
   D. 20 gallons of 20% butterfat and 30 gallons of 15% butterfat

EVALUATION CHART
DIAGNOSTIC MATHEMATICS TEST

Directions: On the following chart, circle the question numbers that you answered incorrectly, and evaluate the results. Then turn to the appropriate topics (listed by chapters), read the explanations, and complete the exercises. Review the other chapters as needed. Finally, complete the Mathematics Progress Tests to check how much you have learned of the California Mathematics Content Standards 6 - Algebra I.

	QUESTIONS	PAGES
Chapter 1: Fractions	7,17	15-29
Chapter 2: Decimals	10,34,37	30-42
Chapter 3: Percents	11,12,13,14	43-57
Chapter 4: Problem Solving and Critical Thinking	35,41,42,61,63,67	58-72
Chapter 5: Integers and Order of Operations	8,19,22,23	73-84
Chapter 6: Exponents and Roots	3,9,16,18	85-96
Chapter 7: Introduction to Algebra	20,21	97-107
Chapter 8: Solving One-Step Equations and Inequalities	70,71,72	108-118
Chapter 9: Solving Multi-Step Equations and Inequalities	70,71,72	119-134
Chapter 10: Algebra Word Problems	80	135-151
Chapter 11: Introduction to Graphing	29,30,31,74,75	152-158
Chapter 12: Graphing and Writing Equations	29,30,31,32,74,75,76,77,78	159-181
Chapter 13: Graphing Inequalities		182-185
Chapter 14: Systems of Equations and Systems of Inequalities	77	186-192
Chapter 15: Polynomials	26,27,69	193-213
Chapter 16: Statistics	1,50,56,57,58,59,60	214-229
Chapter 17: Data Interpretation	2,24,25,28,65	230-236
Chapter 18: Probability	4,5,6,15	237-245
Chapter 19: Patterns and Problem Solving	62,64,66,68,73	246-265
Chapter 20: Measurement	33,38,39	266-274
Chapter 21: Ratios, Proportions, and Scale Drawings	40	275-281
Chapter 22: Plane Geometry	36,43,44,45,47,49,52,53,54,55,79	282-302
Chapter 23: Solid Geometry	46,48	303-319
Chapter 24: Transformations and Plotted Shapes	51	320-328

Chapter 1
Fractions

As part of your California Mathematics Review, you must understand how to add, subtract, multiply, or divide fractions. Answers for fraction problems will be given in a simplified form. In this chapter, you will review and practice all of the skills needed to add, subtract, multiply, and divide fractions as well as to simplify them.

SIMPLIFYING IMPROPER FRACTIONS

EXAMPLE 1: Simplify: $\frac{21}{4} = 21 \div 4 = 5$ remainder 1. The quotient, 5, becomes the whole number portion of the mixed number.

$\frac{21}{4} = 5\frac{1}{4}$

The bottom number of the fraction always remains the same.

The remainder, 1, becomes the top number of the fraction.

EXAMPLE 2: Simplify: $\frac{11}{6}$

Step 1: $\frac{11}{6}$ is the same as $11 \div 6$. $11 \div 6 = 1$ with a remainder of 5.

Step 2: Rewrite as a whole number with a fraction. $1\frac{5}{6}$

Simplify the following improper fractions.

1. $\frac{13}{5} =$ _____
2. $\frac{11}{3} =$ _____
3. $\frac{24}{6} =$ _____
4. $\frac{7}{6} =$ _____
5. $\frac{19}{6} =$ _____
6. $\frac{16}{7} =$ _____
7. $\frac{13}{8} =$ _____
8. $\frac{9}{5} =$ _____
9. $\frac{22}{3} =$ _____
10. $\frac{13}{4} =$ _____
11. $\frac{18}{7} =$ _____
12. $\frac{15}{2} =$ _____
13. $\frac{17}{9} =$ _____
14. $\frac{27}{8} =$ _____
15. $\frac{32}{7} =$ _____
16. $\frac{3}{2} =$ _____
17. $\frac{7}{4} =$ _____
18. $\frac{21}{10} =$ _____

Fractions that have the same denominator (bottom number) can be added quickly. Add the numerators (top numbers), and keep the bottom number the same. Simplify the answer. The first one is done for you.

19. $\frac{2}{9}+\frac{7}{9}+\frac{4}{9} = \frac{13}{9} = 1\frac{4}{9}$
20. $\frac{2}{6}+\frac{4}{6}+\frac{5}{6}$
21. $\frac{3}{8}+\frac{5}{8}+\frac{7}{8}$
22. $\frac{9}{10}+\frac{1}{10}+\frac{3}{10}$
23. $\frac{8}{13}+\frac{6}{13}+\frac{2}{13}$
24. $\frac{6}{7}+\frac{4}{7}+\frac{5}{7}$
25. $\frac{3}{5}+\frac{4}{5}+\frac{3}{5}$
26. $\frac{7}{12}+\frac{5}{12}+\frac{1}{12}$
27. $\frac{4}{11}+\frac{3}{11}+\frac{6}{11}$

Copyright © American Book Company

CHANGING MIXED NUMBERS TO IMPROPER FRACTIONS

EXAMPLE: Change $4\frac{3}{5}$ to an improper fraction.

Step 1: Multiply the whole number (4) by the bottom number of the fraction (5). $4 \times 5 = 20$

Step 2: Add the top number to the product from step 1. $20 + 3 = 23$

Step 3: Put the answer over the bottom number (5).

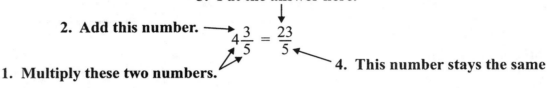

Change the following mixed numbers to improper fractions.

1. $3\frac{1}{2}=$ _____
2. $2\frac{7}{8}=$ _____
3. $9\frac{2}{3}=$ _____
4. $4\frac{3}{5}=$ _____
5. $7\frac{1}{4}=$ _____
6. $8\frac{5}{8}=$ _____
7. $1\frac{2}{7}=$ _____
8. $2\frac{4}{9}=$ _____
9. $6\frac{1}{5}=$ _____
10. $5\frac{2}{7}=$ _____
11. $3\frac{3}{5}=$ _____
12. $9\frac{3}{8}=$ _____
13. $10\frac{4}{5}=$ _____
14. $3\frac{3}{10}=$ _____
15. $4\frac{1}{7}=$ _____
16. $2\frac{5}{6}=$ _____
17. $7\frac{3}{7}=$ _____
18. $6\frac{7}{9}=$ _____
19. $7\frac{2}{5}=$ _____
20. $1\frac{6}{7}=$ _____

Whole numbers become improper fractions when you put them over 1. The first one is done for you.

21. $4=\frac{4}{1}$
22. $10=$ _____
23. $3=$ _____
24. $2=$ _____
25. $15=$ _____
26. $5=$ _____
27. $6=$ _____
28. $11=$ _____
29. $8=$ _____
30. $16=$ _____

GREATEST COMMON FACTOR

To reduce fractions to their simplest form, you must be able to find the greatest common factor.

EXAMPLE: Find the greatest common factor (GCF) of 16 and 24.

To find the **greatest common factor (GCF)** of two numbers, first list the factors of each number.

The factors of 16 are: 1, 2, 4, 8, and 16.
The factors of 24 are: 1, 2, 3, 4, 6, 8, 12, and 24.

What is the **largest** number they both have in common? **8**
8 is the **greatest** (largest number) **common factor**.

Find all the factors and the greatest common factor (GCF) of each pair of numbers below.

	Pairs	Factors	GCF		Pairs	Factors	GCF
1.	10			10.	6		
	15				42		
2.	12			11.	14		
	16				63		
3.	18			12.	9		
	36				51		
4.	27			13.	18		
	45				45		
5.	32			14.	12		
	40				20		
6.	16			15.	16		
	48				40		
7.	14			16.	10		
	42				45		
8.	4			17.	18		
	26				30		
9.	8			18.	15		
	28				25		

REDUCING PROPER FRACTIONS

EXAMPLE: Reduce $\frac{4}{8}$ to lowest terms.

Step 1: First you need to find the greatest common factor of 4 and 8. Think: What is the largest number that can be divided into 4 and 8 without a remainder?

These must be the same number. $?\overline{)4}$ 4 and 8 can both be divided by 4.
$?\overline{)8}$

Step 2: Divide the top and bottom of the fraction by the same number.
$$\frac{4 \div 4}{8 \div 4} = \frac{1}{2}$$

Reduce the following fractions to lowest terms.

1. $\frac{2}{8}$ = _____
2. $\frac{12}{15}$ = _____
3. $\frac{9}{27}$ = _____
4. $\frac{12}{42}$ = _____
5. $\frac{3}{21}$ = _____
6. $\frac{27}{54}$ = _____
7. $\frac{14}{22}$ = _____
8. $\frac{9}{21}$ = _____
9. $\frac{4}{14}$ = _____
10. $\frac{6}{26}$ = _____
11. $\frac{30}{45}$ = _____
12. $\frac{16}{64}$ = _____
13. $\frac{10}{25}$ = _____
14. $\frac{3}{12}$ = _____
15. $\frac{15}{30}$ = _____
16. $\frac{12}{36}$ = _____
17. $\frac{13}{39}$ = _____
18. $\frac{28}{49}$ = _____
19. $\frac{8}{18}$ = _____
20. $\frac{14}{21}$ = _____
21. $\frac{2}{12}$ = _____
22. $\frac{5}{15}$ = _____
23. $\frac{9}{15}$ = _____
24. $\frac{24}{48}$ = _____
25. $\frac{3}{18}$ = _____
26. $\frac{6}{27}$ = _____
27. $\frac{4}{18}$ = _____
28. $\frac{8}{28}$ = _____
29. $\frac{14}{42}$ = _____
30. $\frac{18}{36}$ = _____

MULTIPLYING FRACTIONS

EXAMPLE: Multiply $4\frac{3}{8} \times \frac{8}{10}$

Step 1: Change the mixed numbers in the problem to improper fractions. $\frac{35}{8} \times \frac{8}{10}$

Step 2: When multiplying fractions, you can cancel and simplify terms that have a common factor.

The 8 in the first fraction will cancel with the 8 in the second fraction.

The terms 35 and 10 are both divisible by 5, so 35 simplifies to 7, and 10 simplifies to 2.

Step 3: Multiply the simplified fractions. $\frac{7}{1} \times \frac{1}{2} = \frac{7}{2} = 3\frac{1}{2}$

Step 4: You cannot leave an improper fraction as the answer. You must change it to a mixed number.

Multiply and reduce answers to lowest terms.

1. $3\frac{1}{5} \times 1\frac{1}{2}$

2. $\frac{3}{8} \times 3\frac{3}{7}$

3. $4\frac{1}{3} \times 2\frac{1}{4}$

4. $4\frac{2}{3} \times 3\frac{3}{4}$

5. $1\frac{1}{2} \times 1\frac{2}{5}$

6. $3\frac{3}{7} \times \frac{5}{6}$

7. $3 \times 6\frac{1}{3}$

8. $1\frac{1}{6} \times 8$

9. $6\frac{2}{5} \times 5$

10. $6 \times 1\frac{3}{8}$

11. $\frac{5}{7} \times 2\frac{1}{3}$

12. $1\frac{2}{5} \times 1\frac{1}{4}$

13. $2\frac{1}{2} \times 5\frac{4}{5}$

14. $7\frac{2}{3} \times \frac{3}{4}$

15. $2 \times 3\frac{1}{4}$

16. $3\frac{1}{8} \times 1\frac{3}{5}$

17. $2\frac{3}{4} \times 6\frac{2}{3}$

18. $5\frac{3}{5} \times 1\frac{1}{14}$

19. $3\frac{1}{3} \times 3\frac{3}{4}$

20. $5\frac{1}{9} \times 1\frac{1}{23}$

21. $6\frac{2}{5} \times 1\frac{9}{16}$

22. $2\frac{1}{7} \times 2\frac{1}{10}$

23. $3\frac{3}{10} \times 2\frac{2}{3}$

24. $\frac{3}{5} \times 4\frac{1}{6}$

DIVIDING FRACTIONS

EXAMPLE: $1\frac{3}{4} \div 2\frac{5}{8}$

Step 1: Change the mixed numbers in the problem to improper fractions. $\frac{7}{4} \div \frac{21}{8}$

Step 2: Invert (turn upside down) the **second** fraction and multiply. $\frac{7}{4} \times \frac{8}{21}$

Step 3: Cancel where possible and multiply. $\frac{\cancel{7}^1}{\cancel{4}_1} \times \frac{\cancel{8}^2}{\cancel{21}_3} = \frac{2}{3}$

Divide and reduce answers to lowest terms.

1. $2\frac{2}{3} \div 1\frac{7}{9}$
2. $5 \div 1\frac{1}{2}$
3. $1\frac{5}{8} \div 2\frac{1}{4}$
4. $8\frac{2}{3} \div 2\frac{1}{6}$
5. $2\frac{4}{5} \div 2\frac{1}{5}$
6. $3\frac{2}{3} \div 1\frac{1}{6}$
7. $10 \div \frac{4}{5}$
8. $6\frac{1}{4} \div 1\frac{1}{2}$

9. $\frac{2}{5} \div 2$
10. $4\frac{1}{6} \div 1\frac{2}{3}$
11. $9 \div 3\frac{1}{4}$
12. $5\frac{1}{3} \div 2\frac{2}{5}$
13. $4\frac{1}{5} \div \frac{9}{10}$
14. $2\frac{2}{3} \div 4\frac{4}{5}$
15. $3\frac{3}{8} \div 3\frac{6}{7}$
16. $5\frac{1}{4} \div \frac{3}{4}$

17. $1\frac{2}{3} \div 1\frac{7}{8}$
18. $12 \div \frac{4}{7}$
19. $3\frac{3}{5} \div \frac{4}{7}$
20. $4\frac{1}{3} \div 2\frac{8}{9}$
21. $5\frac{2}{5} \div 3\frac{1}{5}$
22. $1\frac{5}{6} \div \frac{1}{2}$
23. $10\frac{2}{3} \div 1\frac{7}{9}$
24. $4\frac{2}{7} \div 3$

FINDING NUMERATORS

REMEMBER: Any fraction that has the same non-zero Numerator (top number) and Denominator (bottom number) equals 1.

EXAMPLES: $\frac{5}{5} = 1 \qquad \frac{8}{8} = 1 \qquad \frac{12}{12} = 1 \qquad \frac{15}{15} = 1 \qquad \frac{25}{25} = 1$

Any fraction multiplied by 1 in any fraction form remains equal.

EXAMPLE: $\frac{3}{7} \times \frac{4}{4} = \frac{12}{28}$ so $\frac{3}{7} = \frac{12}{28}$

PROBLEM: Find the missing numerator (top number). $\quad \frac{5}{8} = \frac{}{24}$

Step 1: Ask yourself, "What was 8 multiplied by to get 24"? 3 is the answer.

Step 2: The only way to keep the fraction equal is to multiply the top and bottom number by the same number. The bottom number was multiplied by 3, so multiply the top number by 3. $\quad \frac{5}{8} \times \frac{3}{3} = \frac{15}{24}$

Find the missing numerators from the following equivalent fractions.

1. $\frac{2}{6} = \frac{}{18}$
2. $\frac{2}{3} = \frac{}{27}$
3. $\frac{4}{9} = \frac{}{18}$
4. $\frac{7}{15} = \frac{}{45}$
5. $\frac{9}{10} = \frac{}{50}$

6. $\frac{5}{6} = \frac{}{36}$
7. $\frac{1}{4} = \frac{}{36}$
8. $\frac{3}{14} = \frac{}{28}$
9. $\frac{2}{5} = \frac{}{25}$
10. $\frac{4}{11} = \frac{}{33}$

11. $\frac{5}{6} = \frac{}{18}$
12. $\frac{6}{11} = \frac{}{22}$
13. $\frac{8}{15} = \frac{}{45}$
14. $\frac{1}{9} = \frac{}{18}$
15. $\frac{7}{8} = \frac{}{40}$

16. $\frac{1}{12} = \frac{}{48}$
17. $\frac{3}{8} = \frac{}{64}$
18. $\frac{3}{4} = \frac{}{16}$
19. $\frac{2}{7} = \frac{}{49}$
20. $\frac{11}{12} = \frac{}{24}$

21. $\frac{2}{5} = \frac{}{45}$
22. $\frac{4}{5} = \frac{}{15}$
23. $\frac{1}{9} = \frac{}{27}$
24. $\frac{3}{8} = \frac{}{56}$
25. $\frac{2}{13} = \frac{}{26}$

26. $\frac{1}{7} = \frac{}{35}$
27. $\frac{4}{5} = \frac{}{10}$
28. $\frac{3}{10} = \frac{}{40}$
29. $\frac{7}{8} = \frac{}{48}$
30. $\frac{6}{7} = \frac{}{14}$

LEAST COMMON MULTIPLE

Find the least common multiple (LCM) of 6 and 10.

To find the **least common multiple (LCM)** of two numbers, first list the multiples of each number. The multiples of a number are 1 times the number, 2 times the number, 3 times the number, and so on.

The multiples of 6 are: 6, 12, 18, 24, 30 …

The multiples of 10 are: 10, 20, 30, 40, 50…

What is the smallest multiple they both have in common?   **30**
30 is the **least** (smallest number) **common multiple** of 6 and 10.

Find the least common multiple (LCM) of each pair of numbers below.

	Pairs	Multiples	LCM		Pairs	Multiples	LCM
1.	6	6, 12, 18, 24, 30	30	10.	6		
	15	15, 30			7		
2.	12			11.	4		
	16				18		
3.	18			12.	7		
	36				5		
4.	7			13.	30		
	3				45		
5.	12			14.	3		
	8				8		
6.	6			15.	12		
	8				9		
7.	4			16.	5		
	14				45		
8.	9			17.	3		
	6				5		
9.	2			18.	4		
	15				22		

ADDING FRACTIONS

EXAMPLE: Add $3\frac{1}{2} + 2\frac{2}{3}$

Step 1: Rewrite the problem vertically, and find a common denominator. Think: What is the least common multiple of 2 and 3? 6 is the LCM, so 6 will be the common denominator.

$$3\frac{1}{2} = \frac{}{6}$$
$$+2\frac{2}{3} = \frac{}{6}$$

Step 2: To find the numerator for the top fraction, think: What do I multiply 2 by to get 6? 3 is the answer. You must multiply the top and bottom numbers of the fraction by 2 to keep the fraction equal. For the bottom fraction, multiply the top and bottom number by 2.

$$3\frac{1}{2} \begin{smallmatrix}\times 3\\ \times 3\end{smallmatrix} \quad 3\frac{3}{6}$$
$$+2\frac{2}{3} \begin{smallmatrix}\times 2\\ \times 2\end{smallmatrix} \rightarrow +2\frac{4}{6}$$
$$= 5\frac{7}{6}$$

Step 3: Add whole numbers and fractions, and simplify.

$$= 6\frac{1}{6}$$

Add and simplify the answers.

1. $3\frac{5}{9} + 5\frac{2}{3}$

2. $1\frac{1}{4} + 4\frac{2}{5}$

3. $3\frac{3}{4} + 2\frac{3}{5}$

4. $2\frac{1}{4} + 1\frac{7}{8}$

5. $6\frac{5}{6} + 4\frac{1}{3}$

6. $9\frac{1}{5} + 5\frac{5}{6}$

7. $\frac{1}{3} + 7\frac{3}{4}$

8. $9\frac{4}{9} + 3\frac{2}{3}$

9. $4\frac{7}{10} + 8\frac{2}{3}$

10. $5\frac{2}{7} + \frac{1}{2}$

11. $3\frac{3}{11} + 2\frac{3}{4}$

12. $\frac{3}{5} + \frac{4}{9}$

13. $\frac{5}{8} + \frac{2}{3}$

14. $6\frac{4}{7} + 5\frac{1}{8}$

15. $9\frac{1}{3} + 4\frac{7}{8}$

16. $7\frac{3}{4} + 3\frac{5}{12}$

17. $\frac{1}{2} + \frac{2}{3}$

18. $6\frac{1}{2} + 2\frac{7}{8}$

19. $7\frac{2}{3} + 6\frac{3}{8}$

20. $3\frac{7}{8} + 9\frac{4}{5}$

21. $4\frac{7}{12} + \frac{2}{3}$

SUBTRACTING MIXED NUMBERS FROM WHOLE NUMBERS

EXAMPLE: Subtract $15 - 3\frac{3}{4}$

Step 1: Rewrite the problem vertically.

$$\begin{array}{r} 15 \\ -\ 3\frac{3}{4} \\ \hline \end{array}$$

Step 2: You cannot subtract three-fourths from nothing. You must borrow 1 from 15. You will need to put the 1 in fraction form. If you use $\frac{4}{4}$ ($\frac{4}{4} = 1$), you will be ready to subtract.

$$\begin{array}{r} 14\frac{4}{4} \\ -\ 3\frac{3}{4} \\ \hline 11\frac{1}{4} \end{array}$$

Subtract

1. $12 - 3\frac{2}{9}$
2. $3 - 1\frac{4}{7}$
3. $24 - 11\frac{4}{5}$
4. $2 - 1\frac{2}{5}$

5. $4 - 1\frac{5}{8}$
6. $11 - 9\frac{7}{8}$
7. $14 - 9\frac{7}{12}$
8. $8 - 3\frac{1}{3}$

9. $5 - 3\frac{1}{2}$
10. $17 - 13\frac{1}{5}$
11. $3 - 1\frac{5}{11}$
12. $13 - 8\frac{9}{10}$

13. $15 - 6\frac{3}{4}$
14. $6 - 4\frac{8}{9}$
15. $20 - 12\frac{6}{7}$
16. $21 - 1\frac{3}{20}$

17. $9 - 5\frac{2}{3}$
18. $8 - 7\frac{3}{5}$
19. $5 - 4\frac{5}{8}$
20. $14 - 9\frac{1}{7}$

21. $12 - 4\frac{1}{6}$
22. $2 - 1\frac{2}{3}$
23. $42 - 30\frac{2}{9}$
24. $7 - 5\frac{9}{13}$

25. $19 - 13\frac{3}{8}$
26. $14 - 10\frac{5}{9}$
27. $16 - 8\frac{1}{4}$
28. $15 - 3\frac{5}{7}$

SUBTRACTING MIXED NUMBERS WITH BORROWING

EXAMPLE: Subtract $7\frac{1}{4} - 5\frac{5}{6}$

Step 1: Rewrite the problem and find a common denominator.

$$7\frac{1}{4} \times \frac{3}{3} \rightarrow 7\frac{3}{12}$$
$$-5\frac{5}{6} \times \frac{2}{2} \rightarrow -5\frac{10}{12}$$

Step 2: You cannot subtract 10 from 3. You must borrow 1 from the 7. The 1 will be in the fraction form $\frac{12}{12}$ which you must add to the $\frac{3}{12}$ you already have, making $\frac{15}{12}$.

$$\overset{6}{\cancel{7}}\overset{15}{\cancel{\frac{3}{12}}}$$
$$-5\frac{10}{12}$$
$$= 1\frac{5}{12}$$

Step 3: Subtract whole numbers and fractions, and simplify.

Subtract and simplify.

1. $4\frac{1}{3} - 1\frac{5}{9}$

2. $3\frac{4}{9} - 2\frac{5}{6}$

3. $8\frac{4}{7} - 5\frac{1}{3}$

4. $5\frac{2}{5} - 3\frac{1}{2}$

5. $8\frac{2}{5} - 5\frac{3}{10}$

6. $9\frac{2}{5} - 4\frac{3}{4}$

7. $9\frac{3}{4} - 2\frac{1}{3}$

8. $5\frac{1}{7} - \frac{2}{3}$

9. $6\frac{1}{5} - 3\frac{3}{8}$

10. $6\frac{5}{6} - 3\frac{4}{5}$

11. $2\frac{2}{9} - 1\frac{3}{4}$

12. $4\frac{7}{10} - 3\frac{1}{3}$

13. $7\frac{3}{5} - 4\frac{5}{6}$

14. $9\frac{3}{8} - 5\frac{1}{2}$

15. $8\frac{1}{9} - 5\frac{1}{3}$

16. $5\frac{1}{6} - 1\frac{2}{3}$

17. $6\frac{5}{6} - 3\frac{1}{3}$

18. $7\frac{2}{3} - 3\frac{5}{6}$

19. $8\frac{4}{7} - 4\frac{3}{4}$

20. $9\frac{3}{4} - 1\frac{1}{5}$

21. $5\frac{1}{2} - 2\frac{1}{3}$

DEDUCTIONS - FRACTION OFF

Sometimes sale prices are advertised as $\frac{1}{4}$ off or $\frac{1}{3}$ off. To find out how much you will save, just multiply the original price by the fraction off.

EXAMPLE: CD players are on sale for $\frac{1}{3}$ off. How much can you save on a $240 CD player?

$$\frac{1}{\cancel{3}_1} \times \frac{\cancel{240}^{80}}{1} = 80 \quad \text{You can save } \$80.00.$$

Find the amount of savings in the problems below.

J.P. Nichols is having a liquidation sale on all furniture. Sale prices are $\frac{1}{2}$ off the regular price. How much can you save on the following furniture items?

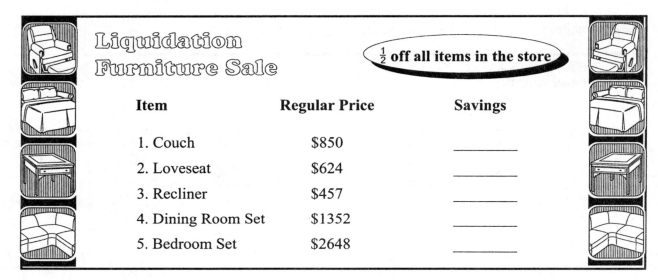

Liquidation Furniture Sale — $\frac{1}{2}$ off all items in the store

Item	Regular Price	Savings
1. Couch	$850	_____
2. Loveseat	$624	_____
3. Recliner	$457	_____
4. Dining Room Set	$1352	_____
5. Bedroom Set	$2648	_____

Buy Rite Computer Store is having a $\frac{1}{3}$ off sale on selected computer items in the store. How much can you save on the following items?

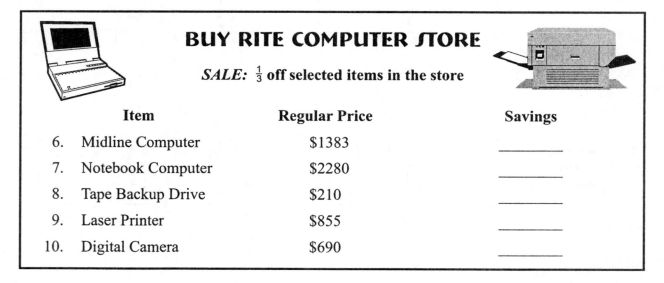

BUY RITE COMPUTER STORE

SALE: $\frac{1}{3}$ off selected items in the store

Item	Regular Price	Savings
6. Midline Computer	$1383	_____
7. Notebook Computer	$2280	_____
8. Tape Backup Drive	$210	_____
9. Laser Printer	$855	_____
10. Digital Camera	$690	_____

FRACTION WORD PROBLEMS

Solve and reduce answers to lowest terms.

1. Sara works for a movie theater and sells candy by the pound. Her first customer bought $1\frac{1}{3}$ pounds of candy, the second bought $\frac{3}{4}$ of a pound, and the third bought $\frac{4}{5}$ of a pound. How many pounds did she sell to the first three customers?

2. Beth has a bread machine that makes a loaf of bread that weighs $1\frac{1}{2}$ pounds. If she makes a loaf of bread for each of her three sisters, how many pounds of bread will she make?

3. A farmer hauled in 120 bales of hay. Each of his cows ate $1\frac{1}{4}$ bales. How many cows did the farmer feed?

4. Juan was competing in a 1000 meter race. He had to pull out of the race after running $\frac{3}{4}$ of it. How many meters did he run?

5. Tad needs to measure where the free-throw line should be in front of his basketball goal. He knows his feet are $1\frac{1}{8}$ feet long and the free-throw line should be 15 feet from the backboard. How many toe-to-toe steps does Tad need to take to mark off 15 feet?

6. A chemical plant takes in $5\frac{1}{2}$ million gallons of water from a local river and discharges $3\frac{2}{3}$ million back into the river. How much water does not go back into the river?

7. In January, Jeff filled his car with $11\frac{1}{2}$ gallons of gas the first week, $13\frac{1}{3}$ gallons the second week, $12\frac{1}{4}$ gallons the third week, and $10\frac{1}{5}$ gallons the fourth week of January. How many gallons of gas did he buy in January?

8. Li Tun makes sandwiches for his family. He has $8\frac{1}{4}$ ounces of sandwich meat. If he divides the meat equally to make $4\frac{1}{2}$ sandwiches, how much meat will each sandwich have?

9. The company water cooler started with $4\frac{1}{3}$ gallons of water. Employees drank $3\frac{3}{4}$ gallons. How many gallons were left in the cooler?

10. Rita bought $\frac{1}{4}$ pound hamburger patties for her family reunion picnic. She bought 50 patties. How many pounds of hamburgers did she buy?

CHAPTER 1 REVIEW

Simplify.

1. $\dfrac{15}{6}$ _____

2. $\dfrac{24}{4}$ _____

3. $\dfrac{20}{15}$ _____

4. $\dfrac{14}{3}$ _____

Reduce.

5. $\dfrac{9}{27}$ _____

6. $\dfrac{4}{16}$ _____

7. $\dfrac{8}{12}$ _____

8. $\dfrac{12}{18}$ _____

Change to an improper fraction.

9. $5\dfrac{1}{10}$ _____

10. 7 _____

11. $3\dfrac{3}{5}$ _____

12. $6\dfrac{2}{3}$ _____

Add and simplify.

13. $\dfrac{5}{9} + \dfrac{7}{9}$ _____

14. $7\dfrac{1}{2} + 3\dfrac{3}{8}$ _____

15. $4\dfrac{4}{15} + \dfrac{1}{5}$ _____

16. $\dfrac{1}{7} + \dfrac{3}{7}$ _____

Subtract and simplify.

17. $10 - 5\dfrac{1}{8}$ _____

18. $3\dfrac{1}{3} - \dfrac{3}{4}$ _____

19. $9\dfrac{3}{4} - 2\dfrac{3}{8}$ _____

20. $6\dfrac{1}{5} - 1\dfrac{3}{10}$ _____

Multiply and simplify.

21. $1\dfrac{1}{3} \times 3\dfrac{1}{2}$ _____

22. $5\dfrac{3}{7} \times \dfrac{7}{8}$ _____

23. $4\dfrac{4}{6} \times 1\dfrac{5}{7}$ _____

24. $\dfrac{2}{3} \times \dfrac{5}{6}$ _____

Divide and simplify.

25. $\dfrac{1}{2} \div \dfrac{4}{5}$ _____

26. $6\dfrac{6}{7} \div 2\dfrac{2}{3}$ _____

27. $3\dfrac{5}{6} \div 11\dfrac{1}{2}$ _____

28. $1\dfrac{1}{3} \div 3\dfrac{1}{5}$ _____

Find the greatest common factor for the following sets of numbers.

29. 9 and 15 _____

30. 12 and 16 _____

31. 10 and 25 _____

32. 8 and 24 _____

Find the least common multiple for the following sets of numbers.

33. 8 and 12 _____

34. 5 and 9 _____

35. 4 and 10 _____

36. 6 and 8 _____

37. Which of the following can be used to compute $\dfrac{1}{8} + \dfrac{1}{6}$?

   A. $\dfrac{1+1}{8+6}$

   B. $\dfrac{1 \cdot 3}{8 \cdot 3} + \dfrac{1 \cdot 4}{6 \cdot 4}$

   C. $\dfrac{1}{8 \cdot 3} + \dfrac{1}{6 \cdot 4}$

   D. $\dfrac{1}{8 \cdot 6} + \dfrac{1}{6 \cdot 6}$

38. Which of the following can be used to compute $\frac{3}{8} - \frac{1}{6}$?

   A. $\frac{3}{8 \cdot 6} - \frac{1}{6 \cdot 8}$
   B. $\frac{1}{8 \cdot 3} - \frac{1}{6 \cdot 4}$
   C. $\frac{3 \cdot 3}{8 \cdot 3} - \frac{1 \cdot 4}{6 \cdot 4}$
   D. $\frac{3 - 1}{8 \cdot 6}$

39. Mrs. Tate brought $5\frac{1}{2}$ pounds of candy to divide among her 22 students. If the candy was divided equally, how many pounds of candy did each student receive?

40. Elenita used $1\frac{1}{5}$ yards of material to recover one dining room chair. How much material would she need to recover all eight chairs?

41. The square tiles in Mr. Cooke's math classroom measure $2\frac{1}{4}$ feet across. The students counted that the classroom was $5\frac{1}{3}$ tiles wide. How wide is Mr. Cooke's classroom?

42. The Vargas family is hiking a $23\frac{1}{3}$ mile trail. The first day, they hiked $10\frac{1}{2}$ miles. How much further do they have to go to complete the trail?

43. Jena walked $\frac{1}{5}$ of a mile to a friend's house, $1\frac{1}{3}$ miles to the store, and $\frac{3}{4}$ of a mile back home. How far did Jena walk?

44. Corey used $2\frac{4}{5}$ gallons of paint to mark one mile of this year's spring road race. How many gallons will he use to mark the entire $6\frac{1}{4}$ mile course?

45.

According to the ad above, how much could you save at Martin's Department Store on a comforter regularly priced at $108?

Chapter 2
Decimals

As you review the curriculum of California Mathematics, you will see exercises that require you to add, subtract, multiply, or divide decimals. You may also have to convert from decimals to fractions or from fractions to decimals. In this chapter, you will review and practice math problems involving decimals.

ADDING DECIMALS

EXAMPLE: Find $0.9 + 2.5 + 63.17$

Step 1:	When you add decimals, first arrange the numbers in columns with the decimal points under each other.	0.9 2.5 $\underline{+\ 63.17}$
Step 2:	Add 0's here to keep your columns straight.	
Step 3:	Start at the right and add each column. Remember to carry when necessary. Bring down the decimal point.	1 0.90 2.50 $\underline{+\ 63.17}$ 66.57

Add. Be sure to write the decimal point in your answer.

1. $5.3 + 6.02 + 0.73$
2. $0.235 + 6.2 + 3.27$
3. $7.542 + 10.5 + 4.57$
4. $\$5.87 + \$7.52$
5. $\$4.68 + \$9.47$

6. $5.08 + 11.2 + 6.075$
7. $5.14 + 2.3 + 5.097$
8. $4.9 + 15.71 + 0.254$
9. $\$3.75 + \$18.90$
10. $\$64.95 + \$4.63$

11. $1.25 + 4.1 + 10.007$
12. $15.4 + 5.074 + 3.15$
13. $45.23 + 9.5 + 0.693$
14. $\$8.63 + \$12.50$
15. $\$6.87 + \$27.23$

16. $0.23 + 5.9 + 12$
17. $8.5784 + 10.03$
18. $85.7 + 205.952$
19. $\$98.45 + \$8.89$
20. $\$7.77 + \$11.19$

SUBTRACTING DECIMALS

EXAMPLE: Find $14.9 - 0.007$

Step 1: When you subtract decimals, arrange the numbers in columns with the decimal points under each other.

$$\begin{array}{r} 14.9 \\ -0.007 \end{array}$$

Step 2: You must fill in the empty places with 0's so that both numbers have the same number of digits after the decimal point.

$$\begin{array}{r} 14.900 \\ -0.007 \end{array}$$

Step 3: Start at the right and subtract each column. Remember to borrow when necessary.

$$\begin{array}{r} 14.900 \\ -0.007 \\ \hline 14.893 \end{array}$$

Subtract. Be sure to write the decimal point in your answer.

1. $5.25 - 4.7$
2. $23.657 - 9.83$
3. $\$56.54 - \$17.92$
4. $\$294.78 - \$80.99$
5. $\$70.00 - \$68.99$

6. $58.6 - 9.153$
7. $405.97 - 7.325$
8. $\$40.09 - \$9.99$
9. $\$115.45 - \$4.79$
10. $\$45.18 - \$23.65$

11. $12.96 - 7.32$
12. $19.2 - 8.63$
13. $8.123 - 5.096$
14. $\$14.32 - \$0.58$
15. $\$30.00 - \$22.95$

16. $15.789 - 6.32$
17. $478.63 - 99.2$
18. $\$15.45 - \$8.58$
19. $102.5 - 1.079$
20. $7.054 - 3.009$

MULTIPLICATION OF DECIMALS

EXAMPLE: 56.2 × 0.17

Step 1: Set up the problem as if you were multiplying whole numbers.

$$\begin{array}{r} 56.2 \\ \times\ 0.17 \end{array}$$

Step 2: Multiply as if you were multiplying whole numbers.

$$\begin{array}{r} {}^{4\,1}\\ 56.2 \\ \times\ 0.17 \\ \hline 13934 \\ 562 \\ \hline 9.554 \end{array}$$

56.2 ← 1 number after the decimal point
× 0.17 ← + 2 numbers after the decimal point
— 3 numbers after the decimal point

Step 3: Count how many numbers are after the decimal points in the problem. In this problem, 2, 1, and 7 come after the decimal points, so the answer must also have three numbers after the decimal point.

Multiply.

1. 15.2 × 3.5
2. 9.54 × 5.3
3. 5.72 × 6.3
4. 4.8 × 3.2
5. 45.8 × 2.2
6. 4.5 × 7.1
7. 0.052 × 0.33
8. 4.12 × 6.8
9. 23.65 × 9.2
10. 1.54 × 0.43
11. 0.47 × 6.1
12. 1.3 × 1.57
13. 16.4 × 0.5
14. 0.87 × 3.21
15. 5.94 × 0.65
16. 7.8 × 0.23

MORE MULTIPLYING DECIMALS

EXAMPLE: Find 0.007 × 0.125

Step 1: Multiply as you would whole numbers.

$$\begin{array}{r} 0.007 \\ \times\ 0.125 \\ \hline 0.000875 \end{array}$$ ← 3 numbers after the decimal point
← + 3 numbers after the decimal point
← 6 numbers after the decimal point

Step 2: Count how many numbers are behind decimal points in the problem. In this case, six numbers come after decimal points in the problem, so there must be six numbers after the decimal point in the answer. In this problem, 0's needed to be written in the answer **in front of** the 8, so there will be 6 numbers after the decimal point.

Multiply. Write in zeros as needed. Round dollar figures to the nearest penny.

1. 0.123 × .45
2. 0.004 × 10.31
3. 1.54 × 1.1
4. 10.05 × 0.45
5. 9.45 × 0.8
6. $6.49 × 0.06
7. 5.003 × 0.009
8. $9.99 × 0.06
9. 6.09 × 5.3
10. $22.00 × 0.075
11. 5.914 × 0.02
12. 4.96 × 0.23
13. 6.98 × 0.02
14. 3.12 × 0.08
15. 7.158 × 0.09
16. 0.0158 × 0.32

DIVISION OF DECIMALS BY WHOLE NUMBERS

EXAMPLE: 52.26 ÷ 6

Step 1: Copy the problem as you would for whole numbers. Copy the decimal point directly above in the place for the answer.

$$6\overline{)52.26}$$

Step 2: Divide the same way as you would with whole numbers.

$$\begin{array}{r} 8.71 \\ 6\overline{)52.26} \\ -48 \\ \hline 4\,2 \\ -4\,2 \\ \hline 6 \\ -6 \\ \hline 0 \end{array}$$

Divide. Remember to copy the decimal point directly above the place for the answer.

1. 42.75 ÷ 3
2. 74.16 ÷ 6
3. 81.50 ÷ 25
4. 82.46 ÷ 14
5. 12.50 ÷ 2
6. 224.64 ÷ 52
7. 183.04 ÷ 52
8. 281.52 ÷ 23
9. 72.36 ÷ 4
10. 379.5 ÷ 15
11. 152.25 ÷ 21
12. 40.375 ÷ 19
13. 102.5 ÷ 5
14. 113.4 ÷ 9
15. 585.14 ÷ 34
16. 93.6 ÷ 24

CHANGING FRACTIONS TO DECIMALS

EXAMPLE: Change $\frac{1}{8}$ to a decimal.

Step 1: To change a fraction to a decimal, simply divide the top number by the bottom number.

$$8\overline{)1}$$

Step 2: Add a decimal point and a 0 after the 1 and divide.

$$\begin{array}{r} 0.1 \\ 8\overline{)1.0} \\ -8 \\ \hline 2 \end{array}$$

Step 3: Continue adding 0's and dividing until there is no remainder.

$$\begin{array}{r} 0.125 \\ 8\overline{)1.000} \\ -8 \\ \hline 20 \\ -16 \\ \hline 40 \\ -40 \\ \hline 0 \end{array}$$

In some problems, the number after the decimal point begins to repeat. Take, for example, the fraction $\frac{4}{11}$. $4 \div 11 = 0.363636$, and the 36 keeps repeating forever. To show that the 36 repeats, simply write a bar above the numbers that repeat, $0.\overline{36}$.

Change the following fractions to decimals.

1. $\frac{4}{5}$
2. $\frac{2}{3}$
3. $\frac{1}{2}$
4. $\frac{5}{9}$
5. $\frac{1}{10}$
6. $\frac{5}{8}$
7. $\frac{5}{6}$
8. $\frac{1}{6}$
9. $\frac{3}{5}$
10. $\frac{7}{10}$
11. $\frac{4}{11}$
12. $\frac{1}{9}$
13. $\frac{7}{9}$
14. $\frac{9}{10}$
15. $\frac{1}{4}$
16. $\frac{3}{8}$
17. $\frac{3}{16}$
18. $\frac{3}{4}$
19. $\frac{8}{9}$
20. $\frac{5}{12}$

CHANGING MIXED NUMBERS TO DECIMALS

If there is a whole number with a fraction, write the whole number to the left of the decimal point. Then change the fraction to a decimal.

EXAMPLES: $4\frac{1}{10} = 4.1$ $\qquad 16\frac{2}{3} = 16.\overline{6} \qquad 12\frac{7}{8} = 12.875$

Change the following mixed numbers to decimals.

1. $5\frac{2}{3}$
2. $8\frac{5}{11}$
3. $15\frac{3}{5}$
4. $13\frac{2}{3}$
5. $30\frac{1}{3}$
6. $3\frac{1}{2}$
7. $1\frac{7}{8}$
8. $4\frac{9}{100}$
9. $6\frac{4}{5}$
10. $13\frac{1}{2}$
11. $12\frac{4}{5}$
12. $11\frac{5}{8}$
13. $7\frac{1}{4}$
14. $12\frac{1}{3}$
15. $1\frac{5}{8}$
16. $2\frac{3}{4}$
17. $10\frac{1}{10}$
18. $20\frac{2}{5}$
19. $4\frac{9}{10}$
20. $5\frac{4}{11}$

CHANGING DECIMALS TO FRACTIONS

EXAMPLE: 0.25

Step 1: Copy the decimal without the point. This will be the top number of the fraction.

$$\frac{25}{\Box}$$

Step 2: The bottom number is a 1 with as many 0's after it as there are digits in the top number.

$$\frac{25}{100} \quad \begin{array}{l}\leftarrow \text{Two digits} \\ \leftarrow \text{Two 0's}\end{array}$$

$$\frac{25}{100} = \frac{1}{4}$$

Step 3: You then need to reduce the fraction.

EXAMPLES: $.2 = \frac{2}{10} = \frac{1}{5}$ $.65 = \frac{65}{100} = \frac{13}{20}$ $.125 = \frac{125}{1000} = \frac{1}{8}$

Change the following decimals to fractions.

1. .55
2. .6
3. .12
4. .9
5. .75
6. .82
7. .3
8. .42
9. .71
10. .42
11. .56
12. .24
13. .35
14. .96
15. .125
16. .375

CHANGING DECIMALS WITH WHOLE NUMBERS TO MIXED NUMBERS

EXAMPLE: Change 14.28 to a mixed number.

Step 1: Copy the portion of the number that is whole.   14

Step 2: Change .28 to a fraction. $14\frac{28}{100}$

Step 3: Reduce the fraction. $14\frac{28}{100} = 14\frac{7}{25}$

Change the following decimals to mixed numbers.

1. 7.125
2. 99.5
3. 2.13
4. 5.1
5. 16.95
6. 3.625
7. 4.42
8. 15.84
9. 6.7
10. 45.425
11. 15.8
12. 8.16
13. 13.9
14. 32.65
15. 17.25
16. 9.82

DIVISION OF DECIMALS BY DECIMALS

EXAMPLE: $374.5 \div 0.07$

Step 1: Copy the problem as you would for whole numbers.

$0.07 \overline{)374.5}$ ← Dividend, with Divisor labeled

Step 2: You cannot divide by a decimal number. You must move the decimal point in the divisor 2 places to the right to make it a whole number. The decimal point in the dividend must also move to the right the same number of places. Notice that in this example, you must add a 0 to the dividend.

$0.0\underline{7.}\overline{)374.5\underline{0.}}$

Step 3: The problem now becomes $37450 \div 7$. Copy the decimal point from the dividend straight above in the place for the answer.

$$007.\overline{)37450.} = 5350.$$

Divide. Remember to move the decimal points.

1. $0.676 \div 0.013$
2. $70.32 \div 0.08$
3. $\$54.60 \div 0.84$
4. $\$10.35 \div 0.45$
5. $18.46 \div 1.3$
6. $14.6 \div 0.002$
7. $\$125.25 \div 0.75$
8. $\$33.00 \div 1.65$
9. $154.08 \div 1.8$
10. $0.4374 \div 0.003$
11. $292.9 \div 0.29$
12. $6.375 \div 0.3$
13. $4.8 \div 0.08$
14. $1.2 \div 0.024$
15. $15.725 \div 3.7$
16. $\$167.50 \div 0.25$

ESTIMATING DIVISION OF DECIMALS

EXAMPLE: The following division of decimals problem has the decimal point missing from the answer. Estimate the answer to determine where the decimal point should go.

$$2489 \div 5.8 = 42913793$$

Step 1: Round off the numbers in the problem to numbers that are divisible without a remainder.

5.8 rounds to 6
2489 rounds to 2400

$2400 \div 6 = 400$

Step 2: The answer should be close to 400, so put the decimal point after the third whole number.

Solution: $2489 \div 5.8 = 429.13793$

For each of the following problems, round off the numbers to determine where the decimal point belongs in the answer.

1. $15.63 \div 4.2 = 3\ 7\ 2\ 1\ 4$
2. $476.3 \div 5.81 = 8\ 1\ 9\ 7\ 9$
3. $7561.5 \div 10.6 = 7\ 1\ 3\ 3\ 4\ 9$
4. $6259 \div 8.1 = 7\ 7\ 2\ 7\ 1\ 6$
5. $11.78 \div .94 = 1\ 2\ 5\ 3\ 1\ 9$
6. $45.69 \div 4.67 = 9\ 7\ 8\ 3\ 7$
7. $768 \div 22.35 = 3\ 4\ 3\ 6\ 2\ 4$
8. $5.16 \div 1.78 = 2\ 8\ 9\ 8\ 8\ 7$
9. $87.32 \div 56.7 = 1\ 5\ 4\ 0\ 0\ 3\ 5$
10. $144.92 \div 12.4 = 1\ 1\ 6\ 8\ 7\ 0\ 9$
11. $456.98 \div 21.5 = 2\ 1\ 2\ 5\ 4\ 8\ 8$
12. $19 \div 8.6 = 2\ 2\ 0\ 9\ 3$
13. $79.19 \div 7.8 = 1\ 0\ 1\ 5\ 2\ 5\ 6$
14. $856.3 \div 8.2 = 1\ 0\ 4\ 4\ 2\ 6\ 8$
15. $11.235 \div .48 = 2\ 3\ 4\ 0\ 6$
16. $9.63 \div 4.1 = 2\ 3\ 4\ 8\ 7\ 8$
17. $96.68 \div 32.56 = 2\ 9\ 6\ 9\ 2\ 8\ 7$
18. $162.3 \div 87.5 = 1\ 8\ 5\ 4\ 8\ 5\ 7$
19. $45.98 \div 2.9 = 1\ 5\ 8\ 5\ 5$
20. $32.65 \div 1.689 = 1\ 9\ 3\ 3\ 0\ 9\ 7$
21. $26.5 \div 5.1 = 5\ 1\ 9\ 6$
22. $6.59 \div 2.147 = 3\ 0\ 6\ 9\ 3\ 9\ 9$
23. $75.26 \div 8.36 = 9\ 0\ 0\ 2\ 3\ 9$
24. $158.4\ 3.09 = 5\ 1\ 2\ 6\ 2$

DECIMAL WORD PROBLEMS

1. Micah can have the oil changed in his car for $19.99, or he can buy the oil and filter and change it himself for $8.79. How much would he save by changing the oil himself?

   _____

2. Megan bought 5 boxes of cookies for $3.75 each. How much did she spend?

   _____

3. Will subscribes to a monthly auto magazine. His one-year subscription costs $29.97. If he pays for the subscription in 3 equal installments, how much is each payment?

   _____

4. Pat purchases 2.5 pounds of hamburger at $0.98 per pound. What is the total cost of the hamburger?

   _____

5. The White family took $650 cash with them on vacation. At the end of their vacation, they had $4.67 left. How much cash did they spend on vacation?

   _____

6. Acer Middle School spent $1443.20 on 55 math books. How much did each book cost?

   _____

7. The Junior Beta Club needs to raise $1513.75 to go to a national convention. If club members decide to sell candy bars at $1.25 profit each, how many will they to need to sell to meet their goal?

   _____

8. Fleta owns a candy store. On Monday, she sold 6.5 pounds of chocolate, 8.34 pounds of jelly beans, 4.9 pounds of sour snaps, and 5.64 pounds of yogurt-covered raisins. How many pounds of candy did she sell total?

   _____

9. Randal purchased a rare coin collection for $1803.95. He sold it at auction for $2700. How much money did he make on the coins?

   _____

10. A leather jacket that normally sells for $259.99 is on sale now for $197.88. How much can you save if you buy it now?

    _____

11. At the movies, Gigi buys 0.6 pounds of candy priced at $2.10 per pound. How much did she spend on candy?

    _____

12. George has $6.00 to buy candy. If each candy bar costs $.60, how many bars can he buy?

    _____

FINDING A PROFIT

EXAMPLE: Tory decided to make dog houses to sell at the hardware store near his home to make money for college. He spent $37 on materials for each doghouse. He sold each doghouse for $60 and gave $5.00 to the hardware store for each doghouse he sold in return for space to display his doghouses. How many doghouses will he have to sell if he wants to earn $1,000 for college?

Step 1: Take the selling price and subtract all the expenses for each doghouse to find the profit on each doghouse. 60 − 37− 5 = **$18 profit on each doghouse.**

Step 2: Take the amount he wants to make and divide by the profit on each doghouse to find out how many he has to sell to make $1,000. 1000 ÷ 18 = 55.5. **He will need to sell 56 doghouses to make $1,000.**

Find the profit for each of the following questions.

1. Troy bought a worm farm and decided to raise worms to make money. The worm farm cost $110.00 and came with 200 worms and 500 worm capsules, each containing about 4 baby worms. He figures he can sell the worms to bait shops for $1.25 per dozen. How many dozen worms will he have to sell to begin to make a profit? Worms eat table scraps, so there are no food expenses.

2. Sharon had a black Labrador retriever with an excellent pedigree which she decided to breed so she could sell the puppies to earn money. Her parents said they would loan her the money she needed to cover her expenses until the puppies were sold. She borrowed $500 for stud fees, $150.00 for vet fees, and $80 for food. Her dog had 8 healthy puppies which she sold for $400.00 each. She paid $25.00 for each puppy for shots and $60.00 for puppy chow, and $25.00 for a heat lamp for the dog house. The ad in the paper cost $12.00/week, and she was happy they all sold in 1 week. After she paid back her parents for all the expenses, how much profit did she make on the puppies?

3. Flores County High School's band needed new uniforms. They decided to sell candy bars to raise money. The candy bar company sent them 1,000 candy bars and a bill for $850.00. The band members sold the bars for $1.50 each. How much was their profit from candy bar sales when all of the bars were sold?

4. Raul and Luis were brothers who decided to go into business together mowing lawns. Their parents told them they could use the family truck but would need to buy their own equipment since the family lawn mower was barely working. They bought two lawn mowers for $150.00 each and had 300 flyers made for 12¢ each. They agreed to work together on every job and split the money evenly. After distributing flyers to all the homes around them, they waited for the calls. They soon had so much business they could hardly keep up. They charged $30 for $\frac{1}{4}$ acre lots and $60 for $\frac{1}{2}$ acre lots. They had to buy a trimmer for $30 and a hedge trimmer for $20. A receipt book and appointment book came to $12. How many $\frac{1}{4}$ acre lots will they have to cut to earn $500 each after they pay their expenses? They estimate they spent about $1.00 on gas for each lawn.

CHAPTER 2 REVIEW

Add.

1. $12.589 + 5.62 + 0.9$ _____
2. $7.8 + 10.24 + 1.903$ _____
3. $152.64 + 12.3 + 0.024$ _____

Subtract.

4. $18.547 - 9.62$ _____
5. $1.85 - 0.093$ _____
6. $45.2 - 37.9$ _____

Multiply.

7. 4.58×0.025 _____
8. 0.879×1.7 _____
9. 30.7×0.0041 _____

Divide.

10. $17.28 \div .054$ _____
11. $174.66 \div 1.23$ _____
12. $2.115 \div 9$ _____

Change to a fraction.

13. 0.55 _____
14. 0.84 _____
15. 0.32 _____

Change to a mixed number.

16. 7.375 _____
17. 9.6 _____
18. 13.25 _____

Change to a decimal.

19. $5\frac{3}{25}$ _____
20. $\frac{7}{100}$ _____
21. $10\frac{2}{3}$ _____

22. Super-X sells tires for $24.56 each. Save-Rite sells the identical tire for $21.97. How much can you save by purchasing a tire from Save-Rite?

_____

23. Gene works for his father sanding wooden rocking chairs. He earns $6.35 per chair. How many chairs does he need to sand in order to buy a portable radio/CD player for $146.05?

_____

24. Margo's Mint Shop has a machine that produces 4.35 pounds of mints per hour. How many pounds of mints are produced in each 8 hour shift?

_____

25. Carter's Junior High track team ran the first leg of a 400 meter relay race in 10.23 seconds, the second leg in 11.4 seconds, the third leg in 10.77 seconds, and the last leg in 9.9 seconds. How long did it take for them to complete the race?

_____

26. Spaulding High School decided to sell boxes of oranges to earn money for new football uniforms. They ordered a truckload of 500 boxes of oranges from a California grower for $16.00 per box. They sold 450 boxes for $19.00 per box. On the last day of the sale, they sold the oranges they had left for $17.00 per box. How much profit did they make?

_____

Chapter 3
Percents

CHANGING PERCENTS TO DECIMALS AND DECIMALS TO PERCENTS

To change a **percent** to a **decimal**, move the **decimal** point two places to the left, and drop the **percent** sign. If there is no decimal point written, it is after the number and before the percent sign. Sometimes you will need to add a "0". (See 5% below.)

EXAMPLES: 14% = 0.14   5% = 0.05   100% = 1   103% = 1.03
(decimal point)

Change the following percents to decimal numbers.

1. 18% = ____
2. 23% = ____
3. 9% = ____
4. 63% = ____
5. 4% = ____
6. 45% = ____
7. 2% = ____
8. 119% = ____
9. 2% = ____
10. 55% = ____
11. 80% = ____
12. 17% = ____
13. 66% = ____
14. 13% = ____
15. 5% = ____
16. 25% = ____
17. 410% = ____
18. 1% = ____
19. 50% = ____
20. 99% = ____
21. 107% = ____

To change a **decimal** number to a **percent**, move the **decimal** point two places to the right, and add a **percent** sign. You may need to add a "0". (See 0.8 below.)

EXAMPLES: 0.62 = 62%   0.07 = 7%   0.8 = 80%   0.166 = 16.6%   1.54 = 154%

Change the following decimal numbers to percents.

22. 0.15 = ____
23. 0.10 = ____
24. 1.53 = ____
25. 0.22 = ____
26. 0.35 = ____
27. 0.375 = ____
28. 0.648 = ____
29. 0.044 = ____
30. 0.58 = ____
31. 0.86 = ____
32. 0.29 = ____
33. 0.05 = ____
34. 0.48 = ____
35. 3.089 = ____
36. 0.042 = ____
37. 0.375 = ____
38. 5.09 = ____
39. 0.75 = ____
40. 0.3 = ____
41. 2.9 = ____
42. 0.06 = ____

CHANGING PERCENTS TO FRACTIONS AND FRACTIONS TO PERCENTS

EXAMPLE: Change 15% to a fraction.

Step 1: Copy the number without the percent sign. **15** is the top number of the fraction.

Step 2: The bottom number of the fraction is 100.

$$15\% = \frac{15}{100}$$

Step 3: Reduce the fraction. $\frac{15}{100} = \frac{3}{20}$

Change the following percents to fractions and reduce.

1. 50%
2. 13%
3. 22%
4. 95%
5. 52%
6. 63%
7. 75%
8. 91%
9. 18%
10. 3%
11. 25%
12. 5%
13. 16%
14. 1%
15. 79%
16. 40%
17. 99%
18. 30%
19. 15%
20. 84%

EXAMPLE: Change $\frac{7}{8}$ to a percent.

Step 1: Divide 7 by 8. Add as many 0's as necessary.

$$\begin{array}{r}.875\\8\overline{)7.000}\\-6\,4\\\hline 60\\-56\\\hline 40\\-40\\\hline 0\end{array}$$

Step 2: Change the decimal answer, .875, to a percent by moving the decimal point 2 places to the right.

$$\frac{7}{8} = .875 = 87.5\%$$

Change the following fractions to percents.

1. $\frac{1}{5}$
2. $\frac{5}{8}$
3. $\frac{7}{16}$
4. $\frac{3}{8}$
5. $\frac{3}{16}$
6. $\frac{19}{100}$
7. $\frac{1}{10}$
8. $\frac{4}{5}$
9. $\frac{15}{16}$
10. $\frac{3}{4}$
11. $\frac{1}{8}$
12. $\frac{5}{16}$
13. $\frac{1}{16}$
14. $\frac{1}{4}$
15. $\frac{4}{100}$
16. $\frac{3}{4}$
17. $\frac{2}{5}$
18. $\frac{16}{25}$

CHANGING PERCENTS TO MIXED NUMBERS AND MIXED NUMBERS TO PERCENTS

EXAMPLE: Change 218% to a fraction.

Step 1: Copy the number without the percent sign. **218** is the top number of the fraction.

Step 2: The bottom number of the fraction is **100**.

$$218\% = \frac{218}{100}$$

Step 3: Reduce the fraction, and convert to a mixed number. $\frac{218}{100} = \frac{109}{50} = 2\frac{9}{50}$

Change the following percents to mixed numbers.

1. 150%
2. 113%
3. 222%
4. 395%
5. 252%
6. 163%
7. 275%
8. 191%
9. 108%
10. 453%
11. 205%
12. 405%
13. 516%
14. 161%
15. 179%
16. 340%
17. 199%
18. 300%
19. 125%
20. 384%

EXAMPLE: Change $5\frac{3}{8}$ to a percent.

Step 1: Divide 3 by 8. Add as many 0's as necessary.

```
       .375
    8)3.000
     -2 4
       60
      -56
       40
      -40
        0
```

Step 2: So, $5\frac{3}{8} = 5.375$. Change the decimal answer to a percent by moving the decimal point 2 places to the right.

$$5\frac{3}{8} = 5.375 = 537.5\%$$

Change the following mixed numbers to percents.

1. $5\frac{1}{2}$
2. $8\frac{3}{4}$
3. $1\frac{5}{8}$
4. $3\frac{1}{4}$
5. $4\frac{7}{8}$
6. $2\frac{3}{100}$
7. $1\frac{3}{10}$
8. $6\frac{1}{5}$
9. $4\frac{7}{10}$
10. $2\frac{13}{25}$
11. $1\frac{1}{8}$
12. $2\frac{5}{16}$
13. $1\frac{3}{16}$
14. $1\frac{1}{16}$
15. $5\frac{17}{100}$
16. $4\frac{4}{5}$
17. $3\frac{2}{5}$
18. $2\frac{17}{100}$

REPRESENTING RATIONAL NUMBERS GRAPHICALLY

You now know how to convert fractions to decimals, decimals to fractions, fractions to percentages, percentages to fractions, decimals to percentages, and percentages to decimals. Study the examples below to understand how fractions, decimals, and percentages can be expressed graphically.

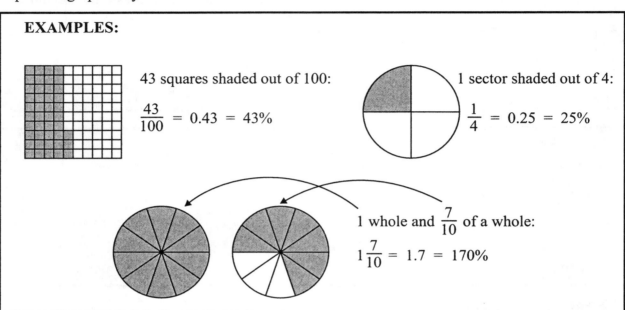

EXAMPLES:

43 squares shaded out of 100: $\frac{43}{100} = 0.43 = 43\%$

1 sector shaded out of 4: $\frac{1}{4} = 0.25 = 25\%$

1 whole and $\frac{7}{10}$ of a whole: $1\frac{7}{10} = 1.7 = 170\%$

Fill in the missing information in the chart below. Shade in the graphic for the problems that are not shaded for you. Reduce all fractions to lowest terms.

Graphic	Fraction	Decimal	Percent	Graphic	Fraction	Decimal	Percent
1.				5.			125%
2.		0.92		6.			
3.				7.			
4.	$\frac{4}{5}$			8.			

46

FINDING THE PERCENT OF THE TOTAL

EXAMPLE: A survey 330 students in the school lunchroom showed that 200 students preferred pizza to chili. What percent of the students preferred chili?

Step 1: First, find how many students preferred chili. Out of 330 students, 200 preferred pizza. That leaves 330 − 200 = 130 students who preferred chili.

Step 2: Divide the number of students who prefer chili by the total number of students.
$\frac{130}{330} = .39 = 39\%$

Answer the following questions.

1. Eighty-five percent of Mrs. Coomer's math class passed her final exam. There were 40 students in her class. How many passed?

2. Fifteen percent of a bag of chocolate candies have a red coating on them. How many red pieces are in a bag of 60 candies?

3. Sixty-eight percent of Valley Creek School students attended this year's homecoming dance. There are 675 students. How many attended the dance?

4. Out of the 4,500 people who attended the rock concert, forty-six percent purchased a T-shirt. How many people bought T-shirts?

5. Nina sold ninety-five percent of her 500 cookies at the bake sale. How many cookies did she sell?

6. Twelve percent of yesterday's customers purchased premium grade gasoline from GasCo. If GasCo had 200 customers, how many purchased premium grade gasoline?

7. The Candy Shack sold 138 pounds of candy on Tuesday. Fifty-two percent of the candy was jelly beans. How many pounds of jelly beans were sold Tuesday?

8. A fund-raiser at the school raised $617.50. Ninety-four percent went to local charities. How much money went to charities?

9. Out of the company's $6.5 million profit, eight percent will be paid to shareholders as dividends. How much will be paid out in dividends?

10. Ted's Toys sold seventy-five percent of its stock of stuffed bean animals on Saturday. If Ted's Toys had 620 originally in stock, how many were sold on Saturday?

TIPS AND COMMISSIONS

Vocabulary

Tip: A **tip** is money given to someone doing a service for you such as a server, hair stylist, porter, cab driver, grocery bagger, etc.

Commission: In many businesses, sales people are paid on **commission** - a percent of the total sales they make.

Problems requiring you to figure a tip, commission, or percent of a total are all done in the same way.

EXAMPLE: Ramon made a 4% commission on an $8,000 pickup truck he sold. How much was his commission?

```
        TOTAL COST          $8,000
   ×  RATE OF COMMISSION   ×  0.04
        COMMISSION          $320.00
```

Solve each of the following problems.

1. Mia makes 12% commission on all her sales. This month she sold $9,000 worth of merchandise. What was her commission?

2. Marcus gives 25% of his income to his parents to help cover expenses. He earns $340 per week. How much money does he give his parents?

3. Jan pays $640 per month for rent. If the rate of inflation is 5%, how much can Jan expect to pay monthly next year?

4. The total bill at Jake's Catfish Place came to $35.80. Palo wanted to leave a 15% tip. How much money will he leave for the tip?

5. Rami makes $2,400 per month and puts 6% in a savings plan. How much does he save per month?

6. Cristina makes $2,550 per month. Her boss promised her a 7% raise. How much more will she make per month?

7. Out of 150 math students, 86% passed. How many students passed math class?

8. Marta sells Sue Ann Cosmetics and gets 20% commission on all her sales. Last month, she sold $560.00 worth of cosmetics. How much was her commission?

FINDING THE AMOUNT OF DISCOUNT

Sale prices are sometimes marked 30% off, or better yet, 50% off. A 30% **discount** means you will pay 30% less than the original price. How much money you save is also known as the amount of the **discount**. Read the example below to learn to figure the amount of a discount.

EXAMPLE: A $179.00 chair is on sale for 30% off. How much can I save if I buy it now?

Step 1: Change 30% to a decimal. 30% = .30

Step 2: Multiply the original price by the discount.

ORIGINAL PRICE	$179.00
× **% DISCOUNT**	× .30
SAVINGS	$ 53.70

Practice finding the amount of the discount. Round off answers to the nearest penny.

1. Tubby Tires is offering a 25% discount on tires purchased on Tuesday. How much can you save if you buy tires on Tuesday regularly priced at $225.00 any other day of the week? _____

2. The regular price for a garden rake is $10.97 at Sly's Super Store. This week, Sly is offering a 30% discount. How much is the discount on the rake? _____

3. Christine bought a sweater regularly priced at $26.80 with a coupon for 20% off any sweater. How much did she save? _____

4. The software that Myoshi needs for her computer is priced at $69.85. If she waits until a store offers it at 20% off, how much will she save? _____

5. Ty purchased jeans that were priced $23.97. He received a 15% employee discount. How much did he save? _____

6. The Bakery Company offers a 60% discount on all bread made the day before. How much can you save on a $2.40 loaf made today if you wait until tomorrow to buy it? _____

7. A furniture store advertises a 40% off liquidation sale on all items. How much would the discount be on a $2530 dining room set? _____

8. Sharta bought a $4.00 nail polish on sale for 30% off. What was the dollar amount of the discount? _____

9. How much is the discount on a $350 racing bike marked 15% off? _____

10. Raymond receives a 2% discount from his credit card company on all purchases made with the credit card. What is his discount on $1575.50 worth of purchases? _____

FINDING THE DISCOUNTED SALE PRICE

To find the discounted sale price, you must go one step further than shown on the previous page. Read the example below to learn how to figure **discount** prices.

EXAMPLE: A $74.00 chair is on sale for 25% off. How much can I save if I buy it now?

Step 1: Change 25% to a decimal.   25% = .25

Step 2: Multiply the original price by the discount.

ORIGINAL PRICE	$74.00
× % DISCOUNT	× .25
SAVINGS	$18.50

Step 3: Subtract the savings amount from the original price to find the sale price.

ORIGINAL PRICE	$74.00
− SAVINGS	− 18.50
SALE PRICE	$55.50

Figure the sale price of the items below. The first one is done for you.

ITEM	PRICE	% OFF	MULTIPLY	SUBTRACT	SALE PRICE
1. pen	$1.50	20%	1.50 × .2 = $0.30	1.50 − 0.30 = 1.20	$1.20
2. recliner	$325	25%			
3. juicer	$55	15%			
4. blanket	$14	10%			
5. earrings	$2.40	20%			
6. figurine	$8	15%			
7. boots	$159	35%			
8. calculator	$80	30%			
9. candle	$6.20	50%			
10. camera	$445	20%			
11. VCR	$235	25%			
12. video game	$25	10%			

SALES TAX

EXAMPLE: The total price of a sofa is $560.00 + 6% **sales tax**. How much is the sales tax? What is the total cost?

Step 1: You will need to change 6% to a decimal.   6% = .06

Step 2: Simply multiply the cost, $560, by the tax rate, 6%.   560 × .06 = 33.6
The answer will be $33.60. (You need to add a 0 to the answer. When dealing with money, there needs to be two places after the decimal point.)

```
      COST       $560
   × 6% TAX      × .06
   SALES TAX    $33.60
```

Step 3: Add the sales tax amount, $33.60, to the cost of the item sold, $560. This is the total cost.

```
      COST       $560.00
   SALES TAX    + 33.60
   TOTAL COST   $593.60
```

NOTE: When the answer to the question involves money, you always need to round off the answer to the nearest hundredth (2 places after the decimal point). Sometimes you will need to add a zero.

Figure the total costs in the problems below. The first one is done for you.

ITEM	PRICE	% SALES TAX	MULTIPLY	ADD PRICE PLUS TAX	TOTAL
1. jeans	$42	7%	$42 × 0.07 = $2.94	42 + 2.94 = 44.94	$44.94
2. truck	$17,495	6%			
3. film	$5.89	8%			
4. T-shirt	$12	5%			
5. football	$36.40	4%			
6. soda	$1.78	5%			
7. 4 tires	$105.80	10%			
8. clock	$18	6%			
9. burger	$2.34	5%			
10. software	$89.95	8%			

FINDING THE PERCENT

EXAMPLE: 15 is what **percent** of 60?

Step 1: To solve these problems, simply divide the smaller amount by the larger amount. You will need to add a decimal point and two 0's.

$$\begin{array}{r} .25 \\ 60\overline{)15.00} \\ -120 \\ \hline 300 \\ -300 \\ \hline 0 \end{array}$$

Step 2: Change the answer, .25, to a percent by moving the decimal point two places to the right.

$$.25 = 25\% \quad 15 \text{ is } 25\% \text{ of } 60.$$

Remember: To change a decimal to a percent, you will sometimes have to add a zero when moving the decimal point two places to the right.

Find the following percents.

1. What percent of 50 is 16?
2. 20 is what percent of 80?
3. 9 is what percent of 100?
4. 19 is what percent of 95?
5. Ruth made 200 cookies for the picnic. Only 25 were left at the end of the day. What percent of the cookies was left?
6. Asad made 116 bird houses to sell at the county fair. The first day he sold 29. What percent of the bird houses did he sell?
7. Eileen planted 90 sweet corn seeds, but only 18 plants came up. What percent of the seeds germinated?
8. Tomika invests $36 of her $240 paycheck in a retirement account. What percent of her pay is she investing?
9. Ray sold a house for $115,000, and his commission was $9,200. What percent commission did he make?
10. Julio was making $16.00 per hour. After one year, he received a $2.00 per hour raise. What percent of a raise did he get?
11. Calvin budgets $235 per month for food. If his salary is $940 per month, what percent of his salary does he budget for food?
12. Katie earned $45 on commission for her sales totaling $225. What percent was her commission?
13. Among the students taking band this year, 2 out of 5 are freshmen. What percent of the band are freshmen?
14. Of the donuts we have left to sell, the ratio of chocolate donuts to non-chocolate is 3 to 5. What percent of the donuts left to sell are chocolate?
15. The school bought 340 new history books for 400 students. What percent of the students got new history books?
16. Of the 48 dogs enrolled in obedience school, 36 successfully completed training. What percent of the dogs completed training?

FINDING THE PERCENT INCREASE AND DECREASE

EXAMPLE 1: Office Supply Co. purchased paper wholesale for $18.00 per case. They sold the paper for $20.00 per case. By what percent did the store increase the price of the paper (or what is the percent markup)?

$$\text{Percent change} = \frac{\text{Amount of change}}{\text{Original Amount}}$$

Step 1: Find the amount of change. In this problem, the price was marked up $2.00. The amount of change is 2.

Step 2: Divide the amount of change, 2, by the wholesale cost, 18. $\frac{2}{18} = .111$

Step 3: Change the decimal, .111, to a percent. .111 = 11.1%

EXAMPLE 2: The price of gas went from $2.40 per gallon to $1.30. What is the percent of decrease in the price of gas?

$$\text{Percent change} = \frac{\text{Amount of change}}{\text{Original Amount}}$$

Step 1: Find the amount of change. In this problem, the price decreased $1.10 The amount of change is 1.10.

Step 2: Divide the amount of change, 1.10, by the original cost, $2.40. $\frac{1.10}{2.40} = .46$

Step 3: Change the decimal, .46, to a percent. .46 = 46% The price of gas has decreased 46%.

Find the percent increase or decrease for each of the problems below.

1. Mary was making $25,000 per year. Her boss gave her a $3,000 raise. What percent increase is that?

2. Last week Matt's total sales were $12,000. This week his total sales were only $2,000. By what percent did his sales for this week decrease?

3. Emil was making $16.00 per hour. After one year, he received a $2.00 per hour raise. What percent raise did he get?

4. Sara owned an office supply store. She marked down pens from $1.50 to $1.20. What percent discount is that?

5. Rosa bought a clock marked $18.00. After sales tax, the total came to $19.08. What percent sales tax did she pay?

6. Cowboys bought boots wholesale for $103.35. They sold the boots in their store for $159. What percent was the markup on the boots?

7. Blakeville has a population of 1600. According to the last census, Blakeville had a population of 1850. What has been the percent decrease in population?

8. Last year, Roswell High School had 680 graduates. This year they graduated 812. What has been the percent increase in graduates?

9. Michi got a new job that pays $52,000 per year. That is $16,000 more than his last job. What percent pay increase is that?

UNDERSTANDING SIMPLE INTEREST

I = PRT is a formula to figure out the **cost of borrowing money** or the **amount you earn** when you **put money in a savings account**. For example, when you want to buy a used truck or car, you go to the bank to borrow the $7,000 you need. The bank will charge you **interest** on the $7,000. If the simple interest rate is 9% for four years, you can figure the cost of the interest with this formula.

First, you need to understand these terms:

> **I** = Interest = The amount charged by the bank or other lender
> **P** = Principal = The amount you borrow
> **R** = Rate = The interest rate the bank is charging you
> **T** = Time = How many years you will take to pay off the loan

EXAMPLE:

> In the problem above: **I = PRT**   This means the **interest** equals the **principal**, times the **rate**, times the **time** in **years**.
>
> I = $7,000 × 9% × 4 years
> I = $7,000 × .09 × 4
> I = $2,520

Use the formula I = PRT to work the following problems:

1. Craig borrowed $1,800 from his parents to buy a stereo. His parents charged him 3% simple interest for 2 years. How much interest did he pay his parents? _____

2. Raul invested $5,000 in a savings account that earned 2% simple interest. If he kept the money in the account for 5 years, how much interest did he earn? _____

3. Bridgette borrowed $11,000 to buy a car. The bank charged 12% simple interest for 7 years. How much interest did she pay the bank? _____

4. A tax accountant invested $25,000 in a money market account for 3 years. The account earned 5% simple interest. How much interest did the accountant make on his investment? _____

5. Linda Kay started a savings account for her nephew with $2,000. The account earned 6% simple interest. How much interest did the account accumulate in 3 years? _____

6. Renada bought a living room set on credit. The set sold for $2,300, and the store charged her 9% simple interest for one year. How much interest did she pay? _____

7. Duane took out a $3,500 loan at 8% simple interest for 3 years. How much interest did he pay for borrowing the $3,500? _____

COMPOUND INTEREST

Vocabulary: **Semiannually** - twice a year or every 6 months
Quarterly - four times a year or every 3 months
Principal - the amount of money you borrow or put in a bank
Rate - the interest rate you pay or receive on your principal
Time - the number of years or parts of a year you will take to pay off the loan.

EXAMPLE: Beth put $1,000 in a savings account that paid 6% compounded semiannually. How much money did she have at the end of a year?

Step 1: Using the I = PRT formula, figure how much Beth had at the end of 6 months.
Principal = $1,000
Rate = 6%
Time = 6 months

$I = \$1,000 \times .06 \times \frac{6}{12}$
I = $30 interest she gained in 6 months

Step 2: Add the $30 she received as interest to the principal she had to start with.
$1,000 + 30 = $1,030.
Now using I=PRT, find the interest she earned for the next six months.
Principal = $1,030
Rate = 6%
Time = 6 months

$I = \$1,030 \times .06 \times \frac{6}{12}$
I = $30.90 interest she gained in 6 months

Step 3: Add interest she gained in the last 6 months to the principal she had at the beginning of the last 6 months.
$1,030 + $30.90 = $1,060.90, amount she had at the end of the year.

Complete the chart below with the missing information.

	Beginning Principal	Interest Rate	Compounded	Duration	Ending Balance
1.	$1,000	9%	Quarterly	Six Months	
2.	$2,500	5%	Semiannually	One Year	
3.	$1,700	6%	Quarterly	Six Months	
4.	$3,200	11%	Quarterly	Six Months	
5.	$6,300	10%	Semiannually	One Year	
6.	$3,700	2%	Monthly	Two Months	
7.	$5,200	3%	Quarterly	Six Months	
8.	$7,500	8%	Yearly	Two Years	
9.	$4,250	4%	Semiannually	One Year	
10.	$5,900	1%	Monthly	Two Months	

CHAPTER 3 REVIEW

Change the following percents to decimals.

1. 45% _____
2. 219% _____
3. 22% _____
4. 1.25% _____

Change the following decimals to percents.

5. 0.52 _____
6. 0.64 _____
7. 1.09 _____
8. 0.625 _____

Change the following percents to fractions.

9. 25% _____
10. 3% _____
11. 68% _____
12. 102% _____

Change the following fractions to percents.

13. $\frac{9}{10}$ _____
14. $\frac{5}{16}$ _____
15. $\frac{1}{8}$ _____
16. $\frac{1}{4}$ _____

17. What is 1.65 written as a percent?

   _____

18. What is $2\frac{1}{4}$ written as a percent?

   _____

19. Change 5.65 to a percent.

   _____

Fill in the equivalent numbers represented by the shaded area.

20. fraction _____
21. decimal _____
22. percent _____

Fill in the equivalent numbers represented by the shaded area.

23. fraction _____
24. decimal _____
25. percent _____

26. Uncle Howard left his only niece 56% of his assets according to his will. If his assets totaled $564,000 when he died, how much did his niece inherit?

   _____

27. Celeste makes 6% commission on her sales. If her sales for a week total $4580, what is her commission?

   _____

28. Peeler's Jewelry is offering a 30% off sale on all bracelets. How much will you save if you buy a $45.00 bracelet during the sale?

   _____

29. How much would an employee pay for a $724.00 stereo if the employee got a 15% discount?

30. Misha bought a CD for $14.95. If sales tax was 7%, how much did she pay total?

31. The Pep band made $640 during a fund-raiser. The band spent $400 of the money on new uniforms. What percent of the total did the band members spend on uniforms?

32. Linda took out a simple interest loan for $7,000 at 11% interest for 5 years. How much interest did she have to pay back?

33. McMartin's is offering a deal on fitness club memberships. You can pay $999 up front for a 3 year membership, or pay $200 down and $30 per month for 36 months. How much would you save by paying up front?

34. Patton, Patton, and Clark, a law firm, won a malpractice law suit for $4,500,000. Sixty-eight percent went to the law firm. How much did the law firm make?

35. Jeneane earned $340.20 commission by selling $5670 worth of products. What percent commission did she earn?

36. Tara put $500 in a savings account that earned 3% simple interest. How much interest did she make after 5 years?

37. Ms. Clark put $3,000 in an account that compounded 5% interest twice a year. If she left the money in for 1 year, how much would she have?

38. Hank got 10 miles to the gallon in his vintage 1966 Mustang. After getting a new carburetor installed in his car, he got 16 miles to the gallon. What percent increase in mileage did he get?

39. A department store is selling all swimsuits for 40% off in August. How much would you pay for a swimsuit that is normally priced at $35.80?

40. High school students voted on where they would go on a field trip. For every 3 students who wanted to see Calaveras Big Trees State Park, 8 students wanted to see Columbia State Historic Park. What percent of the students wanted to go to Columbia State Historic Park?

41. An increase from 20 to 36 is what percent of increase?

42. An auto parts store buys air filters for $12.00 each. The store sells air filters for $18.60 each. What is the percent markup on the air filters?

Chapter 4
Problem Solving and Critical Thinking

MISSING INFORMATION

Problems can only be solved if you are given enough information. Sometimes you are not given enough information to solve a problem.

EXAMPLE: Chuck has worked on his job for 1 year now. At the end of a year, his employer gave him a 12% raise. How much does Chuck make now?

To solve this problem, you need to know how much Chuck made when he began his job one year ago.

Each problem below does not give enough information for you to solve it. Beneath each problem, describe the information you would need to solve the problem.

1. Fourteen percent of the coated chocolate candies in Amin's bag were yellow. At that rate, how many of the candies were yellow?

2. Patrick is putting up a fence around all four sides of his back yard. The fence costs $2.25 per foot, and his yard is 150 feet wide. How much will the fence cost?

3. Yoko worked 5 days last week. She made $6.75 per hour before taxes. What was her total earnings before taxes were taken out?

4. Which is a better buy: a 4 oz. bar of soap for 88¢ or a bath bar for $1.20?

5. Randy bought a used car for $4,568 plus sales tax. What was the total cost of the car?

6. The Portes family ate at a restaurant, and each of their dinners cost $5.95. They left a 15% tip. What was the total amount of the tip?

7. If a kudzu plant grows 3 feet per day, in what month will it be 90 feet long?

8. Bethany traveled by car to her sister's house in Kodiak Bear Lake. She traveled at an average speed of 52 miles per hour. She arrived at 4:00 p.m. How far did she travel?

9. Terrence earns $7.50 per hour plus 5% commission on total sales over $500 per day. Today he sold $6,500 worth of merchandise. How much did he earn for the day?

10. Michelle works at a department store and gets an employee's discount on all of her purchases. She wants to buy a sweater that sells for $38.00. How much will the sweater cost after her discount?

11. Matsu filled his car with 10 gallons of gas and paid for the gas with a $20 bill. How much change did he get back?

12. Olivia budgets $5.00 per work day for lunch. How much does she budget for lunch each month?

13. Joey worked 40 hours and was paid $356.00. His friend Pete worked 38 hours. Who was paid more per hour?

14. A train trip from Columbia to Boston took $18\frac{1}{4}$ hours. How many miles apart are the two cities?

15. Caleb spent 35% of his check on rent, 10% on groceries, and 18% on utilities. How much money did he have left from his check?

16. The Lyons family spent $54.00 per day plus tax on lodging during their vacation. How much tax did they pay for lodging per day?

17. Ricardo bought cologne at a 30% off sale. How much did he save buying the cologne on sale?

18. The bottling machine works 7 days a week and fills 1,000 bottles per hour. How many bottles did it fill last week?

19. Tyler, who works strictly on commission, brought in $25,000 worth of sales in the last 10 days. How much was his commission?

20. Ninety percent of the student body at Parks Middle School bought raffle tickets to help the basketball team buy new uniforms. The main prize was a 25 inch color TV with built-in VCR. How many students bought raffle tickets?

EXACT INFORMATION

Most word problems supply exact information and ask for exact answers. The following problems are the same as those on the previous two pages with the missing information given. Find the exact solution.

1. Fourteen percent of the coated chocolate candies in Amin's bag were yellow. If there were 50 pieces in the bag, how many of the candies were yellow?

2. Patrick is putting up a fence around all four sides of his back yard. The fence costs $2.25 per foot. His yard is 150 feet wide and 200 feet long. How much will the fence cost?

3. Yoko worked 5 days last week, 8 hours each day. She made $6.75 per hour before taxes. How much did she make last week before taxes were taken out?

4. Which is a better buy: a 4 oz. bar of soap for 88¢ or a 6 oz. bath bar for $1.20?

5. Randy bought a used car for $4,568 plus 6% sales tax. What was the total cost of the car?

6. The Portes family ate at a restaurant, and each of the 4 dinners cost $5.95. They left a 15% tip. What was the total amount of the tip?

7. Kudzu is a rapid-growing vine found in southeastern states of the U.S. If a kudzu plant grows 3 feet per day, in what month will it be 90 feet long if it takes root in the middle of May?

8. Bethany traveled from Fairmont by car to her sister's house in Moorhead. She traveled at an average speed of 52 miles per hour. She left at 10:00 a.m. and arrived at 4:00 p.m. How far did she travel?

9. Terrence earns $7.50 per hour plus 5% commission on total sales over $500 per day. Today he sold $6,500 worth of merchandise and worked 7 hours. How much did he earn for the day?

10. Michelle works at a department store and gets a 20% employee's discount on all of her purchases. She wants to buy a sweater that sells for $38.00. How much will the sweater cost after her discount?

11. Matsu filled his car with 10 gallons of gas priced at $1.24 per gallon. He paid for the gas with a $20 bill. How much change did he get back?

12. Olivia budgets $5.00 per work day for lunch. How much does she budget for lunches if she works 21 days this month?

13. Joey worked 40 hours and was paid $356.00. His friend Pete worked 38 hours at $8.70 per hour. Who was paid more per hour?

14. A train trip from Columbia, SC to Boston, MA takes $18\frac{1}{4}$ hours. How many miles apart are the two cities if the train travels at an average speed of 50 miles per hour?

15. Caleb spent 35% of his check on rent, 10% on groceries, and 18% on utilities. How much money did he have left from his $260 check?

16. The Lyons family spent $54.00 per day plus 10% tax on lodging during their vacation. How much tax did they pay per day?

17. Ricardo bought cologne at a 30% off sale. The cologne was regularly priced at $44. How much did he save buying the cologne on sale?

18. The bottling machine works 7 days a week, 14 hours per day and fills 1,000 bottles per hour. How many bottles did it fill last week?

19. Tyler, who works strictly on commission, brought in $25,000 worth of sales in the last 10 days. He earns 15% commission on his sales. How much was his commission?

20. Ninety percent of the total 540 students at Parks High School bought raffle tickets to help the basketball team buy new uniforms. How many students bought raffle tickets?

EXTRA INFORMATION

In each of the following problems, there is extra information given. **Look closely at the question,** and use only the information you need to answer it.

EXAMPLE: Gary was making $6.50 per hour. His boss gave him a 52¢ per hour raise. Gary works 40 hours per week. What percent raise did Gary receive?

Solution: To figure the percent of Gary's raise, you do **not** need to know how many hours per week Gary works. That is extra information not needed to answer the question. To figure the percent increase, simply divide the change in pay, $0.52, by the original wages, $6.50. 0.52 ÷ 6.50 = 0.08

Gary received an 8% raise.

In the following questions, determine what information is needed from the problem to answer the question, and solve.

1. Leah wants a new sound system that is on sale for 15% off the regular price of $420. She has already saved $325 toward the cost. What is the dollar amount of the discount?

2. Praveen bought a shirt for $34.80 and socks for $11.25. He gets $10.00 per week for his allowance. He paid $2.76 sales tax. What was his change from three $20 bills?

3. Marty worked 38 hours this week, and he earned $8.40 per hour. His taxes and insurance deductions amount to 34% of his gross pay. What is his total gross pay?

4. Tamika went shopping and spent $4.80 for lunch. She wants to buy a sweater that is on sale for $\frac{1}{4}$ off the regular $56.00 price. How much will she save?

5. Nick drove an average of 52 miles per hour for 7 hours. His car gets 32 miles per gallon. How far did he travel?

6. The odometer on Melody's car read 45,920 at the beginning of her trip and 46,460 at the end of her trip. Her speed averaged 54 miles per hour, and she used 20 gallons of gasoline. How many miles per gallon did she average?

7. Eighty percent of the eighth graders attended the end-of-the-year class picnic. There are 160 eighth graders and 54% of them ride the bus to school each day. How many students went to the class picnic?

8. Matt has $5.00 to spend on snacks. Tastee Potato Chips cost $2.57 for a one-pound bag at the grocery store. T-Mart sells the same bag of chips for $1.98. How much can he save if he buys the chips at T-Mart?

9. Elaina wanted to make 10 cakes for the band bake sale. She needed $1\frac{3}{4}$ cups of flour and $2\frac{1}{4}$ cups of sugar for each cake. How many cups of flour did she need in all?

ESTIMATED SOLUTIONS

Some problems require an estimated solution. In order to have enough product to complete a job, you often must buy more materials than you actually need. **In the following problems, be sure to round your answer up to the next whole number to find the correct solution.**

1. Endicott Publishing received an order for 550 books. Each shipping box holds 30 books. How many boxes do the packers need to ship the order?

2. Elena's 250 chickens laid 314 eggs in the last 2 days. How many egg cartons holding one dozen eggs would be needed to hold all the eggs?

3. Antoinetta's Italian restaurant uses $1\frac{1}{4}$ quarts of olive oil every day. The restaurant is open 7 days a week. For the month of September, how many gallons should the cooks order to have enough?

4. Eastmont High School is taking 316 students and 22 chaperones on a field trip. Each bus holds 44 persons. How many buses will the school need?

5. Fran volunteered to hem 11 choir robes that came in too long. Each robe is 7 feet around at the bottom. Hemming tape comes three yards to a pack. How many packs will Fran need to buy to go around all the robes?

6. Tonya is making matching vests for the children's choir. Each vest has 5 buttons on it, and there are 23 children in the choir. The button she picked comes 6 buttons to a card. How many cards of buttons does she need?

7. Tiffany is making the bread for the banquet. She needs to make 6 batches with $2\frac{1}{4}$ lb. of flour in each batch. How many 10 lb. bags of flour will she need to buy?

8. The homeless shelter is distributing 250 sandwiches per day to hungry guests. It takes one foot of plastic wrap to wrap each sandwich. There are 150 feet of plastic wrap per box. How many boxes will Mary need to buy to have enough plastic wrap for the week?

9. An advertising company has 15 different kinds of one-page flyers. The company needs 75 copies of each kind of flyer. How many reams of paper will the company need to produce the flyers? One ream equals 500 sheets of paper.

TWO-STEP PROBLEMS

Some problems require two steps to solve.

Read each of the following problems carefully and solve.

1. For a family picnic, Renee bought 10 pounds of hamburger meat and used $\frac{1}{4}$ of a pound of meat to make each hamburger patty. Renee's family ate 32 hamburgers. How many pounds of hamburger meat did she have left?

2. Vic sold 45 raffle tickets. His brother sold twice as many. How many tickets did they sell together?

3. Erin earns $2,200 per month. Her deductions amount to 28% of her paycheck. How much does she take home each month?

4. Matheson Middle School Band is selling T-shirts to raise money for new uniforms. They need to raise $1260. They are selling T-shirts for $12 each. There is a $6 profit for each shirt sold. So far, they have sold 85 T-shirts. How many more T-shirts do they need to sell to raise the $1260?

5. Alphonso was earning $1,860 per month and then got a 12% raise. How much will he make per month now?

6. Barbara and Jeff ate out for dinner. The total came to $15.00. They left a 15% tip. How much was the tip and the meal together?

7. Hillary is bicycling across Montana, taking an 845 mile course. The first week she covered 320 miles. The second week she traveled another 350 miles. How many more miles does she have to travel to complete the course?

8. Jason budgets 30% of his $1,100 income each month for food. How much money does he have to spend for everything else?

9. After Madison makes a 12% down payment on a $2,000 motorcycle, how much will she still owe?

10. Omar bought a pair of shoes for $51, a tie for $18, and a new belt for $23. If the sales tax is 8%, how much sales tax did he pay?

TIME OF TRAVEL

EXAMPLE: Katrina drove 384 miles at an average of 64 miles per hour. How many hours did she travel?

Solution: Divide the number of miles by the miles per hour. $\dfrac{384 \text{ miles}}{64 \text{ miles/hour}} = 6 \text{ hours}$

Katrina traveled 6 hours.

Find the hours of travel in each problem below.

1. Bobbi drove 342 miles at an average speed of 57 miles per hour. How many hours did she drive? _____

2. Jan set her speed control at 55 miles per hour and drove for 165 miles. How many hours did she drive? _____

3. John traveled 2,092 miles in a jet that flew an average of 523 miles per hour. How long was he in the air? _____

4. How long will it take a bus averaging 54 miles per hour to travel 378 miles? _____

5. Kyle drove his motorcycle in a 225 mile race, and he averaged 75 miles per hour. How long did it take for him to complete the race? _____

6. Stacy drove 576 miles at an average speed of 48 miles an hour. How many hours did she drive? _____

7. Kendra flew 250 miles in a glider and averaged 125 miles per hour in speed. How many hours did she fly? _____

8. Travis traveled 496 miles at an average speed of 62 miles per hour. How long did he travel? _____

9. Wanda rode her bicycle an average of 15 miles an hour for 60 miles. How many hours did she ride? _____

10. Rami drove 184 miles at an average speed of 46 miles per hour. How many hours did he drive? _____

11. A train traveled at a constant 85 miles per hour for 425 miles. How many hours did the train travel? _____

12. How long was Amy on the road if she drove 195 miles at an average of 65 miles per hour? _____

Copyright © American Book Company

RATE

EXAMPLE: Laurie traveled 312 miles in 6 hours. What was her average rate of speed?

Solution: Divide the number of miles by the number of hours. $\frac{312 \text{ miles}}{6 \text{ hours}} = 52$ miles/hour

Laurie's average rate of speed was 52 miles per hour (or 52 mph).

Find the average rate of speed in each problem below.

1. A race car went 500 miles in 4 hours. What was its average rate of speed?

2. Carrie drove 124 miles in 2 hours. What was her average speed?

3. After 7 hours of driving, Chad had gone 364 miles. What was his average speed?

4. Anna drove 360 miles in 8 hours. What was her average speed?

5. After 3 hours of driving, Paul had gone 183 miles. What was his average speed?

6. Nicole ran 25 miles in 5 hours. What was her average speed?

7. A train traveled 492 miles in 6 hours. What was its average rate of speed?

8. A commercial jet traveled 1,572 miles in 3 hours. What was its average speed?

9. Jillian drove 195 miles in 3 hours. What was her average speed?

10. Esteban drove 8 hours from his home to a city 336 miles away. At what average speed did he travel?

11. Hammad drove 64 miles in one hour. What was his average speed in miles per hour?

12. After 9 hours of driving, Kate had traveled 405 miles. What speed did she average?

MORE RATES

On the previous page, rate was discussed in terms of miles per hour. Rate can be any measured quantity divided by another measurement such as feet per second, kilometers per minute, mass per unit volume, etc. A rate can be how fast something is done. For example, a bricklayer may lay 80 bricks per hour. Rates can also be used to find measurement such as density. For example, 35 grams of salt in 1 liter of water gives the mixture a density of 35 grams/liter.

EXAMPLE: Nathan entered his snail in a race. His snail went 18 feet in 6 minutes. How fast did his snail move?

Solution: In this problem, the units given are feet and minutes, so the rate will be feet per minute (or feet/minute). You need to find out how far the snail went in one minute.

$$\text{Rate equals } \frac{\text{distance}}{\text{time}} \text{ so } \frac{18 \text{ feet}}{6 \text{ minutes}} = \frac{3 \text{ feet}}{1 \text{ minute}}$$

Nathan's snail went an average of 3 feet per minute or 3 ft./min.

Find the average rate for each of the following problems.

1. Tewanda read a 2000 word news article in 8 minutes. How fast did she read the news article?

2. Mr. Molier is figuring out the semester averages for his history students. He can figure the average for 20 students in an hour. How long does it take him to figure the average for each student?

3. In 1908, John Hurlinger of Austria walked 1400 kilometers from Vienna to Paris on his hands. The journey took 55 days. What was his average speed per day?

4. Nathan added 60 grams of sugar to a 12 cup container of coffee. What is the density of the sugar-coffee mixture?

5. Marcus Page, star receiver for the Big Bulls, was awarded a 5 year contract for 105 million dollars. How much will his annual rate of pay be if he is paid the same amount each year?

6. The new McDonald's in Moscow serves 11,208 customers during a 24 hour period. What is the average number of customers served per hour?

7. Spectators at the Super Circus were amazed to watch a canon shoot a clown 212 feet into a net in 4 seconds. How many feet per second did the clown travel?

8. Duke Delaney scored 28 points during the 4 quarters of the basketball playoffs. What was his average score per quarter?

PRODUCT MEASUREMENTS

Similar to rate problems, measurements can also be expressed as **products**. A product measurement could be the amount of labor and time needed to accomplish a task. For example, it takes 10 person·hours to unload a truck. This means that the **number of people** multiplied by the **number of hours** must equal 10 to complete the task.

EXAMPLE 1: It takes 12 person·hours to assemble a certain piece of equipment. If 3 people work on assembling the equipment, how long will it take to complete the task?

Solution: $\dfrac{12 \text{ person·hours}}{3 \text{ people}} = 4 \text{ hours}$

It will take 3 people 4 hours to assemble the equipment.

EXAMPLE 2: Jana has a printing job that will require 20 press·hours to print. How many presses would she have to use to complete the printing in 5 hours?

Solution: $\dfrac{20 \text{ press·hours}}{5 \text{ hours}} = 4 \text{ presses}$

It will take 4 presses to print the job in 5 hours.

Work the following problems involving product measurements.

1. Chandler wants to mow three large lawns. He estimates it will take 45 person·hours to complete the project. If Chandler has two friends helping him, how many hours will be required to mow the lawns?

2. It once took 400,000 worker·years to build a pyramid in Egypt. If the pharaoh had 20,000 workers, in how many years could a pyramid be built?

3. It takes 45 computer·minutes to decipher a code. If 5 computers worked together to decipher the code, how long would it take?

4. An instruction manual takes 135 copier·seconds to reproduce. How many copier·seconds will it take to make 10 copies of the manual?

5. How much time would it take for 2 copy machines to reproduce the 10 manuals in the previous problem?

6. Your teacher gives your group a science project to complete. She says it should take 15 student·days to complete the project. If you have 3 people in your group, how fast should you be able to complete the project?

DISTANCE

EXAMPLE: Jessie traveled for 7 hours at an average rate of 58 miles per hour. How far did she travel?

Solution: Multiply the number of hours by the average rate of speed.
7 hours × 58 miles/hour = 406 miles

Find the distance in each of the following problems.

1. Miyoko traveled for 9 hours at an average rate of 45 miles per hour. How far did she travel?

2. A tour bus drove 4 hours, averaging 58 miles per hour. How many miles did it travel?

3. Tina drove for 7 hours at an average speed of 53 miles per hour. How far did she travel?

4. Dustin raced for 3 hours, averaging 176 miles per hour. How many miles did he race?

5. Kris drove 5 hours and averaged 49 miles per hour. How far did she travel?

6. Oliver drove at an average of 93 miles per hour for 3 hours. How far did he travel?

7. A commercial airplane traveled 514 miles per hour for 2 hours. How far did it fly?

8. A train traveled at 125 miles per hour for 4 hours. How many miles did it travel?

9. Carl drove a constant 65 miles an hour for 3 hours. How many miles did he drive?

10. Jasmine drove for 5 hours, averaging 40 miles per hour. How many miles did she drive?

11. Javier flew his glider for 2 hours at 87 miles per hour. How many miles did his glider fly?

12. Beth traveled at a constant 65 miles per hour for 4 hours. How far did she travel?

MILES PER GALLON

EXAMPLE: The odometer on Ginger's car read 46,789 before she started her trip. At the end of her trip, it read 47,119. She used 10 gallons of gasoline. How many miles per gallon did she average?

Step 1: Subtract the ending odometer reading from the beginning odometer reading.
47,119 − 46,789 = 330 miles traveled

Step 2: Divide the number of miles traveled by the number of gallons of gasoline used.
$\dfrac{330 \text{ miles}}{10 \text{ gallons}} = 33$ miles per gallon

Compute the number of miles per gallon in each of the following problems.

1. Rachael's odometer read 125,625 at the beginning of her trip. At the end of her trip, it read 125,863. She used 7 gallons of gasoline. How many miles per gallon did she average? _____

2. Tamera traveled 492 miles on 12 gallons of gasoline. What was her average gas mileage? _____

3. The odometer on Blake's car read 3,975 before she started her trip. At the end of her trip, it read 4,625. She used 26 gallons of gasoline. How many miles per gallon did she average? _____

4. Farmer Joe's tractor odometer read 218,754 before tilling his fields. After tilling, it read 218,802. He used 4 gallons of gasoline. How many miles per gallon did his tractor average? _____

5. When Devyn started the week, the odometer on his van read 64,742. At the end of the week, it read 64,984. He used 11 gallons of gasoline. How many miles per gallon did he average? _____

6. Kathryn drove 364 miles on 13 gallons of gasoline. What was her gas mileage? _____

7. Manny drove his jeep to the Pacific Ocean, towing his boat. His odometer read 23,745 before his trip, and it read 24,030 once he arrived at the ocean. His jeep used 15 gallons of gas. How many miles per gallon did he average? _____

8. Bonnie's odometer read 17,846 before she drove to visit her aunt. Once she arrived at her aunt's house, her odometer read 18,726. She used 22 gallons of gas. What was her average gas mileage? _____

9. Ron traveled 74 miles on 2 gallons of gas. How many miles per gallon did he average? _____

10. Before Janet and Bill left for their vacation, their odometer read 87,985. When they arrived back home, their odometer read 88,753. They used 24 gallons of gasoline. What was their average gas mileage? _____

CHAPTER 4 REVIEW

Read each problem carefully and solve using the problem-solving methods you learned in this chapter. If there is not enough information, tell what information is missing.

1. East Point Middle School is taking 321 students, 11 teachers, and 10 parents on a field trip. Each bus holds 45 persons. How many busses will the school need?

2. If six workers are constructing a wall which takes approximately 30 worker·days to complete, how many days will they require?

3. Tyrone bought shoes for $65 and a shirt for $24. He paid 8% sales tax. What was the total cost of the two items?

4. Crystal bought a jacket at a 20% off sale. How much did she save buying the jacket on sale?

5. Laura left her house at 8:30 a.m. in Los Angeles and drove 5 hours to see her brother in Needles. Her trip from Los Angeles to Needles totaled 265 miles. What was her average speed in miles per hour?

6. Someone can drive a vehicle 600 miles on two tanks of gas. If each tank holds 15 gallons, how many miles per gallon does this vehicle get?

7. Li drove 560 miles at an average speed of 70 miles per hour. How many hours did Li drive?

8. Tran worked 4 days last week. She made $6.26 per hour. Her employer deducted 24% of her pay for taxes. How much was Tran's take-home pay?

9. Lori bought a sweater originally marked $32.50 and a belt for $12.25. The sweater was on sale for 20% off. How much money did she save by buying the sweater on sale?

10. Northside High School ordered 556 algebra books from the Galactic Math Company. Thirty-two books fit in a box. How many boxes were needed to ship the order?

11. Cecilia won 1,390 tickets in the arcade. A stuffed bear costs 434 tickets, a stop watch costs 1,650, and a clown wig costs 1,240 tickets. How many more tickets does Cecilia need to win to purchase the stop watch and the stuffed bear?

12. Mrs. Rhodes gave each of her 35 second grade students a Valentine's Day card. The cards that she picked out came 12 to a box. How many boxes did she have to purchase?

13. Seth left his home at 2:00 p.m. He arrived at the airport at 3:30 p.m., $1\frac{1}{2}$ hours before his plane departed. What was Seth's average speed traveling to the airport?

14. Cynthia and her friends, Dwayne and Raquel, worked together on a patchwork quilt in the evenings after their interior decorating class. If the quilt takes approximately 54 worker·hours to complete, how many hours will they need to finish the quilt?

    _____

15. How many people would be required to complete a 42 worker·hour project in three hours?

    _____

16. How much salt is present in a 5 liter container of salt water which has a density of 40 grams/liter?

    _____

17. A vehicle reads 17,804 miles on the odometer at the beginning of a trip on a full tank of gas. When the mileage reads 18,084 miles, the car is filled up again with 8 gallons. How many miles per gallon does this vehicle get?

    _____

18. A powerboat traveled for 5 hours at 48 miles per hour. How far did it move on the water?

    _____

19. A cat, chasing a rabbit, chases at an average rate of 12 miles per hour for three hours. How far did the cat move in its hunt?

    _____

20. David found a motorcycle he wants to buy that is priced at $4,800. If he waits to buy it at the end of the year, it will be on sale for 20% off. How much will it cost then?

    _____

21. Movie tickets sell for $7.50 each. On Wednesday nights, tickets are 30% off the regular price. How much would two tickets cost on Wednesday night?

    _____

22. A living room set can be purchased for $1200 or for $100 down and $120 a month for 12 months. How much less is the cash price than the installment plan?

    _____

23. Janelle drove an average of 50 miles per hour for 275 miles. How many hours did the trip last?

    _____

24. Floor tiles are sold only in boxes of 11 tiles per box. Christy needs 140 tiles for her kitchen floor. How many boxes of tile will she need to buy?

    _____

25. Kyle works 40 hours per week for $11.20 per hour. His total deductions come to $37.80. What is his take-home pay?

    _____

Chapter 5
Integers and Order of Operations

INTEGERS

In elementary school, you learned to use whole numbers.

Whole numbers = { 0, 1, 2, 3, 4, 5 . . . }

For most things in life, whole numbers are all we need to use. However, when a checking account falls below zero or the temperature falls below zero, we need a way to express that. Mathematicians have decided that a negative sign, which looks exactly like a subtraction sign, would be used in front of a number to show that the number is below zero. All the negative whole numbers and positive whole numbers plus zero make up the set of integers.

Integers = { . . . –4, –3, –2, –1, 0, 1, 2, 3, 4 . . . }

ABSOLUTE VALUE

The absolute value of a number is the distance the number is from zero on the number line.

The absolute value of 6 is written | 6 |. | 6 | = 6
The absolute value of –6 is written | –6 |. | –6 | = 6

Both 6 and –6 are the same distance, 6 spaces, from zero so their absolute value is the same, 6.

EXAMPLES:

| –4 | = 4   – | –4 | = –4 | –9 | + 5 = 9 + 5 = 14

| 9 | – | 8 | = 9 – 8 = 1 | 6 | – | –6 | = 6 – 6 = 0 | –5 | + | –2 | = 5 + 2 = 7

Simplify the following absolute value problems.

1. | 9 | = _____
2. – | 5 | = _____
3. | –25 | = _____
4. – | –12 | = _____
5. – | 64 | = _____

6. | –2 | = _____
7. – | –3 | = _____
8. | –4 | – | 3 | = _____
9. | –8 | – | –4 | = _____
10. | 5 | + | –4 | = _____

11. | –2 | + | 6 | = _____
12. | 10 | + | 8 | = _____
13. | –2 | + | 4 | = _____
14. | –3 | + | –4 | = _____
15. | 7 | – | –5 | = _____

ADDING INTEGERS

First, we will see how to add integers on the number line; then, we will learn rules for working the problems without using a number line.

EXAMPLE 1: Add: $(-3) + 7$

Step 1: The first integer in the problem tells us where to start. Find the first integer, -3, on the number line.

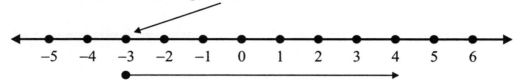

Step 2: $(-3) + 7$ The second integer in the problem, $+7$, tells us the direction to go, positive (toward positive numbers), and how far, 7 places.
$(-3) + 7 = 4$

EXAMPLE 2: Add: $(-2) + (-3)$

Step 1: Find the first integer, (-2), on the number line.

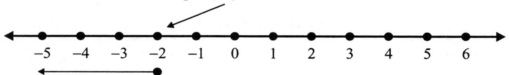

Step 2: $(-2) + (-3)$ The second integer in the problem, (-3), tells us the direction to go, negative (toward the negative numbers), and how far, 3 places.
$(-2) + (-3) = (-5)$

Solve the problems below using this number line.

1. $2 + (-3) =$ _____
2. $4 + (-2) =$ _____
3. $(-3) + 7 =$ _____
4. $(-4) + 4 =$ _____
5. $(-1) + 5 =$ _____
6. $(-1) + (-4) =$ _____
7. $3 + 2 =$ _____
8. $(-5) + 8 =$ _____
9. $3 + (-7) =$ _____
10. $(-2) + (-2) =$ _____
11. $6 + (-7) =$ _____
12. $2 + (-5) =$ _____
13. $(-5) + 3 =$ _____
14. $(-6) + 7 =$ _____
15. $(-3) + (-3) =$ _____
16. $(-8) + 6 =$ _____
17. $(-2) + 6 =$ _____
18. $(-4) + 8 =$ _____
19. $(-7) + 4 =$ _____
20. $(-5) + 8 =$ _____
21. $-2 + (-2) =$ _____
22. $8 + (-6) =$ _____
23. $5 + (-3) =$ _____
24. $1 + (-8) =$ _____

RULES FOR ADDING INTEGERS WITH THE SAME SIGNS

To add integers without using the number line, use these simple rules:

> 1. Add the numbers together
> 2. Give the answer the same sign

EXAMPLE 1: (−2) + (−5) = _____   Both integers are negative. To find the answer, add the numbers together (2 + 5), and give the answer a negative sign.

(−2) + (−5) = (−7)

EXAMPLE 2: 3 + 4 = _____   Both integers are positive, so the answer is positive.

3 + 4 = 7   NOTE: Sometimes positive signs are placed in front of positive numbers. For example 3 + 4 = 7 may be written (+3) + (+4) = +7. Positive signs are optional. If a number has no sign, it is considered positive.

Solve the problems below using the rules for adding integers with the same signs.

1. (−18) + (−4) = _____
2. (−12) + (−3) = _____
3. (−2) + (−7) = _____
4. (22) + (11) = _____
5. (−7) + (−6) = _____
6. 13 + 12 = _____
7. 16 + 11 = _____
8. (−9) + (−8) = _____
9. 8 + 4 = _____
10. (−4) + (−7) = _____
11. (−15) + (−5) = _____
12. (+7) + (+4) = _____
13. (−4) + (−2) = _____
14. (−15) + (−1) = _____
15. (−8) + (−12) = _____
16. 6 + 9 = _____
17. 9 + 7 = _____
18. (−9) + (−7) = _____
19. (−14) + (−6) = _____
20. (6) + (+19) = _____
21. (−11) + (−7) = _____
22. (+8) + (+6) = _____
23. (+5) + 7 = _____
24. (−4) + (−9) = _____
25. (2) + (8) = _____
26. (+18) + 5 = _____
27. 14 + (+7) = _____
28. (−11) + (−19) = _____
29. 13 + (+11) = _____
30. (−8) + (−21) = _____

RULES FOR ADDING INTEGERS WITH OPPOSITE SIGNS

> 1. Ignore the signs and find the difference.
> 2. Give the answer the sign of the larger number.

EXAMPLE 1: $(-4) + 6 =$ _____ To find the difference, take the larger number minus the smaller number. $6 - 4 = 2$ Looking back at the original problem, the larger number, 6, is positive, so the answer is positive.

$(-4) + 6 = 2$

EXAMPLE 2: $3 + (-7) =$ _____ Find the difference. $7 - 3 = 4$ Looking at the problem, the larger number, 7, is a negative number, so the answer is negative.

$3 + (-7) = (-4)$

Solve the problems below using the rules of adding integers with opposite signs.

1. $(-4) + 8 =$ _____
2. $-10 + 12 =$ _____
3. $9 + (-3) =$ _____
4. $(+3) + (-3) =$ _____
5. $+8 + (-7) =$ _____
6. $(-5) + (+12) =$ _____
7. $-14 + (+7) =$ _____
8. $15 + (-3) =$ _____
9. $7 + (-8) =$ _____
10. $6 + (-12) =$ _____
11. $(-11) + 1 =$ _____
12. $(-12) + 8 =$ _____
13. $-14 + 9 =$ _____
14. $14 + (-11) =$ _____
15. $(-20) + 12 =$ _____
16. $-19 + 21 =$ _____
17. $-4 + 18 =$ _____
18. $3 + (-6) =$ _____
19. $4 + (-10) =$ _____
20. $(-2) + 8 =$ _____
21. $-14 + 8 =$ _____
22. $-11 + 15 =$ _____
23. $(-8) + 16 =$ _____
24. $2 + (-15) =$ _____
25. $-2 + 8 =$ _____
26. $(-5) + 15 =$ _____
27. $2 + (-11) =$ _____
28. $-3 + 7 =$ _____
29. $4 + (-12) =$ _____
30. $-12 + 5 =$ _____

Solve the mixed addition problems below using the rules for adding integers.

31. $-7 + 8 =$ _____
32. $5 + 6 =$ _____
33. $(-2) + (-6) =$ _____
34. $3 + (-5) =$ _____
35. $(-7) + (-9) =$ _____
36. $14 + 9 =$ _____
37. $8 + (-5) =$ _____
38. $(-6) + 13 =$ _____
39. $(-9) + (-12) =$ _____
40. $(-7) + (+12) =$ _____
41. $+8 + (-9) =$ _____
42. $(-13) + (-18) =$ _____
43. $(-7) + (+10) =$ _____
44. $(+4) + 11 =$ _____
45. $11 + 6 =$ _____
46. $-4 + (-10) =$ _____
47. $(+6) + (+2) =$ _____
48. $1 + (-17) =$ _____

SUBTRACTING INTEGERS

The easiest way to subtract integers is to change the problem to an addition problem and follow the rules you already know.

RULES FOR SUBTRACTING INTEGERS

> 1. Change the subtraction sign to addition.
> 2. Change the sign of the second number to the opposite sign.

EXAMPLE 1: $-6 - (-2) = $ _____   Change the subtraction sign to addition and -2 to 2. $-6 - (-2) = (-6) + 2$

$(-6) + 2 = (-4)$

EXAMPLE 2: $5 - 6 = $ _____   Change the subtraction sign to addition and 6 to -6. $5 - 6 = 5 + (-6)$

$5 + (-6) = (-1)$

Solve the problems using the rules above.

1. $(-3) - 8 = $ _____
2. $5 - (-9) = $ _____
3. $8 - (-5) = $ _____
4. $(-2) - (-6) = $ _____
5. $8 - (-9) = $ _____
6. $(-4) - (-1) = $ _____
7. $(-5) - (-13) = $ _____
8. $6 - (-7) = $ _____
9. $8 - (-6) = $ _____
10. $(-2) - (-2) = $ _____
11. $(-3) - 7 = $ _____
12. $(-4) - 8 = $ _____
13. $(-7) - 4 = $ _____
14. $1 - (-9) = $ _____
15. $(-5) - 12 = $ _____
16. $(-1) - 9 = $ _____
17. $6 - (-7) = $ _____
18. $(-8) - (-12) = $ _____

Solve the addition and subtraction problems below.

19. $4 - (-2) = $ _____
20. $(-3) + 7 = $ _____
21. $(-4) + 14 = $ _____
22. $(-1) - 5 = $ _____
23. $(-1) + (-4) = $ _____
24. $(-12) + (-2) = $ _____
25. $0 - (-6) = $ _____
26. $2 - (-5) = $ _____
27. $(-5) + 3 = $ _____
28. $(-6) + 7 = $ _____
29. $(-4) + 8 = $ _____
30. $(-4) - 11 = $ _____
31. $(-5) + 8 = $ _____
32. $2 - (-2) = $ _____
33. $(-8) + 9 = $ _____
34. $0 + (-10) = $ _____
35. $30 + (-15) = $ _____
36. $-40 - (-5) = $ _____
37. $25 - 50 = $ _____
38. $-13 + 12 = $ _____
39. $(-21) - (-1) = $ _____
40. $62 - (-3) = $ _____
41. $(-16) + (-2) = $ _____
42. $(-25) + 5 = $ _____

Copyright © American Book Company

MULTIPLYING INTEGERS

You are probably used to seeing multiplication written with a "×" sign, but multiplication can be written two other ways. A "·" between numbers means the same as "×", and parentheses ( ) around a number without a "×" or a "·" also means to multiply.

EXAMPLES: $2 \times 3 = 6$ or $2 \cdot 3 = 6$ or $(2)(3) = 6$

All of these mean the same thing, multiply.

DIVIDING INTEGERS

Division is commonly indicated two ways: with a "÷" or in the form of a fraction.

EXAMPLE: $6 \div 3 = 2$ means the same thing as $\frac{6}{3} = 2$

RULES FOR MULTIPLYING AND DIVIDING INTEGERS

1. If the numbers have the same sign, the answer is positive.
2. If the numbers have different signs, the answer is negative.

EXAMPLES: $6 \times 8 = 48$ $(-6) \times 8 = (-48)$ $(-6) \times (-8) = 48$

$48 \div 6 = 8$ $(-48) \div 6 = (-8)$ $(-48) \div (-6) = 8$

Solve the problems below using the rules of adding integers with opposite signs.

1. $(-4) \div 2 =$ _____
2. $12 \div (-3) =$ _____
3. $\frac{(-14)}{(-2)} =$ _____
4. $-15 \div 3 =$ _____
5. $(-3) \times (-7) =$ _____
6. $(-1) \cdot (5) =$ _____
7. $-1 \times (-4) =$ _____
8. $2(-5) =$ _____
9. $3 \times (-7) =$ _____
10. $(-12) \cdot (-2) =$ _____
11. $\frac{(-18)}{(-6)} =$ _____
12. $21 \div (-7) =$ _____
13. $-5 \times 3 =$ _____
14. $(-6)(7) =$ _____
15. $(-5) \times 8 =$ _____
16. $\frac{-12}{6} =$ _____
17. $8(-4) =$ _____
18. $1 \cdot (-8) =$ _____
19. $(-7) \cdot (-4) =$ _____
20. $(-2) \div (-2) =$ _____
21. $\frac{18}{(-6)} =$ _____

MIXED INTEGER PRACTICE

1. $(-6) + 13 =$ _____
2. $(-3) + (-9) =$ _____
3. $(-4) \times 4 =$ _____
4. $(-18) \div 3 =$ _____
5. $(-1) - 5 =$ _____
6. $(-1) \times (-4) =$ _____
7. $3 + (-5) =$ _____
8. $6 + (-5) =$ _____
9. $(-9) - (-12) =$ _____
10. $2 + (-5) =$ _____
11. $\dfrac{(-24)}{(-6)} =$ _____
12. $(-5) + 3 =$ _____
13. $(-6) - 7 =$ _____
14. $(-33) \div (-11) =$ _____
15. $(-21)(-3) =$ _____
16. $(-7) + (-14) =$ _____
17. $(-5) - 8 =$ _____
18. $1(-8) =$ _____
19. $(-2) \cdot (-2) =$ _____
20. $8 + (-6) =$ _____
21. $\dfrac{-14}{7} =$ _____
22. $(+7) \cdot (-2) =$ _____
23. $(10)(4) =$ _____
24. $24 \div (-4) =$ _____
25. $6(-5) =$ _____
26. $\dfrac{12}{(-3)} =$ _____
27. $36 \div 12 =$ _____

PROPERTIES OF ADDITION AND MULTIPLICATION

The Associative, Commutative, and Distributive properties and the Identity of addition and multiplication are listed below by example as a quick refresher.

Property | **Example**
1. Associative Property of Addition — $(a + b) + c = a + (b + c)$
2. Associative Property of Multiplication — $(a \times b) \times c = a \times (b \times c)$
3. Commutative Property of Addition — $a + b = b + a$
4. Commutative Property of Multiplication — $a \times b = b \times a$
5. Distributive Property — $a \times (b + c) = (a \times b) + (a \times c)$
6. Identity Property of Addition — $0 + a = a$
7. Identity Property of Multiplication — $1 \times a = a$
8. Inverse Property of Addition — $a + (-a) = 0$
9. Inverse Property of Multiplication — $a \times \dfrac{1}{a} = \dfrac{a}{a} = 1 \quad a \neq 0$

In the blanks provided, write the number of the property listed above that describes each of the following statements.

1. $4 + 5 = 5 + 4$ _____
2. $4 + (2 + 8) = (4 + 2) + 8$ _____
3. $10(4 + 7) = (10)(4) + (10)(7)$ _____
4. $(2 \times 3) \times 4 = 2 \times (3 \times 4)$ _____
5. $1 \times 12 = 12$ _____
6. $8\left(\dfrac{1}{8}\right) = 1$ _____
7. $1c = c$ _____
8. $18 + 0 = 18$ _____
9. $9 + (-9) = 0$ _____
10. $p \times q = q \times p$ _____
11. $t + 0 = t$ _____
12. $x(y + z) = xy + xz$ _____
13. $(m)(n \cdot p) = (m \cdot n)(p)$ _____
14. $-y + y = 0$ _____

UNDERSTANDING EXPONENTS

Sometimes it is necessary to multiply a number by itself one or more times. For example, a math problem may need to multiply 3 × 3 or 5 × 5 × 5 × 5. In these situations, mathematicians have come up with a shorter way of writing out this kind of multiplication. Instead of writing 3 × 3, you can write 3^2, or instead of 5 × 5 × 5 × 5, 5^4 means the same thing. The first number is the **base**. The small, raised number is called the **exponent** or **power**. The exponent tells how many times the base should be multiplied by itself.

EXAMPLE 1: 6^3 ← exponent (or power) / base   This means multiply 6 three times: 6 × 6 × 6

EXAMPLE 2: Negative numbers can be raised to exponents also.
An **even** exponent will give a **positive** answer: $(-2)^2 = (-2) \times (-2) = 4$
An **odd** exponent will give a **negative** answer: $(-2)^3 = (-2) \times (-2) \times (-2) = (-8)$

You also need to know two special properties of exponents:

> 1. Any base number raised to the exponent of 1 equals the base number.
> 2. Any base number raised to the exponent of 0 equals 1.

EXAMPLE 3: $4^1 = 4$ $10^1 = 10$ $25^1 = 25$ $4^0 = 1$ $10^0 = 1$ $25^0 = 1$

Rewrite the following problems using exponents.

EXAMPLE: $2 \times 2 \times 2 = 2^3$

1. 7×7×7×7 = _____
2. 10×10 = _____
3. 12×12×12 = _____
4. 4×4×4×4 = _____
5. 9×9×9 = _____
6. 25×25 = _____
7. 15×15×15 = _____
8. 5×5×5×5×5 = _____
9. 2×2×2×2 = _____
10. 14×14 = _____
11. 3×3×3×3×3 = _____
12. 11×11×11 = _____

Use your calculator to figure what product each number with an exponent represents.

EXAMPLE: $2^3 = 2 \times 2 \times 2 = 8$

13. $(-8)^3 =$ _____
14. $12^2 =$ _____
15. $20^1 =$ _____
16. $5^4 =$ _____
17. $15^0 =$ _____
18. $16^2 =$ _____
19. $(-10)^2 =$ _____
20. $3^5 =$ _____
21. $10^4 =$ _____
22. $7^0 =$ _____
23. $4^3 =$ _____
24. $54^1 =$ _____

Express each of the following numbers as a base with an exponent.

EXAMPLE: $4 = 2 \times 2 = 2^2$

25. 9 = _____
26. 16 = _____ or _____
27. 27 = _____
28. 36 = _____
29. 8 = _____
30. 32 = _____
31. 1000 = _____
32. 125 = _____
33. 81 = _____ or _____
34. 64 = _____, _____, or _____
35. 49 = _____
36. 121 = _____

SQUARE ROOT

Just as working with exponents is related to multiplication, so finding square roots is related to division. In fact, the sign for finding the square root of a number looks similar to a division sign. The best way to learn about square roots is to look at examples.

EXAMPLES: This is a square root problem: $\sqrt{64}$

It is asking, "What is the square root of 64?"
It means, "What number multiplied by itself equals 64?"
The answer is 8. $8 \times 8 = 64$.

Find the square root of the following numbers.

$\sqrt{36}$ $6 \times 6 = 36$ so $\sqrt{36} = 6$ $\sqrt{144}$ $12 \times 12 = 144$ so $\sqrt{144} = 12$

Find the square roots of the following numbers.

1. $\sqrt{49}$ _____
2. $\sqrt{81}$ _____
3. $\sqrt{25}$ _____
4. $\sqrt{16}$ _____
5. $\sqrt{121}$ _____

6. $\sqrt{625}$ _____
7. $\sqrt{100}$ _____
8. $\sqrt{289}$ _____
9. $\sqrt{196}$ _____
10. $\sqrt{36}$ _____

11. $\sqrt{4}$ _____
12. $\sqrt{900}$ _____
13. $\sqrt{64}$ _____
14. $\sqrt{9}$ _____
15. $\sqrt{144}$ _____

MIXED PRACTICE

Write with exponents.

1. $2 \times 2 \times 2 \times 2 \times 2$ _____
2. $5 \times 5 \times 5$ _____
3. 3×3 _____
4. $6 \times 6 \times 6 \times 6$ _____
5. 8×8 _____

Find the square root.

6. $\sqrt{1}$ _____
7. $\sqrt{169}$ _____
8. $\sqrt{49}$ _____
9. $\sqrt{256}$ _____
10. $\sqrt{400}$ _____

Write as whole numbers.

11. 16^2 _____
12. 3^4 _____
13. 5^3 _____
14. 10^2 _____
15. 4^3 _____

ORDER OF OPERATIONS

In long math problems with $+$, $-$, $\times$, $\div$, ( ), and exponents in them, you have to know what to do first. Without following the same rules, you could get different answers. If you will memorize the silly sentence, Please Excuse My Dear Aunt Sally, you can easily memorize the order you must follow.

Please   "**P**" stands for parentheses. You must get rid of parentheses first.
Examples: $3(1+4) = 3(5) = 15$
$6(10-6) = 6(4) = 24$

Excuse   "**E**" stands for exponents. You must eliminate exponents next.
Example: $4^2 = 4 \times 4 = 16$

My Dear   "**M**" stands for multiply. "**D**" stands for divide. Start on the left of the equation and perform all multiplications and divisions in the order in which they appear.

Aunt Sally   "**A**" stands for add. "**S**" stands for subtract. Start on the left and perform all additions and subtractions in the order they appear.

EXAMPLE: $12 \div 2(6-3) + 3^2 - 1$

Please	Eliminate **parentheses**. $6-3 = 3$ so now we have	$12 \div 2(3) + 3^2 - 1$
Excuse	Eliminate **exponents**. $3^2 = 9$ so now we have	$12 \div 2(3) + 9 - 1$
My Dear	**Multiply** and **divide** next in order from left to right.	$12 \div 2 = 6$ then $6(3) = 18$
Aunt Sally	Last, we **add** and **subtract** in order from left to right.	$18 + 9 - 1 = 26$

Simplify the following problems.

1. $6 + 9 \times 2 - 4$ = _____
2. $3(4+2) - 6^2$ = _____
3. $3(6-3) - 2^3$ = _____
4. $49 \div 7 - 3 \times 3$ = _____
5. $10 \times 4 - (7-2)$ = _____
6. $2 \times 3 \div 6 \times 4$ = _____
7. $4^3 \div 8(4+2)$ = _____
8. $7 + 8(14-6) \div 4$ = _____
9. $(2+8-12) \times 4$ = _____
10. $4(8-13) \times 4$ = _____
11. $8 + 4^2 \times 2 - 6$ = _____
12. $3^2(4+6) + 3$ = _____
13. $(12-6) + 27 \div 3^2$ = _____
14. $82^0 - 1 + 4 \div 2^2$ = _____
15. $1 - (2-3) + 8$ = _____
16. $12 - 4(7-2)$ = _____
17. $18 \div (6+3) - 12$ = _____
18. $10^2 + 3^3 - 2 \times 3$ = _____
19. $4^2 + (7+2) \div 3$ = _____
20. $7 \times 4 - 9 \div 3$ = _____

When a problem has a fraction bar, simplify the top of the fraction (numerator) and the bottom of the fraction (denominator) separately using the rules for order of operations. You treat the top and the bottom as if they were separate problems. Then reduce the fraction to lowest terms.

EXAMPLE: $\dfrac{2(4-5)-6}{5^2+3(2+1)}$

Please — Eliminate **parentheses**. $(4-5) = -1$ and $(2+1) = 3$ $\quad \dfrac{2(-1)-6}{5^2+3(3)}$

Excuse — Eliminate **exponents**. $5^2 = 25$ $\quad \dfrac{2(-1)-6}{25+3(3)}$

My Dear — **Multiply** and **divide** in the numerator and denominator separately. $3(3) = 9$ and $2(-1) = -1$ $\quad \dfrac{-2-6}{25+9}$

Aunt Sally — **Add** and **subtract** in the numerator and denominator separately. $-2-6 = -8$ and $25+9 = 34$ $\quad \dfrac{-8}{34}$

Now reduce the fraction to lowest terms. $\dfrac{-8}{34} = \dfrac{-4}{17}$

Simplify the following problems.

1. $\dfrac{2^2+4}{5+3(8+1)} =$ _____

2. $\dfrac{8^2-(4+11)}{4^2-3^2} =$ _____

3. $\dfrac{5-2(4-3)}{2(1-8)} =$ _____

4. $\dfrac{10+(2-4)}{4(2+6)-2^2} =$ _____

5. $\dfrac{3^3-8(1+2)}{-10-(3+8)} =$ _____

6. $\dfrac{(9-3)+3^2}{-5-2(4+1)} =$ _____

7. $\dfrac{16-3(10-6)}{(13+15)-5^2} =$ _____

8. $\dfrac{(2-5)-11}{12-2(3+1)} =$ _____

9. $\dfrac{7+(8-16)}{6^2-5^2} =$ _____

10. $\dfrac{16-(12-3)}{8(2+3)-5} =$ _____

11. $\dfrac{-3(9-7)}{7+9-2^3} =$ _____

12. $\dfrac{4-(2+7)}{13+(6-9)} =$ _____

13. $\dfrac{5(3-8)-2^2}{7-3(6+1)} =$ _____

14. $\dfrac{3(3-8)+5}{8^2-(5+9)} =$ _____

15. $\dfrac{6^2-4(7+3)}{8+(9-3)} =$ _____

CHAPTER 5 REVIEW

Simplify the following problems.

1. $(-7) \times (-4) =$ _____
2. $15^0 =$ _____
3. $7 - (-8) =$ _____
4. $9 \div (-3) =$ _____
5. $\sqrt{100} =$ _____
6. $-9 - 7 =$ _____
7. $(-10) \cdot (4) =$ _____
8. $\sqrt{49} =$ _____
9. $(-3)^3 =$ _____
10. $4 + (-4) =$ _____
11. $\dfrac{(-16)}{4} =$ _____
12. $(7)(-7) =$ _____
13. $-5 + (-2) =$ _____
14. $(-12) \div (-3) =$ _____
15. $(-2) \times 6 =$ _____
16. $-4 + (-10) =$ _____
17. $5 - (-5) =$ _____
18. $(9)(-1) =$ _____
19. $\dfrac{(-20)}{(-5)} =$ _____
20. $14 - 25 =$ _____
21. $|-4| + |5| =$ _____
22. $18 - |-6| =$ _____
23. $|-14| + |9| =$ _____
24. $|-2| - |-9| =$ _____
25. $|5| - |-8| =$ _____
26. $|-2| + |7| =$ _____
27. $|12| - |-6| =$ _____
28. $|-3| + 9 =$ _____
29. $|10| - |6| =$ _____
30. $14 - |5| =$ _____

Simplify the following problems using the correct order of operations.

31. $2^3 + (2^2)(5 - 7) =$ _____

32. $10 \div (-1 - 4) + 2 =$ _____

33. $5 + (2)(4 - 1) \div 3 =$ _____

34. $5 - 5^2 + (2 - 4) =$ _____

35. $(8 - 10) \times (5 + 3) - 10 =$ _____

36. $\dfrac{10 + 5^2 - 3}{2^2 + 2(5 - 3)} =$ _____

37. $1 - (3^2 - 1) \div 2 =$ _____

38. $\dfrac{5(3 - 6) + 3^2}{4(2 + 1) - 6} =$ _____

39. $-4(6 + 4) \div (-2) + 1 =$ _____

40. $12 \div (7 - 4) - 2 =$ _____

41. $1 + 4^2 \div (3 + 1) =$ _____

42. $2^3 + (5)(3 - 5) =$ _____

Chapter 6
Exponents and Roots

In the previous chapter on Integers and Order of Operations, you were introduced to exponents and square roots. In this chapter, you will build on that knowledge.

MULTIPLYING EXPONENTS WITH THE SAME BASE

To multiply two expressions with the same base, add the exponents together and keep the base the same.

EXAMPLE 1: $2^3 \times 2^5 = 2^{3+5} = 2^8$

EXAMPLE 2: $3a^2 \times 2a^3 = 6a^{2+3} = 6a^5$ Notice that only the *a*'s are raised to a power and not the 3 or the 2.

Simplify each of the expressions below.

1. $2^3 \times 2^5$
2. $x^5 \times x^3$
3. $2a^3 \times 3a^3$
4. $4^5 \times 4^3$
5. $2x^3 \times x^5$
6. $4b^3 \times 2b^4$
7. $10^5 \times 10^4$
8. $5^2 \times 5^4$
9. $3^3 \times 3^2$
10. $4x \times x^2$
11. $a^2 \times 3a^4$
12. $2^3 \times 2^4$

MULTIPLYING EXPONENTS RAISED TO AN EXPONENT

If a power is raised to another power, multiply the exponents together and keep the base the same.

EXAMPLE 1: $(2^3)^2 = 2^{3 \times 2} = 2^6$

EXAMPLE 2: $(y^4)^3 = y^{4 \times 3} = y^{12}$

Simplify each of the expressions below.

1. $(5^3)^3$
2. $(x^5)^2$
3. $(6^2)^5$
4. $(3^4)^2$
5. $(3^2)^4$
6. $(y^2)^3$
7. $(3^3)^2$
8. $(9^2)^2$
9. $(5^3)^2$
10. $(a^4)^2$
11. $(2^1)^3$
12. $(x^3)^2$

FRACTIONS RAISED TO A POWER

A fraction can be raised to a power also.

EXAMPLE: $\left(\frac{3}{4}\right)^3 = \frac{3^3}{4^3} = \frac{27}{64}$

Simplify the following fractions.

1. $\left(\frac{2}{3}\right)^2$
2. $\left(\frac{7}{8}\right)^3$
3. $\left(\frac{1}{2}\right)^2$
4. $\left(\frac{1}{4}\right)^2$
5. $\left(\frac{2}{3}\right)^3$
6. $\left(\frac{3}{4}\right)^2$
7. $\left(\frac{1}{2}\right)^3$
8. $\left(\frac{5}{7}\right)^2$
9. $\left(\frac{2}{3}\right)^4$
10. $\left(\frac{3}{10}\right)^2$
11. $\left(\frac{4}{5}\right)^2$
12. $\left(\frac{1}{10}\right)^4$

MORE MULTIPLYING EXPONENTS

If a product in parentheses is raised to a power, then each factor is raised to the power when parentheses are eliminated.

EXAMPLE 1: $(2 \times 4)^2 = 2^2 \times 4^2 = 4 \times 16 = 64$

EXAMPLE 2: $(3a)^3 = 3^3 \times a^3 = 27a^3$

EXAMPLE 3: $(7b^5)^2 = 7^2 b^{10} = 49b^{10}$

Simplify each of the expressions below.

1. $(2^3)^2$
2. $(7a^5)^2$
3. $(6b^2)^2$
4. $(3^2)^2$
5. $(3 \times 5)^2$
6. $(3x^4)^2$
7. $(6y^7)^2$
8. $(11w^3)^2$
9. $(3^3)^2$
10. $(3 \times 3)^2$
11. $(2a)^4$
12. $(2^2)^3$
13. $(3 \times 2)^3$
14. $(5x^3)^2$
15. $(4r^7)^3$
16. $(2m^3)^2$
17. $(6 \times 4)^2$
18. $(9a^5)^2$
19. $(7b^5)^2$
20. $(9^2)^2$
21. 4×4^3
22. $(3a)^2$
23. $(2 \times 3)^3$
24. $(5p^4)^3$
25. $(4y^4)^2$
26. $(2b^3)^4$
27. $(5a^2)^2$
28. $(8a^3)^2$
29. $(2 \times 6)^2$
30. $(7^2)^2$
31. $(4 \times 3)^2$
32. $(15c^7)^2$

NEGATIVE EXPONENTS

Expressions can also have negative exponents. Negative exponents do not indicate negative numbers. They indicate **reciprocals**. The **reciprocal** of a number is one divided by that number. For example, the reciprocal of 2 is $\frac{1}{2}$. (A number multiplied by its reciprocal is equal to 1.) If the negative exponent is in the bottom of a fraction, the reciprocal will put the expression on the top of the fraction without the negative sign.

EXAMPLE 1: $2^{-3} = \frac{1}{2^3} = \frac{1}{8}$

EXAMPLE 2: $3a^{-5} = 3 \times \frac{1}{a^5} = \frac{3}{a^5}$ Notice that the 3 is not raised to the -5 power, only the a.

EXAMPLE 3: $\frac{6}{5x^{-2}} = \frac{6x^2}{5}$ Notice that the 5 is not raised to the -2 power, only the x.

Rewrite using only positive exponents.

1. $5m^{-6}$
2. $\frac{5x^{-4}}{7}$
3. $14z^{-8}$
4. $\frac{1}{5s^{-4}}$
5. $14h^{-5}$
6. $\frac{h^{-3}}{5}$
7. $\frac{2y^{-3}}{4}$
8. x^{-4}
9. $-2y^{-2}$
10. $5y^{-5}$
11. $\frac{x^{-3}}{5}$
12. $10z^{-7}$
13. $7x^{-3}$
14. r^{-2}
15. $\frac{m^{-4}}{6}$

MULTIPLYING WITH NEGATIVE EXPONENTS

Multiplying with negative exponents follows the same rules as multiplying with positive exponents.

EXAMPLE 1: $6^2 \cdot 6^{-3} = 6^{2+(-3)} = 6^{-1} = \frac{1}{6}$

EXAMPLE 2: $(5a \times 2)^{-3} = (10a)^{-3} = \frac{1}{(10a)^3} = \frac{1}{1000a^3}$

EXAMPLE 3: $(7a^2)^{-3} = 7^{-3}a^{-6} = \frac{1}{7^3 a^6}$

Simplify the following. Answers should <u>not</u> have any negative exponents.

1. $5^{-2} \cdot 5^5$
2. $(6^3 \cdot 6^{-2})^{-2}$
3. $10^{-4} \cdot 10^2$
4. $11^{-5} \cdot 11^7$
5. $4^7 \cdot 4^{-10}$
6. $20^8 \cdot 20^{-6}$
7. $5^{-8} \cdot 5^4$
8. $(2^{-2} \cdot 2^3)^{-4}$
9. $7^{-2} \cdot 7^{-1}$
10. $(3x^4)^{-3}$
11. $12^{-10} \cdot 12^8$
12. $(10^8 \cdot 10^{-10})^2$
13. $3^{-2} \cdot 2^{-2}$
14. $(8x^5)^{-4}$
15. $(6b^3)^{-6}$
16. $(9y)^{-2}$

DIVIDING WITH EXPONENTS

Exponents that have the same base can also be divided.

EXAMPLE 1: $\dfrac{3^5}{3^3}$ This problem means $3^5 \div 3^3$. Let us look at 2 ways to solve this problem.

Solution 1: $\dfrac{3^5}{3^3} = \dfrac{\cancel{3}\cdot\cancel{3}\cdot\cancel{3}\cdot 3\cdot 3}{\cancel{3}\cdot\cancel{3}\cdot\cancel{3}} = 3\cdot 3 = 9$ First, rewrite the fraction with exponents in expanded form, and then cancel terms and multiply.

Solution 2: $\dfrac{3^5}{3^3} = 3^{5-3} = 3^2 = 9$ A quick way to simplify this same problem is to subtract the exponents. **When dividing exponents with the same base, subtract the exponents.**

EXAMPLE 2: $\dfrac{(4x)^{-3}}{2x^4}$

Step 1: $(4x)^{-3} = \dfrac{1}{(4x)^3} = \dfrac{1}{4^3 x^3}$ Remove the parentheses from the top of the fraction.

Step 2: $\dfrac{1}{4^3 x^3 \cdot 2x^4} = \dfrac{1}{128 x^7}$ The bottom of the fraction remains the same, so put the two together and simplify.

Simplify the problems below. You may be able to cancel. Be sure to follow order of operations. Remove parentheses before canceling.

1. $\dfrac{5^5}{5^3}$

2. $\dfrac{x^2}{x^3}$

3. $\dfrac{(10^2)^4}{10^5}$

4. $\dfrac{3^5}{3^2}$

5. $\dfrac{8^{10}}{8^8}$

6. $\dfrac{5^2}{5}$

7. $\dfrac{(7^2)^3}{7^5}$

8. $\dfrac{(x^3)^4}{x^6}$

9. $\dfrac{4^3}{4^2}$

10. $\dfrac{2}{(2^2)^2}$

11. $\dfrac{(3x)^{-2}}{9x^5}$

12. $\dfrac{(11^4)^4}{(11^7)^2}$

13. $\dfrac{x^3}{(x^2)^3}$

14. $\dfrac{2^2}{2^7}$

15. $\dfrac{6^2}{6}$

16. $\dfrac{9^{11}}{9^9}$

17. $\dfrac{(15)^5}{15^6}$

18. $\dfrac{(x^3)^{-2}}{(x^2)^5}$

19. $\dfrac{12^{-4}}{12^{-2}}$

20. $\dfrac{6^{12}}{6^9}$

21. $\dfrac{8^8}{8^{10}}$

22. $\dfrac{3(x^{-3})^{-2}}{3x^7}$

23. $\dfrac{7^3}{7^5}$

24. $\dfrac{10^3}{10^{-1}}$

SIMPLIFYING SQUARE ROOTS

Square roots can sometimes be simplified even if the number under the square root is not a perfect square. One of the rules of roots is that if a and b are two positive real numbers, then it is always true that $\sqrt{a \cdot b} = \sqrt{a} \cdot \sqrt{b}$. You can use this rule to simplify square roots.

EXAMPLE 1: $\sqrt{100} = \sqrt{4 \cdot 25} = \sqrt{4} \cdot \sqrt{25} = 2 \cdot 5 = 10$

EXAMPLE 2: $\sqrt{200} = \sqrt{100 \cdot 2} = 10\sqrt{2}$ ← Means 10 multiplied by the square root of 2

EXAMPLE 3: $\sqrt{160} = \sqrt{10 \cdot 16} = 4\sqrt{10}$

Simplify.

1. $\sqrt{98}$ _____
2. $\sqrt{600}$ _____
3. $\sqrt{50}$ _____
4. $\sqrt{27}$ _____
5. $\sqrt{8}$ _____
6. $\sqrt{63}$ _____
7. $\sqrt{48}$ _____
8. $\sqrt{75}$ _____
9. $\sqrt{54}$ _____
10. $\sqrt{40}$ _____
11. $\sqrt{72}$ _____
12. $\sqrt{80}$ _____
13. $\sqrt{90}$ _____
14. $\sqrt{175}$ _____
15. $\sqrt{18}$ _____
16. $\sqrt{20}$ _____

ESTIMATING SQUARE ROOTS

EXAMPLE 1: Estimate the value of $\sqrt{3}$.

Step 1: Estimate the value of $\sqrt{3}$ by using the square root of values that you know. $\sqrt{1} = 1$ and $\sqrt{4}$ is 2, so the value of $\sqrt{3}$ is going to be between 1 and 2.

Step 2: To estimate a little closer, try squaring 1.5. $1.5 \times 1.5 = 2.25$, so $\sqrt{3}$ has to be greater than 1.5. If you do further trial and error calculations, you will find that $\sqrt{3}$ is greater than 1.7 ($1.7 \times 1.7 = 2.89$) but less than 1.8 ($1.8 \times 1.8 = 3.24$).

Therefore $\sqrt{3}$ is around 1.75. It is closer to 2 than it is to 1.

EXAMPLE 2: Is $\sqrt{52}$ closer to 7 or 8? Look at the perfect squares above and below 52.

To answer this question, first look at 7^2 which is equal to 49 and 8^2 which is equal to 64. Then ask yourself whether 52 is closer to 49 or to 64? The answer is 49, of course. Therefore, $\sqrt{52}$ is closer to 7 than 8.

Follow the steps above to answer the following questions. Do not use a calculator.

1. Is $\sqrt{66}$ closer to 8 or 9? _____
2. Is $\sqrt{27}$ closer to 5 or 6? _____
3. Is $\sqrt{13}$ closer to 3 or 4? _____
4. Is $\sqrt{78}$ closer to 8 or 9? _____
5. Is $\sqrt{12}$ closer to 3 or 4? _____
6. Is $\sqrt{8}$ closer to 2 or 3? _____
7. Is $\sqrt{20}$ closer to 4 or 5? _____
8. Is $\sqrt{53}$ closer to 7 or 8? _____
9. Is $\sqrt{60}$ closer to 7 or 8? _____
10. Is $\sqrt{6}$ closer to 2 or 3? _____

ADDING AND SUBTRACTING ROOTS

You can add and subtract terms with square roots only if the number under the square root sign is the same.

EXAMPLE 1: $2\sqrt{2} + 3\sqrt{2} = 5\sqrt{2}$
EXAMPLE 2: $12\sqrt{7} - 3\sqrt{7} = 9\sqrt{7}$

Or, look at the following examples where you can simplify the square roots and then add or subtract.

EXAMPLE 3: $2\sqrt{25} + \sqrt{36}$
Step 1: Simplify. You know that $\sqrt{25} = 5$, and $\sqrt{36} = 6$ so the problem simplifies to $2(5) + 6$

Step 2: Solve: $2(5) + 6 = 10 + 6 = 16$

EXAMPLE 4: $2\sqrt{72} - 3\sqrt{2}$

Step 1: Simplify what you know. $\sqrt{72} = \sqrt{36 \cdot 2} = 6\sqrt{2}$
Step 2: Substitute $6\sqrt{2}$ for $\sqrt{72}$ and simplify.
$2(6)\sqrt{2} - 3\sqrt{2} = 12\sqrt{2} - 3\sqrt{2} = 9\sqrt{2}$

Simplify the following addition and subtraction problems.

1. $3\sqrt{5} + 9\sqrt{5}$
2. $3\sqrt{25} + 4\sqrt{16}$
3. $4\sqrt{8} + 2\sqrt{2}$
4. $3\sqrt{32} - 2\sqrt{2}$
5. $\sqrt{25} - \sqrt{49}$
6. $2\sqrt{5} + 4\sqrt{20}$
7. $5\sqrt{8} - 3\sqrt{72}$
8. $\sqrt{27} + 3\sqrt{27}$

9. $3\sqrt{20} - 4\sqrt{45}$
10. $4\sqrt{45} - \sqrt{125}$
11. $2\sqrt{28} + 2\sqrt{7}$
12. $\sqrt{64} + \sqrt{81}$
13. $5\sqrt{54} - 2\sqrt{24}$
14. $\sqrt{32} + 2\sqrt{50}$
15. $2\sqrt{7} + 4\sqrt{63}$
16. $8\sqrt{2} + \sqrt{8}$

17. $2\sqrt{8} - 4\sqrt{32}$
18. $\sqrt{36} + \sqrt{100}$
19. $\sqrt{9} + \sqrt{25}$
20. $\sqrt{64} - \sqrt{36}$
21. $\sqrt{75} + \sqrt{108}$
22. $\sqrt{81} + \sqrt{100}$
23. $\sqrt{192} - \sqrt{75}$
24. $3\sqrt{5} + \sqrt{245}$

MULTIPLYING ROOTS

You can also multiply square roots. To multiply square roots, you just multiply the numbers under the square root sign and then simplify. Look at the examples below.

EXAMPLE 1: $\sqrt{2} \times \sqrt{6}$

Step 1: $\sqrt{2} \times \sqrt{6} = \sqrt{2 \times 6} = \sqrt{12}$ Multiply the numbers under the square root sign.

Step 2: $\sqrt{12} = \sqrt{4 \times 3} = 2\sqrt{3}$ Simplify.

EXAMPLE 2: $3\sqrt{3} \times 5\sqrt{6}$

Step 1: $(3 \times 5)\sqrt{3 \times 6} = 15\sqrt{18}$ Multiply the numbers in front of the square root, and multiply the numbers under the square root sign.

Step 2: $15\sqrt{18} = 15\sqrt{2 \times 9}$
$15 \times 3\sqrt{2} = 45\sqrt{2}$ Simplify.

EXAMPLE 3: $\sqrt{14} \times \sqrt{42}$ For this more complicated multiplication problem, use the rule of roots that you learned on the top of page 89, $\sqrt{a \cdot b} = \sqrt{a} \cdot \sqrt{b}$.

Step 1: $\sqrt{14} = \sqrt{7} \times \sqrt{2}$ and
$\sqrt{42} = \sqrt{2} \times \sqrt{3} \times \sqrt{7}$ Instead of multiplying 14 by 42, divide these numbers into their roots.

$\sqrt{14} \times \sqrt{42} = \sqrt{7} \times \sqrt{2} \times \sqrt{2} \times \sqrt{3} \times \sqrt{7}$

Step 2: Since you know that $\sqrt{7} \times \sqrt{7} = 7$, and $\sqrt{2} \times \sqrt{2} = 2$, the problem simplifies to $(7 \times 2)\sqrt{3} = 14\sqrt{3}$

Simplify the following multiplication problems.

1. $\sqrt{5} \times \sqrt{7}$
2. $\sqrt{32} \times \sqrt{2}$
3. $\sqrt{10} \times \sqrt{14}$
4. $2\sqrt{3} \times 3\sqrt{6}$
5. $4\sqrt{2} \times 2\sqrt{10}$

6. $\sqrt{5} \times 3\sqrt{15}$
7. $\sqrt{45} \times \sqrt{27}$
8. $5\sqrt{21} \times \sqrt{7}$
9. $\sqrt{42} \times \sqrt{21}$
10. $4\sqrt{3} \times 2\sqrt{12}$

11. $\sqrt{56} \times \sqrt{24}$
12. $\sqrt{11} \times 2\sqrt{33}$
13. $\sqrt{13} \times \sqrt{26}$
14. $2\sqrt{2} \times 5\sqrt{5}$
15. $\sqrt{6} \times \sqrt{12}$

DIVIDING ROOTS

When dividing a number or a square root by another square root, you cannot leave the square root sign in the denominator (the bottom number) of a fraction. You must simplify the problem so that the square root is not in the denominator. Look at the examples below.

EXAMPLE 1: $\dfrac{\sqrt{2}}{\sqrt{5}}$

Step 1: $\dfrac{\sqrt{2}}{\sqrt{5}} \times \dfrac{\sqrt{5}}{\sqrt{5}}$ ⬅ The fraction $\dfrac{\sqrt{5}}{\sqrt{5}}$ is equal to 1, and multiplying by 1 does not change the value of a number.

Step 2: $\dfrac{\sqrt{2 \times 5}}{5} = \dfrac{\sqrt{10}}{5}$ Multiply and simplify. Since $\sqrt{5} \times \sqrt{5}$ equals 5, you no longer have a square root in the denominator.

EXAMPLE 2: $\dfrac{6\sqrt{2}}{2\sqrt{10}}$ In this problem, the numbers outside of the square root will also simplify.

Step 1: $\dfrac{6}{2} = 3$ so you have $\dfrac{3\sqrt{2}}{\sqrt{10}}$

Step 2: $\dfrac{3\sqrt{2}}{\sqrt{10}} \times \dfrac{\sqrt{10}}{\sqrt{10}} = \dfrac{3\sqrt{2 \times 10}}{10} = \dfrac{3\sqrt{20}}{10}$

Step 3: $\dfrac{3\sqrt{20}}{10}$ will further simplify because $\sqrt{20} = 2\sqrt{5}$, so you then have $\dfrac{3 \times 2\sqrt{5}}{10}$ which reduces to $\dfrac{3 \times \cancel{2}\sqrt{5}}{\cancel{10}_{5}}$ or $\dfrac{3\sqrt{5}}{5}$

Simplify the following division problems.

1. $\dfrac{9\sqrt{3}}{\sqrt{5}}$

2. $\dfrac{16}{\sqrt{8}}$

3. $\dfrac{24\sqrt{10}}{12\sqrt{3}}$

4. $\dfrac{\sqrt{121}}{\sqrt{6}}$

5. $\dfrac{\sqrt{40}}{\sqrt{90}}$

6. $\dfrac{33\sqrt{15}}{11\sqrt{2}}$

7. $\dfrac{\sqrt{32}}{\sqrt{12}}$

8. $\dfrac{\sqrt{11}}{\sqrt{5}}$

9. $\dfrac{\sqrt{2}}{\sqrt{6}}$

10. $\dfrac{2\sqrt{7}}{\sqrt{14}}$

11. $\dfrac{5\sqrt{2}}{4\sqrt{8}}$

12. $\dfrac{4\sqrt{21}}{7\sqrt{7}}$

13. $\dfrac{9\sqrt{22}}{2\sqrt{2}}$

14. $\dfrac{\sqrt{35}}{2\sqrt{14}}$

15. $\dfrac{\sqrt{40}}{\sqrt{15}}$

SCIENTIFIC NOTATION

Mathematicians use **scientific notation** to express very large and very small numbers. **Scientific notation** expresses a number in the following form:

$$x.xx \times 10^x$$

only one digit before the decimal ⟶

multiplied by a multiple of ten

remaining digits not ending in zeros after the decimal ⟶

USING SCIENTIFIC NOTATION FOR LARGE NUMBERS

Scientific notation simplifies very large numbers that have many zeros. For example, Pluto averages a distance of 5,900,000,000 kilometers from the sun. In scientific notation, a decimal is inserted after the first digit (5.); the rest of the digits are copied except for the zeros at the end (5.9), and the result is multiplied by 10^9. The exponent = the total number of digits in the original number minus 1 or the number of spaces the decimal point moved.

$5,900,000,000 = 5.9 \times 10^9$ The following are more examples:

EXAMPLES: $32,560,000,000 = 3.256 \times 10^{10}$ $5,060,000 = 5.06 \times 10^6$

decimal moves 10 spaces to the left   decimal moves 6 spaces to the left

Convert the following numbers to scientific notation.

1. 4,230,000,000 = _____
2. 64,300,000 = _____
3. 951,000,000,000 = _____
4. 12,300 = _____
5. 20,350,000,000 = _____
6. 9,000 = _____
7. 450,000,000,000 = _____
8. 6,200 = _____
9. 87,000,000 = _____
10. 105,000,000 = _____
11. 1,083,000,000,000 = _____
12. 304,000 = _____

To convert a number written in scientific notation back to conventional form, reverse the steps.

EXAMPLE: $4.02 \times 10^5 = 4.02000 = 402,000$ Move the decimal 5 spaces to the right and add zeros.

Convert the following numbers from scientific notation to conventional numbers.

13. 6.85×10^8 = _____
14. 1.3×10^{10} = _____
15. 4.908×10^4 = _____
16. 7.102×10^6 = _____
17. 2.5×10^3 = _____
18. 9.114×10^5 = _____
19. 5.87×10^7 = _____
20. 8.047×10^8 = _____
21. 3.81×10^5 = _____
22. 9.5×10^{12} = _____
23. 1.504×10^6 = _____
24. 7.3×10^9 = _____

USING SCIENTIFIC NOTATION FOR SMALL NUMBERS

Scientific notation also simplifies very small numbers that have many zeros. For example, the diameter of a helium atom is 0.000000000244 meters. It can be written in scientific notation as 2.44×10^{-10}. The first number is always greater than 0, and the first number is always followed by a decimal point. The negative exponent indicates how many digits the decimal point moved to the right. The exponent is negative when the original number is less than 1. To convert small numbers to scientific notation, follow the examples below.

EXAMPLES: $0.00058 = 5.8 \times 10^{-4}$ $\qquad$ $0.00003059 = 3.059 \times 10^{-5}$

decimal point moves 4 spaces to the right — negative exponent indicates the original number is less than 1. — decimal moves 5 spaces to the right

Convert the following numbers to scientific notation.

1. $0.00000254 =$ _____
2. $0.00000000508 =$ _____
3. $0.000008004 =$ _____
4. $0.00047 =$ _____
5. $0.000000005478 =$ _____
6. $0.00000059 =$ _____
7. $0.00000004712 =$ _____
8. $0.00025 =$ _____
9. $0.0000000501 =$ _____
10. $0.0000006 =$ _____
11. $0.0000000000875 =$ _____
12. $0.00004 =$ _____

Now convert small numbers written in scientific notation back to conventional form.

EXAMPLE: $3.08 \times 10^{-5} = 00003.08 = 0.0000308$ $\qquad$ Move the decimal 5 spaces to the left, and add zeros.

Convert the following numbers from scientific notation to conventional numbers.

13. $1.18 \times 10^{-7} =$ _____
14. $2.3 \times 10^{-5} =$ _____
15. $6.205 \times 10^{-9} =$ _____
16. $4.1 \times 10^{-6} =$ _____
17. $7.632 \times 10^{-4} =$ _____
18. $5.48 \times 10^{-10} =$ _____
19. $2.75 \times 10^{-8} =$ _____
20. $4.07 \times 10^{-7} =$ _____
21. $5.2 \times 10^{-3} =$ _____
22. $7.01 \times 10^{-6} =$ _____
23. $4.4 \times 10^{-5} =$ _____
24. $3.43 \times 10^{-2} =$ _____

CHAPTER 6 REVIEW

Simplify the following expressions. Reduce to simplest form. Make all exponents positive.

1. $5^2 \times 5^3$
2. $(4^4)^5$
3. $(4y^3)^3$
4. $6x^{-3}$
5. $(3a^2)^{-2}$
6. $(b^3)^{-4}$
7. $\dfrac{4^6}{4^4}$
8. $\left(\dfrac{3}{5}\right)^2$
9. $\dfrac{(3a^2)^3}{a^3}$
10. $(2x)^{-4}$
11. $3^3 \times 3^2$
12. $(2^4)^2$
13. $5^7 \times 5^{-4}$
14. $x^3 \cdot x^{-7}$
15. $(4^2)^{-2}$
16. $(5^{-9} \cdot 5^7)^{-2}$
17. $\dfrac{(2^3)^2}{2^4}$
18. $\dfrac{y^{-2}}{3y^4}$
19. $\dfrac{(2^3)^2}{2^4}$
20. $(4d^5)^{-3}$

Simplify the following square root expressions.

21. $\sqrt{50}$
22. $\sqrt{44}$
23. $\sqrt{12}$
24. $\sqrt{18}$
25. $\sqrt{8}$
26. $\sqrt{48}$
27. $\sqrt{75}$
28. $\sqrt{200}$
29. $\sqrt{32}$
30. $\sqrt{20}$
31. $\sqrt{63}$
32. $\sqrt{80}$

Estimate the following square root solutions.

33. Is $\sqrt{5}$ closer to 2 or 3?
34. Is $\sqrt{52}$ closer to 7 or 8?
35. Is $\sqrt{130}$ closer to 11 or 12?
36. Is $\sqrt{619}$ closer to 24 or 25?
37. Is $\sqrt{79}$ closer to 8 or 9?
38. Is $\sqrt{106}$ closer to 10 or 11?
39. Is $\sqrt{160}$ closer to 12 or 13?
40. Is $\sqrt{29}$ closer to 5 or 6?

Simplify the following square root problems.

41. $5\sqrt{27} + 7\sqrt{3}$
42. $\sqrt{40} - \sqrt{10}$
43. $\sqrt{64} + \sqrt{81}$
44. $8\sqrt{50} - 3\sqrt{32}$
45. $14\sqrt{5} + 8\sqrt{80}$
46. $\sqrt{63} \times \sqrt{28}$
47. $\dfrac{\sqrt{56}}{\sqrt{35}}$
48. $\sqrt{8} \times \sqrt{50}$
49. $\dfrac{\sqrt{20}}{\sqrt{45}}$
50. $5\sqrt{40} \times 3\sqrt{20}$
51. $2\sqrt{48} - \sqrt{12}$
52. $\dfrac{2\sqrt{5}}{\sqrt{30}}$
53. $\dfrac{3\sqrt{22}}{2\sqrt{3}}$
54. $\sqrt{72} \times 3\sqrt{27}$
55. $4\sqrt{5} + 8\sqrt{45}$

Convert the following numbers to scientific notation.

56. $5{,}340{,}000 =$ _____
57. $0.00000005874 =$ _____
58. $1{,}451 =$ _____
59. $0.0000041 =$ _____
60. $0.0004178 =$ _____
61. $105{,}000 =$ _____
62. $705{,}000{,}000 =$ _____
63. $0.0000747 =$ _____
64. $0.08 =$ _____
65. $105 =$ _____
66. $0.0048754 =$ _____
67. $62{,}400 =$ _____

Convert the following numbers from scientific notation to conventional numbers.

68. $5.204 \times 10^{-5} =$ _____
69. $1.02 \times 10^{7} =$ _____
70. $8.1 \times 10^{5} =$ _____
71. $2.0078 \times 10^{-4} =$ _____
72. $4.7 \times 10^{-3} =$ _____
73. $7.75 \times 10^{-8} =$ _____
74. $9.795 \times 10^{9} =$ _____
75. $3.51 \times 10^{2} =$ _____
76. $6.32514 \times 10^{3} =$ _____
77. $1.584 \times 10^{-6} =$ _____
78. $7.041 \times 10^{4} =$ _____
79. $4.09 \times 10^{-7} =$ _____

Chapter 7
Introduction to Algebra

ALGEBRA VOCABULARY

Vocabulary Word	Example	Definition
variable	$4x$ (x is the variable)	a letter that can be replaced by a number
coefficient	$4x$ (4 is the coefficient)	a number multiplied by a variable or variables
term	$5x^2 + x - 2$ terms	numbers or variables separated by $+$ or $-$ signs
constant	$5x + 2y + 4$ constant	a term that does not have a variable
numerical expression	$2^3 + 6 - 5$	two or more terms using only constants (numbers)
algebraic expression	$2x + 5^2 - 7$	two or more terms that include one or more variables
sentence	$2x = 7$ or $5 \leq x$	two algebraic expressions connected by $=, \neq, <, >, \leq, \geq,$ or $\approx$
equation	$4x = 8$	a sentence with an equal sign
inequality	$7x < 30$ or $x \neq 6$	a sentence with one of the following signs: $\neq, <, >, \leq,$ or $\geq$
base	6^3 ←— base	the number used as a factor
exponent	6^3 ←— exponent	the number of times the base is multiplied by itself

SUBSTITUTING NUMBERS FOR VARIABLES

These problems may look difficult at first glance, but they are very easy. Simply replace the variable with the number the variable is equal to, and solve the problems.

EXAMPLE 1: In the following problems, substitute 10 for *a*.

	PROBLEM	CALCULATION	SOLUTION
1.	$a + 1$	Simply replace the *a* with 10. $10 + 1$	11
2.	$17 - a$	$17 - 10$	7
3.	$9a$	This means multiply. 9×10	90
4.	$\frac{30}{a}$	This means divide. $30 \div 10$	3
5.	a^3	$10 \times 10 \times 10$	1000
* 6.	$5a + 6$	$(5 \times 10) + 6$	56

* **Note:** Be sure to do all multiplying and dividing before adding and subtracting.

EXAMPLE 2: In the following problems, let $x = 2$, $y = 4$, and $z = 5$.

	PROBLEM	CALCULATION	SOLUTION
1.	$5xy + z$	$5 \times 2 \times 4 + 5$	45
2.	$xz^2 + 5$	$2 \times 5^2 + 5 = 2 \times 25 + 5$	55
3.	$\frac{yz}{x}$	$(4 \times 5) \div 2 = 20 \div 2$	10

In the following problems, $t = 7$. Solve the problems.

1. $t + 3 =$ _____
2. $18 - t =$ _____
3. $\frac{21}{t} =$ _____
4. $3t - 5 =$ _____
5. $t^2 + 1 =$ _____
6. $2t - 4 =$ _____
7. $9t \div 3 =$ _____
8. $\frac{t^2}{7} =$ _____
9. $5t + 6 =$ _____

In the following problems $a = 4$, $b = -2$, $c = 5$, and $d = 10$. Solve the problems.

10. $4a + 2c =$ _____
11. $3bc - d =$ _____
12. $\frac{ac}{d} =$ _____
13. $d - 2a =$ _____
14. $a^2 - b =$ _____
15. $abd =$ _____
16. $5c - ad =$ _____
17. $cd + bc =$ _____
18. $\frac{6b}{a} =$ _____
19. $9a + b =$ _____
20. $5 + 3bc =$ _____
21. $d^2 + d + 1 =$ _____

UNDERSTANDING ALGEBRA WORD PROBLEMS

The biggest challenge to solving word problems is figuring out whether to add, subtract, multiply, or divide. Below is a list of key words and their meanings. This list does not include every situation you might see, but it includes the most common examples.

Words Indicating Addition	Example	Add
and	6 **and** 8	$6 + 8$
increased	The original price of $15 **increased** by $5.	$15 + 5$
more	3 coins and 8 **more**	$3 + 8$
more than	Josh has 10 points. Will has 5 **more than** Josh.	$10 + 5$
plus	8 baseballs **plus** 4 baseballs	$8 + 4$
sum	the **sum** of 3 and 5	$3 + 5$
total	the **total** of 10, 14, and 15	$10 + 14 + 15$

Words Indicating Subtraction	Example	Subtract
decreased	$16 **decreased** by $5	$16 - 5$
difference	the **difference** between 18 and 6	$18 - 6$
less	14 days **less** 5	$14 - 5$
less than	Jose completed 2 laps **less than** Mike's 9.	*$9 - 2$
left	Ray sold 15 out of 35 tickets. How many did he have **left**?	*$35 - 15$
lower than	This month's rainfall is 2 inches **lower than** last month's rainfall of 8 inches.	*$8 - 2$
minus	15 **minus** 6	$15 - 6$

* In subtraction word problems, you cannot always subtract the numbers in the order that they appear in the problem. Sometimes the first number should be subtracted from the last. You must read each problem carefully.

Words Indicating Multiplication	Example	Multiply
double	Her $1000 profit **doubled** in a month.	1000×2
half	**Half** of the $600 collected went to charity.	$\frac{1}{2} \times 600$
product	the **product** of 4 and 8	4×8
times	Li scored 3 **times** as many points as Ted who only scored 4.	3×4
triple	The bacteria **tripled** its original colony of 10,000 in just one day.	$3 \times 10,000$
twice	Ron has 6 Cd's. Tom has **twice** as many.	2×6

Words Indicating Division	Example	Divide
divide into, by, or among	The group of 70 **divided into** 10 teams.	$70 \div 10$ or $\frac{70}{10}$
quotient	the **quotient** of 30 and 6	$30 \div 6$ or $\frac{30}{6}$

Match the phrase on the left with the correct algebraic expression on the right. The answers on the right will be used more than once.

1. _____ 2 more than y A. $y - 2$
2. _____ 2 divided into y B. $2y$
3. _____ 2 less than y C. $y + 2$
4. _____ twice y D. $\frac{y}{2}$
5. _____ the quotient of y and 2   E. $2 - y$
6. _____ y increased by 2
7. _____ 2 less y
8. _____ the product of 2 and y
9. _____ y decreased by 2
10. _____ y doubled
11. _____ 2 minus y
12. _____ the total of 2 and y

Now practice writing parts of algebraic expressions from the following word problems.

EXAMPLE: the product of 3 and a number, t Answer: $3t$

13. 3 less than x _____
14. y divided among 10 _____
15. the sum of t and 5 _____
16. n minus 14 _____
17. 5 times k _____
18. the total of z and 12 _____
19. double the number b _____
20. x increased by 1 _____
21. the quotient of t and 4 _____
22. half of a number y _____

23. bacteria culture, b, doubled _____
24. triple John's age, y _____
25. a number, n, plus 4 _____
26. quantity, t, less 6 _____
27. 18 divided by a number, x _____
28. n feet lower than 10 _____
29. 3 more than p _____
30. the product of 4 and m _____
31. a number, y, decreased by 20 _____
32. 5 times as much as x _____

If a word problem contains the word "sum" or "difference," put the numbers that "sum" or "difference" refer to in parentheses to be added or subtracted first. Do not separate them. Look at the examples below.

EXAMPLES:

	RIGHT	**WRONG**
sum of 2 and 4, times 5	$5(2+4) = 30$	$2 + 4 \times 5 = 22$
the sum of 4 and 6, divided by 2	$\frac{(4+6)}{2} = 5$	$4 + \frac{6}{2} = 7$
4 times the difference between 10 and 5	$4(10-5) = 20$	$4 \times 10 - 5 = 35$
20 divided by the difference between 4 and 2	$\frac{20}{(4-2)} = 10$	$20 \div 4 - 2 = 3$
the sum of x and 4, multiplied by 2	$2(x+4) = 2x + 8$	$x + 4 \times 2 = x + 8$

Change the following phrases into algebraic expressions.

1. 5 times the sum of x and 6

2. the difference between 5 and 3, divided by 4

3. 30 divided by the sum of 2 and 3

4. twice the sum of 10 and x

5. the difference between x and 9, divided by 10

6. 7 times the difference between x and 4

7. 9 multiplied by the sum of 3 and 4

8. the difference between x and 5, divided by 6

9. x divided by the sum of 4 and 9

10. x minus 5, times 10

11. 100 multiplied by the sum of x and 6

12. twice the difference between 3 and x

13. 4 times the sum of 5 and 1

14. 5 times the difference between 4 and 2

15. 12 divided by the sum of 2 and 4

16. four minus x, multiplied by 2

Look at the examples below for more phrases that may be used in algebra word problems.

EXAMPLES:

one-half of the sum of x and 4	$\frac{1}{2}(x+4)$ or $\frac{x+4}{2}$
six more than four times a number, x	$6 + 4x$
100 decreased by the product of a number, x, and 5	$100 - 5x$
ten less than the product of 3 and x	$3x - 10$

Change the following phrases into algebraic expressions.

1. one-third of the sum of x and 5

2. three more than the product of a number, x, and 7

3. ten less than the sum of t and 4

4. the product of 4 and n, minus 3

5. 15 less the sum of 3 and x

6. the difference of 10, and 3 times a number, n

7. one-fifth of t

8. the product of 3 and x, minus 14

9. x times the difference between 4 and x

10. five plus the quotient of x and 6

11. the sum of 5 and k, divided by 2

12. one less than the product of 3 and x

13. 5 increased by one-half of a number, n

14. 10 more than twice x

15. six subtracted from four times m

16. 8 times x, subtracted from 20

SETTING UP ALGEBRA WORD PROBLEMS

So far, you have seen only the first part of algebra word problems. To complete an algebra problem, an equal sign must be added. The words "**is**" or "**are**" as well as "**equal(s)**" signal that you should add an equal sign.

EXAMPLE: Double Jake's age, x, minus 4 is 22.

$$2x - 4 = 22$$

Translate the following word problems into algebra problems. DO NOT find the solutions to the problems yet.

1. Triple the original number, n, is 2,700.

2. The product of a number, y, and 5 is equal to 15.

3. Four times the difference of a number, x, and 2 is 20.

4. The total, t, divided into 5 groups is 45.

5. The number of parts in inventory, p, minus 54 parts sold today is 320.

6. One-half an amount, x, added to $50 is $262.

7. One hundred seeds divided by 5 rows equals n number of seeds per row.

8. A number, y, less than 50 is 82.

9. His base pay of $200 increased by his commission, x, is $500.

10. Seventeen more than half a number, h, is 35.

11. This month's sales of $2,300 are double January's sales, x.

12. The quotient of a number, w, and 4 is 32.

13. Six less a number, d, is 12.

14. Four times the sum of a number, y, and 10 is 48.

15. We started with x number of students. When 5 moved away, we had 42 left.

16. A number, b, divided into 36 parts is 12.

MATCHING ALGEBRAIC EXPRESSIONS

Match each set of algebraic expressions with the correct phrase underneath them.

1. ____ $2x + 5$
2. ____ $2(x + 5)$
3. ____ $2x - 5$
4. ____ $2(x - 5)$

A. twice the sum of x and 5
B. five less than the product of 2 and x
C. five more than the product of 2 and x
D. two times the difference of x and 5

5. ____ $4(y - 2)$
6. ____ $\frac{y-2}{4}$
7. ____ $4y - 2$
8. ____ $\frac{y}{4} - 2$

A. two less than the product of y and 4
B. the difference of y and 2 divided by 4
C. two less than one-fourth of y
D. four times the difference of y and 2

9. ____ $5y + 8$
10. ____ $5(y + 8)$
11. ____ $8y + 5$
12. ____ $8(y + 5)$

A. eight times the sum of y and 5
B. eight more than the product of 5 and y
C. five more than eight times y
D. five multiplied by the sum of y and 8

13. ____ $9 - x = 7$
14. ____ $x - 9 = 7$
15. ____ $9 - 7 = x$

A. nine less than x is 7
B. nine less x is 7
C. the difference between 9 and 7 is x

16. ____ $\frac{n+5}{2} = 10$
17. ____ $\frac{n}{2} + 5 = 10$
18. ____ $\frac{1}{2}n - 5 = 10$

A. one-half the sum of n and 5 is 10
B. five less than half of n is 10
C. five added to half of n is 10

19. ____ $x + \frac{4}{5} = 8$
20. ____ $\frac{x}{5} + 4 = 8$
21. ____ $\frac{x+4}{5} = 8$

A. the sum of x and 4, divided by 5 is 8
B. x added to the quotient of 4 and 5 is 8
C. four more than x divided by 5 is 8

22. ____ $7t + 1 = 5$
23. ____ $7(t + 1) = 5$
24. ____ $7t = 5$

A. one more than seven times t is 5
B. seven times the sum of t and 1 is 5
C. the product of seven and t is 5

CHANGING ALGEBRA WORD PROBLEMS TO ALGEBRAIC EQUATIONS

EXAMPLE: There are 3 people who have a total weight of 595 pounds. Sally weighs 20 pounds less than Jessie. Rafael weighs 15 pounds more than Jessie. How much does Jessie weigh?

Step 1: Notice everyone's weight is given in terms of Jessie. Sally weighs 20 pounds less than Jessie. Rafael weighs 15 pounds more than Jessie. First, we write everyone's weight in terms of Jessie, j.

$$\text{Jessie} = j$$
$$\text{Sally} = j - 20$$
$$\text{Rafael} = j + 15$$

Step 2: We know that all three together weigh 595 pounds. We write the sum of everyone's weight equal to 595.

$$j + j - 20 + j + 15 = 595$$

We will learn to solve these problems in the next chapter.

Change the following word problems to algebraic equations.

1. Fluffy, Spot, and Shampy have a combined age in dog years of 91. Spot is 14 years younger than Fluffy. Shampy is 6 years older than Fluffy. What is Fluffy's age, f, in dog years?

2. Jerry Marcosi puts 5% of the amount he makes per week into a retirement account, r. He is paid $11.00 per hour and works 40 hours per week for a certain number of weeks, w. Write an equation to help him find out how much he puts into his retirement account.

3. A furniture store advertises a 40% off liquidation sale on all items. What would the sale price (p) be on a $2530 dining room set?

4. Kyle Thornton buys an item which normally sells for a certain price, x. Today the item is selling for 25% off the regular price. A sales tax of 6% is added to the equation to find the final price, f.

5. Tamika Francois runs a floral shop. On Tuesday, Tamika sold a total of $600 worth of flowers. The flowers cost her $100, and she paid an employee to work 8 hours for a given wage, w. Write an equation to help Tamika find her profit, p, on Tuesday.

6. Sharice is a waitress at a local restaurant. She makes an hourly wage, $3.50, plus she receives tips. On Monday, she worked 6 hours and received tip money, t. Write an equation showing what Sharice made on Monday, y.

7. Jenelle buys x shares of stock in a company at $34.50 per share. She later sells the shares at $40.50 per share. Write an equation to show how much money, m, Jenelle has made.

CHAPTER 7 REVIEW

Solve the following problems, using $x = 2$.

1. $3x + 4 =$ ____
2. $\dfrac{6x}{4} =$ ____
3. $x^2 - 5 =$ ____
4. $\dfrac{x^3 + 8}{2} =$ ____
5. $12 - 3x =$ ____
6. $x - 5 =$ ____
7. $-5x + 4 =$ ____
8. $9 - x =$ ____
9. $2x + 2 =$ ____

Solve the following problems. Let $w = -1$, $y = 3$, $z = 5$.

10. $5w - y =$ ____
11. $wyz + 2 =$ ____
12. $z - 2w =$ ____
13. $\dfrac{3z + 5}{wz} =$ ____
14. $\dfrac{6w}{y} + \dfrac{z}{w} =$ ____
15. $25 - 2yz =$ ____
16. $-2y + 3 =$ ____
17. $4w - (yw) =$ ____
18. $7y - 5z =$ ____

Write out the algebraic expression given in each word problem.

19. three less the sum of x and 5   _____

20. double Amy's age, a _____

21. the number of bacteria, b, tripled   _____

22. five less than the product of 5 and y _____

23. half of a number, n, less 15   _____

24. the quotient of a number, x, and 6   _____

For questions 25-27, write an equation to match each problem.

25. Calista earns $450 per week for a 40 hour work week plus $16.83 per hour for each hour of overtime after 40 hours. Write an equation that would be used to determine her weekly wages where w is her wages, and v is the number of overtime hours worked.

    _____

26. Daniel purchased a 1 year CD, c, from a bank. He bought it at an annual interest rate of 6%. After 1 year, Daniel cashes in the CD. What is the total amount it is worth?

    _____

27. Omar is a salesman. He earns an hourly wage of $8.00 per hour plus he receives a commission of 7% on the sales he makes. Write an equation which would be used to determine his weekly salary, w, where x is the number of hours worked, and y is the amount of sales for the week.

    _____

28. Tom earns $500 per week before taxes are taken out. His employer takes out a total of 33% for state, federal, and Social Security taxes. Which expression below will help Tom figure his net pay check?

   A.   500 − .33
   B.   500 ÷ .33
   C.   500 + .33(500)
   D.   500 − .33(500)

29. Rosa has to pay $100 of her medical expenses in a year before she qualifies for her insurance company to begin paying. After paying the $100 "deductible," her insurance company will pay 80% of her medical expenses. This year, her total medical expenses came to $960.00. Which expression below shows how much her insurance company will pay?

   A.   .80(960 − 100)
   B.   100 + (960 ÷ .80)
   C.   960(100 − .80)
   D.   .80(960 + 100)

30. A plumber charges $45.00 per hour plus a $25.00 service charge. If a represents his total charges in dollars, and b represents the number of hours worked, which formula below could the plumber use to calculate his total charges?

   A. $a = 45 + 25b$
   B. $a = 45 + 25 + b$
   C. $a = 45b + 25$
   D. $a = (45)(25) + b$

31. In 1999, Bell Computers announced to its sales force to expect a 2.6% price increase on all computer equipment in the year 2000. A certain sales representative wanted to see how much the increase would be on a computer, c, that sold for $2,200 in 1999. Which expression below will help him find the cost of the computer in the year 2000?

   A.   .26(2200)
   B.   2200 − .026(2200)
   C.   2200 + .026(2200)
   D.   .026(2200) − 2200

32. Juan sold a boat that he bought 5 years ago. He sold it for 60% less than he originally paid for it. If the original cost is b, write an expression that shows how much he sold the boat for.

   _____

33. Toshi is going to get a 7% raise after he works at his job for 1 year. If s represents his starting salary, write an expression that shows how much he will make after his raise.

   _____

Chapter 8
Solving One-Step Equations and Inequalities

ONE-STEP ALGEBRA PROBLEMS WITH ADDITION AND SUBTRACTION

You have been solving algebra problems since second grade by filling in blanks. For example, $5 + \_\_\_ = 8$. The answer is 3. You can solve the same kind of problems using algebra. The problems only look a little different because the blank has been replaced with a letter. The letter is called a **variable**.

EXAMPLE: Arithmetic $\quad 5 + \_\_\_ = 14$
Algebra $\quad 5 + x = 14$

The goal in any algebra problem is to move all the numbers to one side of the equal sign and have the letter (called a **variable**) on the other side. In this problem, the 5 and the "x" are on the same side. The 5 is added to x. To move it, do the **opposite** of **add**. The **opposite** of **add** is **subtract**, so subtract 5 from both sides of the equation. Now the problem looks like this:

$5 + x = 14$ $\quad$ To check your answer, put 9 in the place of x
$-5 \quad\quad -5$ $\quad$ in the original problem. Does $5 + 9 = 14$?
$\quad x = 9$ $\quad$ Yes, it does.

EXAMPLE: $y - 16 = 27$ $\quad$ Again, the 16 has to move. To move it to the other side of the equation, we do the **opposite** of **subtract**. We **add** 16 to both sides.

$y - 16 = 27$
$+16 \quad +16$ $\quad$ Check by putting 43 in place of the y in the original problem.
$y \quad\quad = 43$ $\quad$ Does $43 - 16 = 27$? Yes.

Solve the problems below.

1. $n + 9 = 27$
2. $12 + y = 55$
3. $51 + v = 67$
4. $f + 16 = 31$
5. $5 + x = 23$
6. $15 + x = 24$
7. $w - 14 = 89$
8. $t - 26 = 20$
9. $m - 12 = 17$
10. $c - 7 = 21$
11. $k - 5 = 29$
12. $a + 17 = 45$
13. $d + 26 = 56$
14. $15 + x = 56$
15. $y + 19 = 32$
16. $t - 16 = 28$
17. $m + 14 = 37$
18. $y - 21 = 29$
19. $f + 7 = 31$
20. $h - 12 = 18$
21. $r - 12 = 37$
22. $h - 17 = 22$
23. $x - 37 = 46$
24. $r - 11 = 28$
25. $t - 5 = 52$

MORE ONE-STEP ALGEBRA PROBLEMS

Sometimes the answer to the algebra problem is a negative number. The problems work the same. Study the **examples** below.

EXAMPLES:

$$\begin{array}{r} n + 8 = 6 \\ -8 \quad -8 \\ \hline n \quad = (-2) \end{array} \qquad \begin{array}{r} x - 10 = -14 \\ +10 \quad +10 \\ \hline x \quad = (-4) \end{array} \qquad \begin{array}{r} y - (-6) = (-2) \\ +(-6) \quad +(-6) \\ \hline y \quad = (-8) \end{array}$$

Solve the problems below.

1. $w + 4 = (-6)$
2. $q - 8 = (-9)$
3. $j + 7 = 1$
4. $y + 14 = 6$
5. $k - 5 = (-8)$
6. $h - 7 = (-2)$
7. $7 + d = (-3)$
8. $(-7) + k = (-4)$
9. $(-4) + h = 8$
10. $(-5) + g = -2$
11. $y - (-8) = (-5)$
12. $x + 21 = (-2)$
13. $w - (-3) = 8$
14. $q - (-6) = 12$
15. $z + 8 = 3$
16. $(-4) + m = 4$
17. $(-10) + r = (-2)$
18. $(-6) + b = 9$
19. $p - (-7) = (-2)$
20. $q - 5 = (-11)$
21. $x + 17 = (-4)$
22. $(-9) + x = 14$
23. $(-17) + r = -12$
24. $(-3) + y = 19$
25. $t - (-2) = (-8)$
26. $d - 9 = (-16)$
27. $x + 16 = -3$
28. $w + 14 = (-9)$
29. $q + 12 = (-5)$
30. $j + (-7) = (-1)$
31. $y - (-9) = 8$
32. $x + 12 = 6$

Copyright © American Book Company

ONE-STEP ALGEBRA PROBLEMS WITH MULTIPLICATION AND DIVISION

Solving one-step algebra problems with multiplication and division is just as easy as solving addition and subtraction problems. Again, you perform the **opposite** operation. If the problem is a **multiplication** problem, you **divide** to find the answer. If it is a **division** problem, you **multiply** to find the answer. Carefully read the examples below, and you will see how easy they are.

EXAMPLE 1: $4x = 20$ ($4x$ means **4 times** x. 4 is the **coefficient** of x.)

The goal is to get the numbers on one side of the equal sign and the variable x on the other side. In this problem, the 4 and x are on the same side of the equal sign. The 4 has to be moved over. $4x$ means 4 times x. The opposite of **multiply** is **divide**. If we divide both sides of the equation by 4, we will find the answer.

$4x = 20$ We need to divide both sides by 4.

This means divide by 4. $\dfrac{4x}{4} = \dfrac{20}{4}$ We see that $1x = 5$ so $x = 5$

When you put 5 in place of x in the original problem, it is correct. $4 \times 5 = 20$

EXAMPLE 2: $\dfrac{y}{4} = 2$ This problem means y divided by 4 is equal to 2. In this case, the opposite of **divide** is **multiply**. We need to multiply both sides of the equation by 4.

$4 \times \dfrac{y}{4} = 2 \times 4$ so $y = 8$

When you put 8 in place of y in the original problem, it is correct. $\dfrac{8}{4} = 2$

Solve the problems below.

1. $2x = 14$
2. $\dfrac{w}{5} = 11$
3. $3h = 45$
4. $10y = 30$

5. $5a = 60$
6. $\dfrac{x}{3} = 9$
7. $6d = 66$
8. $\dfrac{w}{9} = 3$

9. $7r = 98$
10. $\dfrac{y}{3} = 2$
11. $\dfrac{x}{4} = 36$
12. $\dfrac{r}{4} = 7$

13. $8t = 96$
14. $\dfrac{z}{2} = 15$
15. $\dfrac{n}{9} = 5$
16. $4z = 24$

17. $6d = 84$
18. $\dfrac{t}{3} = 3$
19. $\dfrac{m}{6} = 9$
20. $9p = 72$

Sometimes the answer to the algebra problem is a **fraction**. Read the example below, and you will see how easy it is.

> **EXAMPLE**
>
> $4x = 5$ Problems like this are solved just like the problems on the previous page. The only difference is that the answer is a **fraction**.
>
> In this problem, the 4 is **multiplied** by x. To solve, we need to divide both sides of the equation by 4.
>
> $4x = 5$ Now **divide** by 4. $\frac{4x}{4} = \frac{5}{4}$ Now cancel. $\frac{\cancel{4}x}{\cancel{4}} = \frac{5}{4}$ so $x = \frac{5}{4}$
>
> When you put $\frac{5}{4}$ in place of x in the original problem, it is correct.
>
> $4 \times \frac{5}{4} = 5$ Now cancel. ⟶ $\cancel{4} \times \frac{5}{\cancel{4}} = 5$ so $5 = 5$

Solve the problems below. Some of the answers will be fractions. Some answers will be integers.

1. $2x = 3$
2. $4y = 5$
3. $5t = 2$
4. $12b = 144$
5. $9a = 72$
6. $8y = 16$
7. $7x = 21$

8. $4z = 64$
9. $7x = 126$
10. $6p = 10$
11. $2n = 9$
12. $5x = 11$
13. $15m = 180$
14. $5h = 21$

15. $3y = 8$
16. $2t = 10$
17. $3b = 2$
18. $5c = 14$
19. $4d = 3$
20. $5z = 75$
21. $9y = 4$

22. $7d = 12$
23. $2w = 13$
24. $9g = 81$
25. $6a = 18$
26. $2p = 16$
27. $15w = 3$
28. $5x = 13$

MULTIPLYING AND DIVIDING WITH NEGATIVE NUMBERS

EXAMPLE 1: $-3x = 15$ In the problem, -3 is **multiplied** by x. To find the solution, we must do the opposite. The opposite of **multiply** is **divide**. We must **divide** both sides of the equation by -3.

$$\frac{-3x}{-3} = \frac{15}{-3}$$ Then cancel. $\frac{-3x}{-3} = \frac{15}{-3}$ $x = -5$

EXAMPLE 2: $\frac{y}{-4} = -20$ In this problem, y is **divided** by -4. To find the answer, do the opposite. **Multiply** both sides by -4.

$$-4 \times \frac{y}{-4} = (-20) \times (-4)$$ so $y = 80$

EXAMPLE 3: $-6a = 2$ The answer to an algebra problem can also be a negative fraction.

$$\frac{-6a}{-6} = \frac{2}{-6}$$ ← reduce to get $a = \frac{1}{-3}$ or $-\frac{1}{3}$

Note: A negative fraction can be written several different ways.

$$\frac{1}{-3} = \frac{-1}{3} = -\frac{1}{3} = -\left(\frac{1}{3}\right)$$

All mean the same thing.

Solve the problems below. Reduce any fractions to lowest terms.

1. $2z = -6$
2. $\frac{y}{-5} = 20$
3. $-6k = 54$
4. $4x = -24$
5. $\frac{t}{7} = -4$

6. $\frac{r}{-2} = -10$
7. $9x = -72$
8. $\frac{x}{-6} = 3$
9. $\frac{w}{-11} = 5$
10. $5y = -35$

11. $\frac{x}{-4} = -9$
12. $7t = -49$
13. $-14x = -28$
14. $\frac{m}{3} = -12$
15. $-8z = 32$

16. $-15w = -60$
17. $\frac{y}{-9} = -4$
18. $\frac{d}{8} = -7$
19. $-12v = 36$
20. $\frac{c}{-6} = -6$

21. $-4x = -3$ 26. $\dfrac{b}{-2} = -14$ 31. $-9y = -1$ 36. $-8d = -12$

22. $-12y = 7$ 27. $-24x = -6$ 32. $\dfrac{d}{5} = -10$ 37. $-24w = 9$

23. $\dfrac{a}{-2} = 22$ 28. $-6p = 42$ 33. $\dfrac{z}{-13} = -2$ 38. $\dfrac{y}{-9} = -6$

24. $-18b = 6$ 29. $\dfrac{x}{-23} = -1$ 34. $-5c = 45$ 39. $-9a = -18$

25. $13a = -36$ 30. $7x = -7$ 35. $2d = -3$ 40. $\dfrac{p}{-2} = 15$

VARIABLES WITH A COEFFICIENT OF NEGATIVE ONE

The answer to an algebra problem should not have a negative sign in front of the variable. For example, the problem $-x = 5$ is not completely simplified. Study the examples below to learn how to finish simplifying this problem.

EXAMPLE 1: $-x = 5$ $-x$ means the same thing as $-1x$ or -1 times x. To simplify this problem, **multiply** by -1 on both sides of the equation.

$$(-1)(-1x) = (-1)(5) \quad \text{so} \quad x = -5$$

EXAMPLE 2: $-y = -3$ Solve the same way.

$$(-1)(-y) = (-1)(-3) \quad \text{so} \quad y = 3$$

Simplify the following equations.

1. $-w = 14$ 4. $-x = -25$ 7. $-p = -34$ 10. $-v = -9$

2. $-a = 20$ 5. $-y = -16$ 8. $-m = 81$ 11. $-k = 13$

3. $-x = -15$ 6. $-t = 62$ 9. $-w = 17$ 12. $-q = 7$

GRAPHING INEQUALITIES

An inequality is a sentence that contains a ≠, <, >, ≤, or ≥ sign. Look at the following graphs of inequalities on a number line.

NUMBER LINE

$x < 3$ is read "x is less than 3."

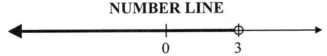

There is no line under the < sign, so the graph uses an **open** endpoint to show x is less than 3 but does not include 3.

$x \leq 5$ is read "x is less than or equal to 5."

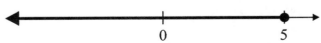

If you see a line under < or > (≤ or ≥), the endpoint is filled in. The graph uses a **closed** circle because the number 5 **is** included in the graph.

$x > -2$ is read "x is greater than −2."

$x \geq 1$ is read "x is greater than or equal to 1."

There can be more than one inequality sign. For example:

$-2 \leq x < 4$ is read "−2 is less than or equal to x, and x is less than 4."

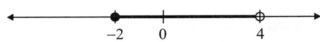

$x < 1$ or $x \geq 4$ is read "x is less than 1, or x is greater than or equal to 4."

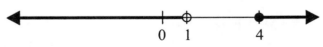

Graph the solution sets of the following inequalities.

1. $x > 8$
2. $x \leq 5$
3. $-5 < x < 1$
4. $x > 7$
5. $1 \leq x < 4$
6. $x < -2$ and $x > 1$
7. $x \geq 10$
8. $x < 4$
9. $x \leq 3$ and $x \geq 5$
10. $x < -1$ and $x > 1$

Give the inequality represented by each of the following number lines.

11. (closed at 0, extending right) _____
12. (open at −4, closed at 10) _____
13. (closed at 2 and 4) _____
14. (open at 8, extending left) _____
15. (open at −10, open at −4) _____
16. (closed at −1, open at 3) _____
17. (closed at −5, extending right) _____
18. (open at 6, extending left) _____
19. (open at −6 and −4) _____
20. (closed at 4, open at 7) _____

SOLVING INEQUALITIES BY ADDITION AND SUBTRACTION

If you add or subtract the same number to both sides of an inequality, the inequality remains the same. It works just like an equation.

EXAMPLE: Solve and graph the solution set for $x - 2 \leq 5$.

Step 1: Add 2 to both sides of the inequality.

$$\begin{array}{r} x - 2 \leq 5 \\ +2 \ +2 \\ \hline x \ \leq 7 \end{array}$$

Step 2: Graph the solution set for the inequality.

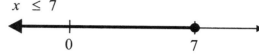

Solve and graph the solution set for the following inequalities.

1. $x + 5 > 3$
2. $x - 10 < 5$
3. $x - 2 \leq 1$
4. $9 + x \geq 7$
5. $x - 4 > -2$
6. $x + 11 \leq 20$
7. $x - 3 < -12$
8. $x + 6 \geq -3$
9. $x + 12 \leq 8$
10. $15 + x > 5$
11. $x - 6 < -2$
12. $x + 7 \geq 4$
13. $14 + x \leq 8$
14. $x - 8 > 24$
15. $x + 1 \leq 12$
16. $11 + x \geq 11$
17. $x - 3 < 17$
18. $x + 9 > -4$
19. $x + 6 \leq 14$
20. $x - 8 \geq 19$

SOLVING INEQUALITIES BY MULTIPLICATION AND DIVISION

If you multiply or divide both sides of an inequality by a **positive** number, the inequality symbol stays the same. However, if you multiply or divide both sides of an inequality by a **negative** number, **you must reverse the direction of the inequality symbol**.

EXAMPLE 1: Solve and graph the solution set for $4x \leq 20$.

Step 1: Divide both sides of the inequality by 4. $\dfrac{\cancel{4}x}{\cancel{4}} \leq \dfrac{\cancel{20}^5}{\cancel{4}}$

Step 2: Graph the solution. $x \leq 5$

EXAMPLE 2: Solve and graph the solution set for $6 > -\dfrac{x}{3}$.

Step 1: Multiply both sides of the inequality by -3, and **reverse the direction of the symbol**.

$$(-3) \times 6 < \dfrac{x}{\cancel{-3}} \times (\cancel{-3})$$

Step 2: Graph the solution. $-18 < x$

Solve and graph the following inequalities.

1. $\dfrac{x}{5} > 4$

2. $2x \leq 24$

3. $-6x \geq 36$

4. $\dfrac{x}{10} > -2$

5. $-\dfrac{x}{4} > 8$

6. $-7x \leq -49$

7. $-3x > 18$

8. $-\dfrac{x}{7} \geq 9$

9. $9x \leq 54$

10. $\dfrac{x}{8} > 1$

11. $-\dfrac{x}{9} \leq 3$

12. $-4x < -12$

13. $-\dfrac{x}{2} \geq -20$

14. $10x \leq 30$

15. $\dfrac{x}{12} > -4$

16. $-6x < 24$

CHAPTER 8 REVIEW

Solve the following one-step algebra problems.

1. $5y = -25$
2. $x + 4 = 24$
3. $d - 11 = 14$
4. $\dfrac{a}{6} = -8$
5. $-t = 2$
6. $-14b = 12$
7. $\dfrac{c}{-10} = -3$
8. $z - 15 = -19$
9. $-13d = 4$
10. $\dfrac{x}{-14} = 2$

11. $-4k = -12$
12. $y + 13 = 27$
13. $15 + h = 4$
14. $14p = 2$
15. $\dfrac{b}{4} = 11$
16. $p - 26 = 12$
17. $x + (-2) = 5$
18. $m + 17 = 27$
19. $\dfrac{k}{-4} = 13$
20. $-18a = -7$

21. $21t = -7$
22. $z - (-9) = 14$
23. $23 + w = 28$
24. $n - 35 = -16$
25. $-a = 26$
26. $-19 + f = -9$
27. $\dfrac{w}{11} = 3$
28. $-7y = 28$
29. $x + 23 = 20$
30. $z - 12 = -7$

31. $-16 + g = 40$
32. $\dfrac{m}{-3} = -9$
33. $d + (-6) = 17$
34. $-p = 47$
35. $k - 16 = 5$
36. $9y = -3$
37. $-2z = -36$
38. $10h = 12$
39. $w - 16 = 4$
40. $y + 10 = -8$

Graph the solution sets of the following inequalities.

41. $x \leq -3$

42. $x > 6$

43. $x < -2$

44. $x \geq 4$

Give the inequality represented by each of the following number lines.

45. (closed dot at 3, arrow right) _____

46. (closed dot at 5, open dot at 9) _____

47. (open dot at -2, closed dot at 0, arrow right) _____

48. (open dot at 10, arrow right) _____

Solve and graph the solution set for the following inequalities.

49. $x - 2 > 8$

50. $4 + x < -1$

51. $6x \geq 54$

52. $-2x \leq 8$

53. $\frac{x}{2} > -1$

54. $-x < -9$

55. $-\frac{x}{3} \leq 5$

56. $x + 10 \leq 4$

57. $x - 6 \geq -2$

58. $7x < -14$

59. $-3x > -12$

60. $-\frac{x}{6} \leq -3$

Chapter 9 Solving Multi-Step Equations and Inequalities

TWO-STEP ALGEBRA PROBLEMS

In the following two-step algebra problems, **additions** or **subtractions** are performed **first** and *then* **divisions** or **multiplications**.

EXAMPLE 1: $-4x + 7 = 31$

Step 1: Subtract 7 from both sides.

$$-4x + 7 = 31$$
$$\; -7 \;\; -7$$
$$-4x = 24$$

Step 2: Divide both sides by -4.

$$\frac{-4x}{-4} = \frac{24}{-4} \quad \text{so} \quad x = -6$$

EXAMPLE 2: $-8 - y = 12$

Step 1: Add 8 to both sides.
$$-8 - y = 12$$
$$+8 \quad\;\; +8$$
$$-y = 20$$

Step 2: **REMEMBER:** To finish solving an algebra problem with a negative sign in front of the variable, multiply both sides by -1. The variable needs to be positive in the answer.

$$(-1)(-y) = (-1)(20) \quad \text{so} \quad y = -20$$

Solve the two-step algebra problems below.

1. $6x - 4 = -34$
2. $5y - 3 = 32$
3. $8 - t = 1$
4. $10p - 6 = -36$
5. $11 - 9m = -70$

6. $4x - 12 = 24$
7. $3x - 17 = -41$
8. $9d - 5 = 49$
9. $10h + 8 = 78$
10. $-6b - 8 = 10$

11. $-g - 24 = -17$
12. $-7k - 12 = 30$
13. $9 - 5r = 64$
14. $6y - 14 = 34$
15. $12f + 15 = 51$

16. $21t + 17 = 80$
17. $20y + 9 = 149$
18. $15p - 27 = 33$
19. $22h + 9 = 97$
20. $-5 + 36w = 175$

TWO-STEP ALGEBRA PROBLEMS WITH FRACTIONS

An algebra problem may contain a fraction. Study the following example to understand how to solve algebra problems that contain a fraction.

EXAMPLE: $\frac{x}{2} + 4 = 3$

Step 1: $\quad \frac{x}{2} + 4 = 3$ 	Subtract 4 from both sides.
$\quad\quad\quad\quad\;\; -4 \quad -4$

Step 2: $\quad \frac{x}{2} = -1$ 	Now this looks like the one-step algebra problems you solved in Chapter 8. Multiply both sides by 2 to solve for x.

$\frac{x}{\cancel{2}} \times \cancel{2} = -1 \times 2 \quad x = -2$

Simplify the following algebra problems.

1. $4 + \frac{y}{3} = 7$
2. $\frac{a}{2} + 5 = 12$
3. $\frac{w}{5} - 3 = 6$
4. $\frac{x}{9} - 9 = -5$
5. $\frac{b}{6} + 2 = -4$
6. $7 + \frac{z}{2} = -13$
7. $\frac{x}{2} - 7 = 3$
8. $\frac{c}{5} + 6 = -2$

9. $3 + \frac{x}{11} = 7$
10. $16 + \frac{m}{6} = 14$
11. $\frac{p}{3} + 5 = -2$
12. $\frac{t}{8} + 9 = 3$
13. $\frac{v}{7} - 8 = -1$
14. $5 + \frac{h}{10} = 8$
15. $\frac{k}{7} - 9 = 1$
16. $\frac{y}{4} + 13 = 8$

17. $15 + \frac{z}{14} = 13$
18. $\frac{b}{6} - 9 = -14$
19. $\frac{d}{3} + 7 = 12$
20. $10 + \frac{v}{6} = 4$
21. $2 + \frac{p}{4} = -6$
22. $\frac{t}{7} - 9 = -5$
23. $\frac{a}{10} - 1 = 3$
24. $\frac{a}{8} + 16 = 9$

MORE TWO-STEP ALGEBRA PROBLEMS WITH FRACTIONS

Study the following example to understand how to solve algebra problems that contain a different type of fraction.

EXAMPLE: $\dfrac{x+2}{4} = 3$ In this example, "$x + 2$" is divided by 4, and not *just* the x or the 2.

Step 1: $\dfrac{x+2}{\cancel{4}} \times \cancel{4} = 3 \times 4$ First, multiply both sides by 4 to eliminate the fraction.

Step 2: $\quad x + 2 = 12$ Next, subtract 2 from both sides.
$\qquad\qquad \underline{-2 \ -2}$
$\qquad\qquad \ \ x = 10$

Solve the following problems.

1. $\dfrac{x+1}{5} = 4$

2. $\dfrac{z-9}{2} = 7$

3. $\dfrac{b-4}{4} = -5$

4. $\dfrac{y-9}{3} = 7$

5. $\dfrac{d-10}{-2} = 12$

6. $\dfrac{w-10}{-8} = -4$

7. $\dfrac{x-1}{-2} = -5$

8. $\dfrac{c+40}{-5} = -7$

9. $\dfrac{13+h}{2} = 12$

10. $\dfrac{k-10}{3} = 9$

11. $\dfrac{a+11}{-4} = 4$

12. $\dfrac{x-20}{7} = 6$

13. $\dfrac{t+2}{6} = -5$

14. $\dfrac{b+1}{-7} = 2$

15. $\dfrac{f-9}{3} = 8$

16. $\dfrac{4+w}{6} = -6$

17. $\dfrac{3+t}{3} = 10$

18. $\dfrac{x+5}{5} = -3$

19. $\dfrac{g+3}{2} = 11$

20. $\dfrac{k+1}{-6} = 5$

21. $\dfrac{y-14}{2} = -8$

22. $\dfrac{z-4}{-2} = 13$

23. $\dfrac{w+2}{15} = -1$

24. $\dfrac{3+h}{3} = 6$

COMBINING LIKE TERMS

In an algebra problem, **terms** are separated by $+$ and $-$ signs. The expression $5x - 4 - 3x + 7$ has 4 terms: $5x$, 4, $3x$, and 7. Terms having the same variable can be combined (added or subtracted) to simplify the expression. In this case, $5x - 4 - 3x + 7$ simplifies to $2x + 3$.

$$5x - 3x \quad -4 + 7$$

Simplify the following expressions.

1. $7x + 12x =$ _____
2. $8y - 5y + 8 =$ _____
3. $4 - 2c + 9 =$ _____
4. $11a - 16 - a =$ _____
5. $9w + 3w + 3 =$ _____
6. $-5x + x + 2x =$ _____
7. $w - 15 + 9w =$ _____
8. $21 - 10t + 9 - 2t =$ _____
9. $-3 + x - 4x + 9 =$ _____
10. $7b + 12 + 4b =$ _____
11. $4h - h + 2 - 5 =$ _____
12. $-6k + 10 - 4k =$ _____
13. $2a + 12a - 5 + a =$ _____
14. $5 + 9c - 10 =$ _____
15. $-d + 1 + 2d - 4 =$ _____
16. $-8 + 4h + 1 - h =$ _____
17. $12x - 4x + 7 =$ _____
18. $10 + 3z + z - 5 =$ _____
19. $14 + 3y - y - 2 =$ _____
20. $11p - 4p + p =$ _____
21. $11m + 2 - m + 1 =$ _____

SOLVING EQUATIONS WITH LIKE TERMS

When an equation has two or more like terms on the same side of the equation, like terms should be combined as the **first** step in solving the equation.

EXAMPLE: $7x + 2x - 7 = 21 + 8$

Step 1: Combine like terms on both sides of the equation.

$$7x + 2x - 7 = 21 + 8$$
$$9x - 7 = 29$$

Step 2: Solve the two-step algebra problem as explained previously.

$$\begin{array}{r} +7 \quad +7 \\ \hline \dfrac{9x}{9} \quad \dfrac{36}{9} \\ x = 4 \end{array}$$

Solve the equations below, combining like terms first.

1. $3w - 2w + 4 = 6$
2. $7x + 3 + x = 16 + 3$
3. $5 - 6y + 9y = -15 + 5$
4. $-14 + 7a + 2a = -5$
5. $-2t + 4t - 7 = 9$
6. $9d + d - 3d = 14$
7. $-6c - 4 - 5c = 10 + 8$
8. $15m - 9 - 6m = 9$
9. $-4 - 3x - x = -16$
10. $9 - 12p + 5p = 14 + 2$
11. $10y + 4 - 7y = -17$
12. $-8a - 15 - 4a = 9$

If the equation has like terms on both sides of the equation, you must get all of the terms with a **variable** on one side of the equation and all of the **integers** on the other side of the equation.

EXAMPLE: $3x + 2 = 6x - 1$

Step 1: $\quad 3x + 2 = 6x - 1$ 　　Subtract $6x$ from both sides to move all the **variables**
$\quad\quad\quad\quad\; \underline{-6x \quad\quad -6x}$ 　　to the left side.

Step 2: $\quad -3x + 2 = -1$ 　　Subtract 2 from both sides to move all the **integers** to
$\quad\quad\quad\quad\; \underline{\quad\quad -2 \;\; -2}$ 　　the right side.

Step 3: $\quad \dfrac{-3x}{-3} = \dfrac{-3}{-3}$ 　　Divide by -3 to solve for x.

$\quad\quad\quad\quad\;\; x = 1$

Solve the following problems.

1. $3a + 1 = a + 9$
2. $2d - 12 = d + 3$
3. $5x + 6 = 14 - 3x$
4. $15 - 4y = 2y - 3$
5. $9w - 7 = 12w - 13$
6. $10b + 19 = 4b - 5$
7. $-7m + 9 = 29 - 2m$
8. $5x - 26 = 13x - 2$
9. $19 - p = 3p - 9$
10. $-7p - 14 = -2p + 11$

11. $16y + 12 = 9y + 33$
12. $13 - 11w = 3 - w$
13. $-17b + 23 = -4 - 8b$
14. $k + 5 = 20 - 2k$
15. $12 + m = 4m + 21$
16. $7p - 30 = p + 6$
17. $19 - 13z = 9 - 12z$
18. $8y - 2 = 4y + 22$
19. $5 + 16w = 6w - 45$
20. $-27 - 7x = 2x + 18$

21. $-12x + 14 = 8x - 46$
22. $27 - 11h = 5 - 9h$
23. $5t + 36 = -6 - 2t$
24. $17y + 42 = 10y + 7$
25. $22x - 24 = 14x - 8$
26. $p - 1 = 4p + 17$
27. $4d + 14 = 3d - 1$
28. $7w - 5 = 8w + 12$
29. $-3y - 2 = 9y + 22$
30. $17 - 9m = m - 23$

Sometimes an equation has two variables, and you may be asked to solve for one of the variables.

EXAMPLE 1: If $5x + y = 19,$ then $y =$

Solution: The goal is to have only y on one side of the equation and the rest of the terms on the other side of the equation. Follow order of operations to solve.

$$5x + y - 5x = 19 - 5x \quad \text{Subtract } 5x \text{ from each side of the equation.}$$
$$y = 19 - 5x$$

EXAMPLE 2: If $7m + n = 30,$ then $m =$

Solution: The goal is to have only m on one side of the equation and the rest of the terms on the other side of the equation. Follow order of operations to solve.

$$7m + n = 30 \quad \text{Subtract } n \text{ from both sides of the equation.}$$
$$7m + n - n = 30 - n$$
$$\frac{7m}{7} = \frac{30 - n}{7} \quad \text{Divide both sides of the equation by 7.}$$
$$m = \frac{30 - n}{7}$$

Solve each of the equations below for the variable indicated. Be sure to follow order of operations.

1. If $5a + b = 14,$ then $a =$

2. If $7c - d = 20,$ then $d =$

3. If $4m - n = 10,$ then $m =$

4. If $4r + 2s = 20,$ then $r =$

5. If $15m - 9n - 6m = 9n,$ then $m =$

6. If $-4y - 3x - x = -16y,$ then $x =$

7. If $-2t + 4t - 7s = 9s,$ then $s =$

8. If $5x - 4y + 9y = -15x + 5,$ then $y =$

9. If $-14b + 7a + 2a = -5b,$ then $a =$

10. If $7x + 3y + x = 16y + 3,$ then $x =$

REMOVING PARENTHESES

In this chapter, you will use the distributive principle to remove parentheses in problems with a variable (letter).

EXAMPLE 1: $2(a + 6)$ You multiply 2 by each term inside the parentheses. $2 \times a = 2a$ and $2 \times 6 = 12$. The 12 is a positive number, so use a plus sign between the terms in the answer.

$$2(a + 6) = 2a + 12$$

EXAMPLE 2: $7(2b - 5)$ You multiply 7 by each term inside the parentheses. $7 \times 2b = 14b$ and $7 \times -5 = -35$. The -35 is a negative number, so use a minus sign between the terms in the answer.

$$7(2b - 5) = 14b - 35$$

EXAMPLE 3: $4(-5c + 2)$ The first term inside the parentheses could be negative. Multiply in exactly the same way as in the examples above. $4 \times (-5c) = -20c$ and $4 \times 2 = 8$

$$4(-5c + 2) = -20c + 8$$

Remove parentheses in the problems below.

1. $7(n + 6)$
2. $8(2g - 5)$
3. $11(5z - 2)$
4. $6(-y - 4)$
5. $3(-3k + 5)$
6. $4(d - 8)$
7. $2(-4x + 6)$
8. $7(4 + 6p)$
9. $5(-4w - 8)$
10. $6(11x + 2)$
11. $10(9 - y)$
12. $9(c - 9)$
13. $12(-3t + 1)$
14. $3(4y + 9)$
15. $8(b + 3)$
16. $5(8a + 7)$
17. $3(2b - 4)$
18. $2(-9x - 7)$
19. $4(8 - 7v)$
20. $10(3c + 5)$
21. $5(2x - 9)$
22. $11(y + 3)$
23. $9(7t + 4)$
24. $6(8 - g)$

The number in front of the parentheses can also be negative. Remove these parentheses the same way.

EXAMPLE: $-2(b-4)$ First, multiply $-2 \times b$. $-2 \times b = -2b$
Second, multiply -2×-4. $-2 \times -4 = 8$

Copy the two products. The second product is a positive number, so put a plus sign between the terms in the answer.

$-2(b-4) = -2b + 8$

Remove the parentheses in the following problems.

1. $-7(x+2)$
2. $-5(4-y)$
3. $-4(2b-2)$
4. $-2(8c+6)$
5. $-5(-w-8)$
6. $-3(4x-2)$

7. $-2(-z+2)$
8. $-4(7p+7)$
9. $-9(t-6)$
10. $-10(2w+4)$
11. $-3(9-7p)$
12. $-9(-k-3)$

13. $-1(7b-9)$
14. $-6(-5t-2)$
15. $-7(-v+4)$
16. $-3(-x-5)$
17. $-11(4y+2)$
18. $-1(-c+100)$

19. $-5(-2t-4)$
20. $-2(7z-12)$
21. $-45(y-1)$
22. $-100(a+1)$
23. $-6(-x-11)$
24. $-12(-2b+1)$

MULTI-STEP ALGEBRA PROBLEMS

You can now use what you know about removing parentheses, combining like terms, and solving simple algebra problems to solve problems that involve three or more steps. Study the examples below to see how easy it is to solve multi-step problems.

EXAMPLE 1: $3(x + 6) = 5x - 2$

Step 1: $3x + 18 = 5x - 2$ — Use the distributive property to remove parentheses.

Step 2: $\underline{-5x \qquad -5x}$
$-2x + 18 = -2$ — Subtract $5x$ from each side to move the terms with variables to the left side of the equation.

Step 3: $\underline{\qquad -18 \quad -18}$
$\dfrac{-2x}{-2} = \dfrac{-20}{-2}$ — Subtract 18 from each side to move the integers to the right side of the equation.

Step 4: Divide both sides by -2 to solve for x.

$x = 10$

EXAMPLE 2: $\dfrac{3(x - 3)}{2} = 9$

Step 1: $\dfrac{3x - 9}{2} = 9$ — Use the distributive property to remove parentheses.

Step 2: $\dfrac{2(3x - 9)}{2} = 2(9)$ — Multiply both sides by 2 to eliminate the fraction.

Step 3: $3x - 9 = 18$
$\underline{\quad +9 \quad +9}$ — Add 9 to both sides, and combine like terms.

Step 4: $\dfrac{3x}{3} = \dfrac{27}{3}$ — Divide both sides by 3 to solve for x.

$x = 9$

Solve the following multi-step algebra problems.

1. $2(y - 3) = 4y + 6$
2. $\dfrac{2(a + 4)}{2} = 12$
3. $\dfrac{10(x - 2)}{5} = 14$
4. $\dfrac{12y - 18}{6} = 4y + 3$
5. $2x + 3x = 30 - x$
6. $\dfrac{2a + 11}{3} = a + 5$
7. $5(b - 4) = 3b - 6$
8. $-8(y + 4) = 10y + 4$
9. $\dfrac{x + 4}{-3} = 6 - x$

10. $\dfrac{4(n+3)}{5} = n - 3$

11. $3(2x - 5) = 8x - 9$

12. $7 - 10a = 9 - 9a$

13. $7 - 5x = 10 - (6x + 7)$

14. $4(x - 3) - x = x - 6$

15. $4a + 4 = 3a - 4$

16. $-3(x - 4) + 5 = -2x - 2$

17. $5b - 11 = 13 - b$

18. $\dfrac{-4x + 3}{2x} = \dfrac{7}{2x}$

19. $-(x + 1) = -2(5 - x)$

20. $4(2c + 3) - 7 = 13$

21. $6 - 3a = 9 - 2(2a + 5)$

22. $-5x + 9 = -3x + 11$

23. $3y + 2 - 2y - 5 = 4y + 3$

24. $3y - 10 = 4 - 4y$

25. $-(a + 3) = -2(2a + 1) - 7$

26. $5m - 2(m + 1) = m - 10$

27. $\dfrac{1}{2}(b - 2) = 5$

28. $-3(b - 4) = -2b$

29. $4x + 12 = -2(x + 3)$

30. $\dfrac{7x + 4}{3} = 2x - 1$

31. $9x - 5 = 8x - 7$

32. $7x - 5 = 4x + 10$

33. $\dfrac{4x + 8}{2} = 6$

34. $2(c + 4) + 8 = 10$

35. $y - (y + 3) = y + 6$

36. $4 + x - 2(x - 6) = 8$

MULTI-STEP INEQUALITIES

Remember that adding and subtracting with inequalities follow the same rules as with equations. When you multiply or divide both sides of an inequality by the same positive number, the rules are also the same as for equations. However, when you multiply or divide both sides of an inequality by a **negative** number, you must **reverse** the inequality symbol.

EXAMPLE 1:
$-x > 4$
$(-1)(-x) < (-1)(4)$
$x < 4$

EXAMPLE 2:
$-4x < 2$
$\dfrac{-4x}{-4} > \dfrac{2}{-4}$
$x > -\dfrac{1}{2}$

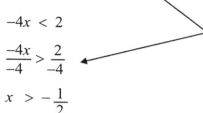

Reverse the symbol when you multiply or divide by a negative number.

When solving multi-step inequalities, first add and subtract to isolate the term with the variable. Then multiply and divide.

EXAMPLE 3:

$2x - 8 > 4x + 1$
$\underline{-4x \qquad -4x}$
$-2x - 8 > 1$
$\underline{\;\;\;+8 \;\; +8}$
$\dfrac{-2x}{-2} < \dfrac{9}{-2}$
$x < -\dfrac{9}{2}$

Step 1: Subtract $4x$ from both sides.

Step 2: Add 8 to both sides.

Step 3: Divide by -2. Remember to change the direction of the inequality sign.

Solve each of the following inequalities.

1. $8 - 3x \leq 7x - 2$

2. $3(2x - 5) \geq 8x - 5$

3. $4 + 2(3 - 2y) \leq 6y - 20$

4. $7 + 3y > 2y - 5$

5. $3a + 5 < 2a - 6$

6. $3(a - 2) > -5a - 2(3 - a)$

Solve each of the following inequalities.

7. $2x - 7 \geq 4(x - 3) + 3x$

8. $6x - 2 \leq 5x + 5$

9. $-\frac{x}{4} > 12$

10. $-\frac{2x}{3} \leq 6$

11. $3b + 5 < 2b - 8$

12. $4x - 5 \leq 7x + 13$

13. $4x + 5 \leq -2$

14. $2y - 4 > 7$

15. $\frac{1}{3}b - 2 > 5$

16. $-4c + 6 \leq 8$

17. $-\frac{1}{2}x + 2 > 9$

18. $\frac{1}{4}y - 3 \leq 1$

19. $-3x + 4 > 5$

20. $\frac{y}{2} - 2 \geq 10$

21. $7 + 4c < -2$

22. $2 - \frac{a}{2} > 1$

23. $10 + 4b \leq -2$

24. $-\frac{1}{2}x + 3 > 4$

25. $12d - 8 > 28$

26. $\frac{3}{4}f + 21 \geq 18$

27. $-7m + 14 < 70$

28. $-15p + 19 \geq -6p + 13$

29. $\frac{z}{5} + 3 < 43$

30. $3t - 13 \leq 5t$

31. $12 - \frac{5}{6}g > 0$

32. $8s + 20 \geq 3s - 25$

33. $7p - 6 < 8p + 9$

SOLVING EQUATIONS AND INEQUALITIES WITH ABSOLUTE VALUES

When solving equations and inequalities which involve variables placed in absolute values, remember that there will be two or more numbers that will work as correct answers. This is because the absolute value variable will signify both positive and negative numbers as answers.

EXAMPLE 1: $5 + 3|k| = 8$ Solve as you would any equation.
Step 1: $3|k| = 3$ Subtract 5 from each side.
Step 2: $|k| = 1$ Divide by 3 on each side.
Step 3: $k = 1$ and $k = -1$ Because k is an absolute value, the answer can be 1 or -1.

EXAMPLE 2: $2|x| - 3 < 7$ Solve as you normally would an inequality.
Step 1: $2|x| < 10$ Add 3 to both sides.
Step 2: $|x| < 5$ Divide by 2 on each side.
Step 3: $x < 5$ and $x > -5$ Because x is an absolute value, the answer is a set of both
           or $-5 < x < 5$ positive and negative numbers.
           Therefore, the following set of numbers is the solution:
           $\{-4, -3, -2, -1, 0, 1, 2, 3, 4\}$

Read each problem, and write the number or set of numbers which solves each equation or inequality.

1. $7 + 2|y| = 15$
2. $4|x| - 9 < 3$
3. $6|k| + 2 = 14$
4. $10 - 4|n| > -14$
5. $-3 = 5|z| + 12$
6. $-4 + 7|m| < 10$
7. $5|x| - 12 > 13$

8. $21|g| + 7 = 49$
9. $-9 + 6|x| = 15$
10. $12 - 6|w| > -12$
11. $31 > 13 + 9|r|$
12. $-30 = 21 - 3|t|$
13. $9|x| - 19 < 35$
14. $-13|c| + 21 \geq -31$

15. $5 - 11|k| < -17$
16. $-42 + 14|p| = 14$
17. $15 < 3|b| + 6$
18. $9 + 5|q| = 29$
19. $-14|y| - 38 < -45$
20. $36 = 4|s| + 20$
21. $20 \leq -60 + 8|e|$

MORE SOLVING EQUATIONS AND INEQUALITIES WITH ABSOLUTE VALUE

Now, look at the following examples in which numbers and variables are added or subtracted in the absolute value symbols (| |).

EXAMPLE 1: $|3x - 5| = 10$ Remember, an equation with the absolute value symbols has two solutions.

Step 1:
$3x - 5 = 10$
$3x - 5 + 5 = 10 + 5$
$\dfrac{3x}{3} = \dfrac{15}{3}$
$x = 5$

To find the first solution, remove the absolute value symbol and solve the equation.

Step 2:
$-(3x - 5) = 10$
$-3x + 5 = 10$
$-3x + 5 - 5 = 10 - 5$
$-3x = 5$
$x = \dfrac{-5}{3}$

To find the second solution, solve the equation for the negative of the expression in absolute value symbols.

Solutions: $x = 5$ and $x = \dfrac{-5}{3}$

EXAMPLE 2: $|5z - 10| < 20$

Step 1:
$5z - 10 < 20$
$5z - 10 + 10 < 20 + 10$
$\dfrac{5z}{5} < \dfrac{30}{5}$
$z < 6$

Remove the absolute value symbols and solve the inequality.

Step 2:
$-(5z - 10) < 20$
$-5z + 10 < 20$
$-5z + 10 - 10 < 20 - 10$
$\dfrac{-5z}{-5} < \dfrac{10}{5}$
$z > -2$

Next, solve the equation for the negative of the expression in absolute value symbols.

Solutions: $-2 < z < 6$, so the solution set is $\{-2, -1, 0, 1, 2, 3, 4, 5, 6\}$

EXAMPLE 3: $|4y + 7| - 5 > 18$

Step 1: $4y + 7 - 5 + 5 > 18 + 5$ Remove the absolute value symbols and solve the
$4y + 7 > 23$ inequality.
$4y + 7 - 7 > 23 - 7$
$4y > 16$
$y > 4$

Step 2: $-(4y + 7) - 5 > 18$ Solve the equation for the negative of the expression
$-4y - 7 - 5 + 5 > 18 + 5$ in absolute value symbols.
$-4y - 7 + 7 > 23 + 7$
$-4y > 30$
$y < -7\frac{1}{2}$

Solutions: $y > 4$ or $y < -7\frac{1}{2}$

Solve the following equations and inequalities below.

1. $-4 + |2x + 4| = 14$
2. $|4b - 7| + 3 > 12$
3. $6 + |12e + 3| < 39$
4. $-15 + |8f - 14| > 35$
5. $70 + |5z + 12| = 18$
6. $|4g - 9| - 3 < 26$
7. $22 + |5a + 21| > 45$
8. $|11c - 25| + 13 = 21$
9. $|-9b + 13| - 12 = 10$
10. $-25 + |7b + 11| < 35$
11. $|7w + 2| - 60 > 30$
12. $63 + |3d - 12| = 21$
13. $|-23 + 8x| - 12 > +37$
14. $|61 + 20x| + 32 > 51$
15. $|21 + 11y| + 18 < 32$

16. $|4a + 13| + 31 = 50$
17. $4 + |4k - 32| < 51$
18. $8 + |4x + 3| = 21$
19. $|28 + 7v| - 28 < 77$
20. $35 + |5c - 12| > 48$
21. $89 > 62 + |9d - 8|$
22. $83 < 31 + |9e + 7|$
23. $|13x + 3| + 6 = 35$
24. $|62p + 31| + 43 = 136$
25. $18 - |6v + 22| < 22$
26. $12 = 4 + |42 + 10m|$
27. $53 < 18 + |12e + 31|$
28. $38 > -39 + |7j + 14|$
29. $9 = |14 + 15u| + 7$
30. $-73 + |24b - 16| > -61$

31. $11 - |2j + 50| > 45$
32. $|35 + 6i| - 3 = 14$
33. $|26 - 8r| - 9 > 41$
34. $|25 + 6z| - 21 = 28$
35. $|9u - 24| + 15 = 36$
36. $54 > 25 + |6f - 83|$
37. $|3g - 41| - 20 < 18$
38. $28 > |31 + 6k| - 15$
39. $12 < |2t + 6| - 14$
40. $50 > |9q - 10| + 6$
41. $12 + |8v - 18| > 26$
42. $-38 + |16i - 33| = 41$
43. $|-14 + 6p| - 9 < 7$
44. $28 > |25 - 5f| - 12$
45. $-41 + |10c - 32| = 67$

CHAPTER 9 REVIEW

Solve each of the following equations.

1. $4a - 8 = 28$
2. $-7 + 23w = 108$
3. $5 + \dfrac{x}{8} = -4$
4. $\dfrac{c}{3} - 13 = 5$
5. $\dfrac{y - 8}{6} = 7$
6. $\dfrac{b + 9}{12} = -3$

Simplify the following expressions by combining like terms.

7. $-4a + 8 + 3a - 9$
8. $14 + 3z - 8 - 5z$
9. $-7 - 7x - 2 - 9x$

Solve.

10. $19 - 8d = d - 17$
11. $6 + 16x = -2x - 12$
12. $7w - 8 = -4w - 30$

Simplify the following expression by removing parentheses.

13. $3(-4x + 7)$
14. $11(2y + 5)$
15. $6(8 - 9b)$
16. $-8(-2 + 3a)$
17. $-2(5c - 3)$
18. $-5(7y - 1)$

Solve each of the following equations and inequalities.

19. $6(b - 4) = 8b - 18$
20. $\dfrac{4x - 16}{2} = 7x + 2$
21. $\dfrac{-11c - 35}{4} = 4c - 2$
22. $5 + x - 3(x + 4) = -17$
23. $-9b - 3 = -3(b + 2)$
24. $7a - 5 = 2(2a - 13)$
25. $4(2x + 3) \geq 2x$
26. $3x - 5 + 3(x + 3) = 10$
27. $7(2x + 6) < -28$

Write the number or set of numbers which solves each equation or inequality.

28. $-11|k| < -22$
29. $3(x + 2) < 7x - 10$
30. $7x < 4(3x + 1)$
31. $-\dfrac{y}{2} > 14$
32. $-\dfrac{3}{4}x \leq 6$
33. $2(3x - 1) \geq 3x - 7$
34. $\dfrac{5(n + 4)}{3} = n - 8$
35. $18 - 3|w| > -18$
36. $\dfrac{t}{5} + 2 > 8$
37. $21 = -4 + |5x + 5|$
38. $39 + |10x - 8| > 41$
39. $|3x + 2| - 4 \geq -2$

Chapter 10
Algebra Word Problems

An equation states that two mathematical expressions are equal. In working with word problems, the words that mean equal are **equals, is, was, is equal to, amounts to,** and other expressions with the same meaning. To translate a word problem into an algebraic equation, use a variable to represent the unknown or unknowns you are looking for.

In the following example, let n be the number you are looking for.

EXAMPLE: Four more than twice a number is two less than three times the number.

Step 1: Translation: $4 + 2n = 3n - 2$

Step 2: Now solve:
$$\begin{aligned} 4 + 2n &= 3n - 2 \\ -2n \quad &\quad -2n \\ \hline 4 &= n - 2 \\ +2 &\quad +2 \\ \hline 6 &= n \end{aligned}$$

The number is 6.
Substitute the number back into the original equation to check.

Translate the following word problems into equations and solve.

1. Seven less than twice a number is eleven. Find the number.

2. Four more than three times a number is one less than four times the number. What is the number?

3. The sum of three times a number and the number is 24. What is the number?

4. Negative 16 is the sum of five and a number. Find the number.

5. Negative 20 is equal to ten minus the product of six and a number. What is the number?

6. Two less than twice a number equals the number plus 15. What is the number?

7. The difference between three times a number and 21 is three. What is the number?

8. Eighteen is fifteen less than the product of a number and three. What is the number?

9. Six more than twice a number is four times the difference between three and the number. What is the number?

10. Four less than twice a number is five times the sum of one and the number. What is the number?

Copyright © American Book Company

GEOMETRY WORD PROBLEMS

The perimeter of a geometric figure is the distance around the outside of the figure.

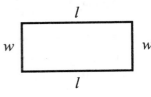

perimeter = $2l + 2w$

perimeter = $a + b + c$

EXAMPLE 1: The perimeter of a rectangle is 44 feet. The length of the rectangle is 6 feet more than the width. What is the measure of the width?

Step 1: Let the variable be the length of the unknown side.
width = w length = $6 + w$

Step 2: Use the equation for the perimeter of a rectangle as follows:
$2l + 2w$ = perimeter of a rectangle.
$2(w + 6) + 2w = 44$

Step 3: Solve for w.

Solution: width = 8 feet

EXAMPLE 2: The perimeter of a triangle is 26 feet. The second side is twice as long as the first. The third side is 1 foot longer than the second side. What is the length of the 3 sides?

Step 1: Let x = first side $2x$ = second side $2x + 1$ = third side

Step 2: Use the equation for perimeter of a triangle as follows:
sum of the length of the sides = perimeter of a triangle.
$x + 2x + 2x + 1 = 26$

Step 3: Solve for x. $5x + 1 = 26$ so $x = 5$

Solution: first side $x = 5$ second side $2x = 10$ third side $2x + 1 = 11$

Solve the following geometry word problems.

1. The length of a rectangle is 6 times longer than the width. The perimeter is 42. What is the width?

2. The length of a rectangle is 4 more than twice the width. The perimeter is 28. What is the length?

3. The perimeter of a triangle is 24 feet. The second side is two feet longer than the first. The third side is two feet longer than the second. What are the lengths of the sides?

136   Copyright © American Book Company

4. In an isosceles triangle, two sides are equal. The third side is two less than twice the length of the two equal sides. The perimeter is 38. What are the lengths of the three sides?

5. The sum of the measures of the angles of a triangle is 180°. The second angle is twice the measure of the first angle. The third angle is three times the measure of the second angle. Find the measure of each angle.

6. The sum of the measures of the angles of a triangle is 180°. The second angle of a triangle is twice the measure of the first angle. The third angle is 4 more than 5 times the first. What are the measures of the three angles?

AGE PROBLEMS

EXAMPLE: Tara is twice as old as Gwen. Their sister, Amy, is 5 years older than Gwen. If the sum of their ages is 29 years, find each of their ages.

Step 1: We want to find each of their ages so there are three unknowns. Tara is twice as old as Gwen, and Amy is older than Gwen, so Gwen is the youngest. Let x be Gwen's age. From the problem we can see that:

$$\left. \begin{array}{l} \text{Gwen} = x \\ \text{Tara} = 2x \\ \text{Amy} = x + 5 \end{array} \right\} \text{ The sum of their ages is 29.}$$

Step 2: Set up the equation, and solve for x.

$$x + 2x + x + 5 = 29$$
$$4x + 5 = 29$$
$$4x = 29 - 5$$
$$x = \frac{24}{4}$$
$$x = 6$$

Solutions: Gwen's age $(x) = 6$
Tara's age $(2x) = 12$
Amy's age $(x + 5) = 11$

Solve the following age problems.

1. Carol is 25 years older than her cousin Amanda. Cousin Bill is 3 times as old as Amanda. The sum of their ages is 90. Find each of their ages.

2. Derrick is 5 less than twice as old as Brandon. The sum of their ages is 43. How old are Derrick and Brandon?

3. Beth's mom is 6 times older than Beth. Beth's dad is 7 years older than Beth's mom. The sum of their ages is 72. How old are each of them?

4. Delores is 2 years more than three times as old as her son, Raul. If the difference between their ages is 26, how old are Delores and Raul?

5. Eileen is 6 years older than Karen. John is three times as old as Karen. The sum of their ages is 56. How old are Eileen, Karen and John?

6. Taylor is 18 years younger than Jim. Andrew is twice as old as Taylor. The sum of their ages is 26. How old are Taylor, Jim, and Andrew?

The following problems work in the same way as the age problems. There are two or three items of different weight, distance, number, or size. You are given the total and asked to find the amount of each item.

7. Three boxes have a total weight of 640 pounds. Box A weighs twice as much as Box B. Box C weighs 30 pounds more than Box A. How much do each of the boxes weigh?

8. There are 158 students registered for American History classes. There are twice as many students registered in second period as first period. There are 10 less than three times as many students in third period as in first period. How many students are in each period?

9. Mei earns $2 less than three times as much as Olivia. Shane earns twice as much as Mei. Together they earn $594 per week. How much does each person earn per week?

10. Ellie, the elephant, eats 4 times as much as Popcorn, the pony. Zac, the zebra, eats twice as much as Popcorn. Altogether, they eat 490 kilograms of feed per week. How much feed does each of them require each week?

11. The school cafeteria served three kinds of lunches today to 225 students. The students chose the cheeseburgers three times more often than the grilled cheese sandwiches. There were twice as many grilled cheese sandwiches sold as fish sandwiches. How many of each lunch were served?

12. Three friends drove west into Illlinois. Kyle drove half as far as Jamaal. Conner drove 4 times as far as Kyle. Altogether, they drove 357 miles. How far did each friend drive?

13. Bianca is taking collections for this year's Feed the Hungry Project. So far she has collected $200 more from Company A than from Company B and $800 more from Company C than from Company A. Until now, she has collected $3,000. How much did Company C give?

14. For his birthday, Torin got $25.00 more from his grandmother than from his uncle. His uncle gave him $10.00 less than his cousin. Torin received $290.00 in total. How much did he receive from his cousin?

15. Cassidy loves black and yellow jelly beans. She noticed when she was counting them that she had 8 less than three times as many black jelly beans as she had yellow jelly beans. In total, she counted 348 jelly beans. How many black jelly beans did she have?

16. Mrs. Vargus planted a garden with red and white rose bushes. Because she was studying to be a botanist, she counted the number of blossoms on each bush. She counted 5 times as many red blossoms as white blossoms. In total, she counted 1,680 blossoms. How many red blossoms did she count?

MIXTURE WORD PROBLEMS

When a coffee manufacturer buys coffee from two different sources at two different prices and then combines them to make a blend, he or she can use algebra to design a mixture.

The formula for mixture problems is $V = AC$.

V = Value of an ingredient
A = Amount of an ingredient
C = Cost per unit of the ingredient

EXAMPLE: A coffee manufacturer bought some beans from Columbia for $5.00 per pound and some beans from Jamaica for $3.00 per pound. He wants to make 10 pounds of a mixture that will cost him $4.75 per pound. How many pounds of each coffee should he use?

Let x = The amount of $5.00 coffee from Columbia
$10 - x$ = The amount of $3.00 coffee from Jamaica
$x + (10 - x) = 10$ pounds of coffee blend

Step 1: Multiply each amount of coffee by it's unit cost.

	Amount	×	Unit Cost	=	Value
$5 coffee	x	×	$5	=	$5x$
$3 coffee	$10 - x$	×	$3	=	$3(10 - x)$
$4.75 coffee	10	×	$4.75	=	$4.75(10)$

The first two rows added together equals the third row.

The Value of the $5.00 coffee + Value of the $3.00 coffee = Value of the $4.75 blend.

In algebra, it looks like this: $5x + 3(10 - x) = 4.75(10)$

Step 2: Solve: $5x + 3(10 - x) = 4.75(10)$

$5x + 30 - 3x = 47.5$
$2x + 30 - 30 = 47.5 - 30$
$\dfrac{2x}{2} = \dfrac{17.5}{2}$
$x = 8.75$

Solution: Substitute the value for x, 8.75, in the equation above. The manufacturer must use 8.75 pounds of $5.00 coffee and 1.25 pounds of $3.00 coffee to get 10 pounds of coffee at $4.75 per pound.

Solve the following mixture problems.

1. A meat distributor paid $2.50 per pound for hamburger and $4.50 per pound for ground sirloin. How many pounds of each did he use to make 100 pounds of meat mixture that will cost $3.24 per pound?

2. How many pounds of walnuts which cost $4 per pound must be mixed with 25 pounds of almonds costing $7.50 per pound to make a mixture which will cost $6.50 per pound?

3. In the gourmet cheese shop, employees grated cheese costing $5.20 per pound to mix with 10 pounds of cheese which cost $3.60 per pound to make grated cheese topping costing $4.80 per pound. How many pounds of cheese costing $5.20 did they use?

4. A 200 pound bin of animal feed sells for $1.25 per pound. How many pounds of feed costing $2.50 per pound should be mixed with it to make a mixture which costs $2.00 per pound?

5. A grocer mixed grape juice which costs $2.25 per gallon with cranberry juice which costs $1.75 per gallon. How many gallons of each should be used to make 200 gallons of cranberry/grape juice which will cost $2.10 per gallon?

6. A nut merchant bought peanuts for $.75 per pound. She also bought 20 pounds of peanuts from another farmer for $1.10 per pound. How many pounds of $.75 peanuts must be combined with 20 pounds of $1.10 peanuts to make a mixture that will cost $.90 per pound?

PERCENT MIXTURE PROBLEMS

EXAMPLE: A goldsmith has 12 grams of a 40% gold alloy (alloy means it has been mixed with other metals). How many grams of pure gold should be added to make an alloy which is 65% gold?

Step 1: Make a chart for the amount of gold, A, and the percent of gold concentration, r. $Ar = Q$. Q is the quantity of the substance. Let the unknown quantity of pure gold equal x.

	Amount, A	×	Concentration, r	=	Quantity, Q
100% gold	x	×	1.00	=	$1.00\,x$
40% gold	12	×	0.40	=	0.40 (12)
65% gold	$x + 12$	×	0.65	=	0.65 $(x + 12)$

The first two rows added together equals the third.

Step 2: Solve: $x + 0.4\,(12) = 0.65\,(x + 12)$
$x + 4.8 = .65x + 7.8$
$.35x = 3$
$x = 8.57$ grams (round to 2 decimal places)

Solution: 8.57 grams of 100% pure gold must be added to make a 65% gold alloy.

Check: Replace x in the chart above to check.
100% gold: $8.57\,(1.00) = 8.57$
40% gold: $12\,(.40) = 4.8$
65% gold: $0.65\,(8.57 + 12) = 13.37$

equation: $8.57 + 4.8 = 13.37$

Solve the following percent mixture problems.

1. How many gallons of a 10% ammonia solution should be mixed with 50 gallons of a 30% ammonia solution to make a 15% ammonia solution?

2. How many gallons of a 25% alcohol solution must be mixed with 10 gallons of a 50% alcohol solution to make 30 gallons of a 40% alcohol solution?

3. Debbie is mixing orange juice concentrate for her restaurant. One juice concentrate is 64% real orange juice. The other is only 48% real orange juice. How many ounces of 48% real orange juice should she use to make 160 ounces of 58% real juice?

4. How many gallons of 60% antifreeze should be mixed with 40% antifreeze to make 80 gallons of 45% antifreeze?

5. Hank has some 60% maple syrup and some 100% maple syrup in his restaurant. How many ounces of each should he use to make 100 ounces of 85% maple syrup?

6. A butcher has some hamburger which is 4% fat and some hamburger which is 20% fat. How much of each will he need to make 120 pounds of hamburger which is 10% fat?

COIN AND STAMP PROBLEMS

Coin problems are a type of mixture problem.

EXAMPLE: Emily has 15 more dimes than quarters. If the total value of the coins is $3.95, how many dimes and quarters does she have?

Step 1: Let x = number of quarters.
$x + 15$ = number of dimes

We know the value of the quarters + the value of the dimes = $3.95.

Step 2: Make a chart showing the value of each coin.

	Value of coin	×	Number of coins	=	Value
Quarters	$0.25	×	x	=	$0.25 x$
Dimes	$0.10	×	$x + 15$	=	$0.10 (x + 15)$

Step 3: Set the total value equal to $3.95 and solve.
$0.25x + 0.10 (x + 15) = 3.95$
$x = 7$

Solution: $x = 7$, so there are 7 quarters.
Substitute 7 for x in the chart above, and we have $7 + 15 = 22$ dimes.

Check: $0.25 (7) + 0.10 (7 + 15) = \$3.95$
$\$3.95 = \$3.95$

Solve the following coin problems.

1. Brent has 8 more nickels than quarters. Altogether he has $6.10. How many of each coin does he have?

2. A child's piggy bank has 3 times as many dimes as nickels. Altogether she has $3.85. How many dimes and nickels does she have?

3. A coin bank contains 60 coins in dimes and quarters. The total amount of money is $12.30. How many quarters are there?

4. Mr. Weng has a stamp collection of 3¢ and 5¢ stamps. He has 2 less than 5 times as many 3¢ stamps as 5¢ stamps. The face value of the stamps totals $6.14. How many 3¢ stamps does he have?

5. A stamp collection consists of 6¢ stamps and 15¢ stamps. The number of 6¢ stamps is half the number of 15¢ stamps. The total face value of the stamps is $15.84. How many 6¢ stamps are there?

6. There are 50 bills in Dan's top drawer. Some of the bills are $1 bills, and the rest are $5 bills. The total amount of cash is $198.00. How many of each bill does he have?

UNIFORM MOTION PROBLEMS

A car which travels constantly at 60 miles per hour is in **uniform motion**. The formula used in uniform motion problems is $d = rt$.

d = distance traveled r = rate of travel (speed) t = time spent traveling

EXAMPLE: A train leaves a station, traveling at 50 miles per hour. Two hours later, a second train leaves on a parallel track, traveling the same direction at 60 miles per hour. In how many hours will the second train catch up to the first train? (Assume there are no stops, and both trains travel at a constant speed.)

Step 1: Record data in a chart. When the second train starts, the first train is already two hours ahead, so let t = time for second train and $t + 2$ = time for first train.

	Rate (r)	×	Time (t)	=	Distance (d)
1st Train	50	×	$t + 2$	=	$50(t + 2)$
2nd Train	60	×	t	=	$60t$

Step 2: The two trains will have traveled the same distance, so set the two distances equal to each other. Solve for t.

$50(t + 2) = 60t$
$t = 10$

Solution: The trains will meet in 10 hours.

Solve the following uniform motion problems.

1. A cyclist, riding at 21 miles per hour, leaves town. Four hours later, another cyclist leaves from the same starting point, traveling in the same direction at an average of 33 miles per hour. How long did it take the second cyclist to catch up to the first cyclist?

2. A jet plane, traveling an average of 480 miles per hour, passes a propeller plane that took off from the same airport two hours before. The propeller plane is traveling at an average of 160 miles per hour. How far from the airport does the jet plane catch up to the propeller plane?

3. Jamal left home and was driving to his brother's house in Colorado at an average speed of 45 miles per hour. One hour later, his sister left from the same house and traveled at an average of 54 miles per hour to catch up with him. She traveled by the same roads as Jamal. How long did it take her to catch up?

4. A train leaves Mt. Carmel, going west at 35 miles per hour. Two hours later, a second train leaves Mt. Carmel, traveling west at 45 miles per hour on a track parallel to the first train. How far from Mt. Carmel will the two trains meet?

5. An airplane took off from Seattle to Los Angeles, traveling at an average of 600 miles per hour. One hour later, a plane on the same route takes off, traveling an average of 800 miles per hour. How long will the first plane be in the air before it is passed by the second plane?

6. A man on a bicycle averages 10 miles per hour, riding down the road. A car leaves from the same point two hours later, traveling at an average of 50 miles per hour. How many miles down the road will the car pass the cyclist?

RETURN TRIP MOTION PROBLEMS

EXAMPLE: Felipe and Mateo took a bike ride. They averaged 12 miles per hour on the trip out into the country and 9 miles per hour on the trip back by the same roads. How far out can they go in 7 hours?

Step 1: Let the time riding out be t and the time riding back be $7 - t$.

Make a chart for rate, time, and distance using the formula $d = rt$.

	Rate (r)	×	Time (t)	=	Distance (d)
out	12	×	t	=	$12t$
back	9	×	$7 - t$	=	$9(7 - t)$

Step 2: The distance out and back are equal, so set the two distances equal to each other. Solve for t.

$12t = 9(7 - t)$
$t = 3$, so it takes Felipe and Mateo 3 hours to ride out into the country.

Solution: Substitute 3 for *t* in the chart to find the distance they can go.

$12(3) = 36$ miles   Felipe and Mateo can travel out 36 miles and get back in 7 hours.

Solve the following round trip problems.

1. A pilot flew an average of 300 miles per hour on a flight out. On the return flight to the same airport, he flew at an average speed of 500 miles per hour. The total flight time was 8 hours. How far did he fly each way?

2. Keenan went out in his sailboat on Lake Tahoe one Sunday afternoon. He sailed at 5 miles per hour for the trip out. He sailed twice as fast on the trip back. The entire trip took 6 hours. How far out did he go on the sailboat?

3. Carrie drove to the mountains, averaging 48 miles per hour. Coming back by the same roads, she averaged 40 miles per hour. The total driving time was 11 hours. How far away did she travel?

4. One airplane pilot flew at 270 miles per hour to her destination. On her return trip, she flew at 360 miles per hour. Her total flying time was 7 hours. How far was it from one airport to the other?

5. Mike and Jeff put their canoe in the river and paddled downstream at 9 miles per hour. Then, they turned the canoe around and paddled upstream at 3 miles per hour to return to their car. How far did they go downstream if the whole trip took 6 hours?

6. Cathy and Tina drove to the beach at an average speed of 40 miles per hour. They returned by the same route at an average speed of 25 miles per hour. The total driving time was 13 hours. How far away was the beach?

OPPOSITE DIRECTION MOTION PROBLEMS

EXAMPLE: Two planes take off from the same airport at the same time, heading in opposite directions. One plane is traveling 200 miles per hour faster than the other. In five hours, they are 5,500 miles apart. Find the average rate of speed of each plane.

Step 1: Let r = the rate of the first plane.
$r + 200$ = rate of the second plane
Make a chart using the formula $d = rt$.

	Rate (r) ×	Time (t)	=	Distance (d)
1st plane	r ×	5	=	$5r$
2nd plane	$r + 200$ ×	5	=	$5(r + 200)$

} sum equals 5,500 miles

Step 2: In this problem, the sum of the distances equals 5,500 miles, so the equation is
$5r + 5(r + 200) = 5,500$.
$r = 450$ miles per hour

Solution: The first plane was traveling at 450 mph.
The second plane was traveling 200 mph faster, so it was traveling at 650 mph.

Check: First plane: $5(450) = 2250$ miles
Second plane: $5(450 + 200) = 3250$ miles
$2250 + 3250 = 5500$
$5500 = 5500$

Solve the following opposite direction problems.

1. One plane took off from San Francisco, traveling east. At the same time, another plane took off from the same airport, traveling west. The plane traveling east was going 120 miles per hour faster than the plane traveling west. After three hours, the planes were 2,400 miles apart. How fast was each plane going?

2. Two trains left the station at the same time, traveling in opposite directions. After five hours, they were 595 miles apart. One train was traveling 15 miles per hour faster than the other. How fast was the faster train going?

3. Two cyclists travel from the same point in opposite directions on a course. One is traveling at an average of 13 miles per hour and the other at an average of 15 miles per hour. After three hours, how far apart are they?

4. Two buses left Bakersfield at the same time. One headed east, and one headed west. The one headed east was going 12 miles per hour slower than the westbound bus. After 6 hours, the buses were 696 miles apart. How fast was each bus going?

5. Two long distance runners started at the same point, at the same time, running in opposite directions. One runner ran an average of 2 miles per hour faster than the other. After 1.5 hours, they were 21 miles apart. How fast was the faster runner running?

6. A truck driver started driving north. At the same time, another truck driver started driving south from the same starting point. The first driver drove 10 miles per hour faster than the second driver. After 8 hours, they were 912 miles apart. What was the average speed of each truck?

WORKING TOGETHER PROBLEMS

EXAMPLE 1: If Barbara can do a certain job in 4 hours, and Kelly can do the same job in 6 hours, how long would it take them to do the job if they worked together?

Caution: At first glance, you may want to just average the two times together and conclude that Barbara and Kelly could do the job together in 5 hours. But, think about this problem carefully. If Barbara can do the job by herself in 4 hours, the job would be done even faster than 4 hours if she had Kelly's help, even if Kelly doesn't work as fast as Barbara. So, the answer you are looking for should be fewer than 4 hours.

Step 1: Let x = the amount of time it takes them to do the job together.
Barbara can do $\frac{1}{4}$ the job in 1 hour, and Kelly can do $\frac{1}{6}$ of the job in 1 hour.
Make a chart of what you know:

	Barbara	+	Kelly	=	Together
Number of tasks	1 job		1 job		1 job
Time	4 hours		6 hours		x hours

Step 2: Write an equation and solve: $\frac{1 \text{ job}}{4 \text{ hours}} + \frac{1 \text{ job}}{6 \text{ hours}} = \frac{1 \text{ job}}{x \text{ hours}}$

Find a common denominator: $\frac{3}{12} + \frac{2}{12} = \frac{1}{x} \rightarrow \frac{5}{12} = \frac{1}{x}$

Solve by cross multiplying: $5x = 12$, so $x = \frac{12}{5}$ or $2\frac{2}{5}$

Solution: Kelly and Barbara can do the job together in $2\frac{2}{5}$ hours.

EXAMPLE 2: Jerome and Kristen are conducting a hand recount of city election ballots for the office of mayor. Jerome can count 240 ballots per hour, while Kristen can count 320 ballots per hour. How long would it take for Kristen and Jerome to count 840 votes if they worked together?

Let x = the number of hours the job takes if they work together.

	Jerome	+	Kristen	=	Together
Number of tasks	240 ballots		320 ballots		840 ballots
Time	1 hour		1 hour		x hours

$$\frac{240 \text{ ballots}}{1 \text{ hour}} + \frac{320 \text{ ballots}}{1 \text{ hour}} = \frac{840 \text{ ballots}}{x \text{ hours}} \rightarrow \frac{560}{1} = \frac{840}{x}$$

$560x = 840 \quad x = \frac{840}{560}$ or $1\frac{1}{2}$

Solution: Jerome and Kristen can count 840 ballots in $1\frac{1}{2}$ hours.

---

1. Simone can assemble a radio in 4 hours, and Sheila can assemble the same model radio in 7 hours. How long would it take them to assemble one radio if they worked together?

2. John can change the front brake pads on a two axle car in 3 hours. Manuel can change the front brake pads in 2 hours. How long would it take John and Manuel to change the brake pads on a car if they worked together?

3. Jessica can type 4 pages per hour. Her friend, Alejandro, can type 6 pages per hour. How long would it take to type a 14 page paper if they worked together?

4. Sandra and Michael both work at a restaurant as servers. Sandra is able to serve six tables of people per hour, while Michael is able to serve 4 tables per hour. If both of the servers work on a private party at the restaurant serving 13 tables of people, how long will it take them to serve the tables?

5. Diedra can paint a wall mural every three months. Lucy can paint a wall mural every five months. How long would it take them to paint one wall mural if they worked together?

6. Phillip can bake 12 dozen cookies per hour. Brian can bake one dozen cookies in the same amount of time. How long would it take to bake 30 dozen cookies if they both worked together?

CONSECUTIVE INTEGER PROBLEMS

Consecutive integers follow each other in order

Examples:
1, 2, 3, 4
−3, −4, −5, −6

Algebraic notation:
$n, n+1, n+2, n+3$

Consecutive **even** integers:

2, 4, 6, 8, 10
−12, −14, −16, −18

$n, n+2, n+4, n+6$

Consecutive **odd** integers:

3, 5, 7, 9
−5, −7, −9, −11

$n, n+2, n+4, n+6$

EXAMPLE 1: The sum of three consecutive odd integers is 63. Find the integers.

Step 1: Represent the three odd integers:
Let n = the first odd integer
$n + 2$ = the second odd integer
$n + 4$ = the third odd integer

Step 2: The sum of the integers is 63, so the algebraic equation is
$n + n + 2 + n + 4 = 63$. Solve for n.
$n = 19$

Solution: the first odd integer = 19
the second odd integer = 21
the third odd integer = 23

Check: Does $19 + 21 + 23 = 63$? Yes, it does.

EXAMPLE 2: Find three consecutive odd integers such that the sum of the first and second is three less than the third.

Step 1: Represent the three odd integers just like above:
Let n = the first odd integer
$n + 2$ = the second odd integer
$n + 4$ = the third odd integer

Step 2: In this problem, the sum of the first and second integers is three less than the third integer, so the algebraic equation is written as follows:
$n + n + 2 = n + 4 - 3$
$n = -1$

Solution: the first odd integer = -1
the second odd integer = 1
the third odd integer = 3

Check: Is the sum of -1 and 1 three less than 3?
$-1 + 1 = 3 - 3$ or $0 = 0$ Yes, it is.

Solve the following problems.

1. Find three consecutive odd integers whose sum is 141.
2. Find three consecutive integers whose sum is -21.
3. The sum of three consecutive even integers is 48. What are the numbers?
4. Find two consecutive even integers such that six times the first equals five times the second.
5. Find two consecutive odd integers such that seven times the first equals five times the second.
6. Find two consecutive odd numbers whose sum is forty-four.

INEQUALITY WORD PROBLEMS

Inequality word problems involve staying under a limit or having a minimum goal one must meet.

EXAMPLE: A contestant on a popular game show must earn a minimum of 800 points by answering a series of questions worth 40 points each per category in order to win the game. The contestant will answer questions from each of four categories. Her results for the first three categories are as follows: 160 points, 200 points, and 240 points. Write an inequality which describes how many points, (p), the contestant will need on the last category in order to win.

Step 1: Add to find out how many points she already has. $160 + 200 + 240 = 600$
Step 2: Subtract the points she already has from the minimum points she needs.
$800 - 600 = 200$. She must get at least 200 points in the last category to win. If she gets more than 200 points, that is okay, too. To express the number of points she needs, use the following inequality statement:

$p \geq 200$ The points she needs must be greater than or equal to 200.

Solve each of the following problems using inequalities.

1. Stella wants to place her money in a high interest money market account. However, she needs at least $1000 to open an account. Each month, she sets aside some of her earnings in a savings account. In January through June, she added the following amounts to her savings: $121, $206, $138, $212, $109, and $134. Write an inequality which describes the amount of money she can set aside in July to qualify for the money market account.

2. A high school band program will receive $2000.00 for selling $10,000.00 worth of coupon books. Six band classes participate in the sales drive. Classes 1–5 collect the following amounts of money: $1,400, $2,600, $1,800, $2,450, and $1,550. Write an inequality which describes the amount of money the sixth class must collect so that the band will receive $2,000.

3. A small elevator has a maximum capacity of 1,000 pounds before the cable holding it in place snaps. Six people get on the elevator. Five of their weights follow: 146, 180, 130, 262, and 135. Write an inequality which describes the amount the sixth person can weigh without snapping the cable.

4. A small high school class of 9 students were told they would receive a pizza party if their class average was 92% or higher on the next exam. Students 1–8 scored the following on the exam: 86, 91, 98, 83, 97, 89, 99, and 96. Write an inequality which describes the score the ninth student must make for the class to qualify for the pizza party.

5. Raymond wants to spend his entire credit limit on his credit card. His credit limit is $2000. He purchases items costing $600, $800, $50, $168, and $3. Write an inequality which describes the amounts Raymond can put on his credit card for his next purchases.

CHAPTER 10 REVIEW

Solve each of the following problems.

1. Terrell drove to the mountains at an average of 60 miles per hour. His return trip, by the same roads, averaged 40 miles per hour. His total driving time was 12 hours. How far did he drive one way?

2. Leah and Ryanne went out for a bike ride into the country. They averaged 24 miles per hour on the way out and 18 miles per hour on the way back, by the same roads. Their total travel time was 7 hours. How far did they go?

3. Deanna is five more than six times older than Ted. The sum of their ages is 47. How old is Ted?

4. Ross is six years older than twice his sister Holly's age. The difference in their ages is 18 years. How old is Holly?

5. Annie has quarters and dimes in her bank. There are eight less than four times as many dimes as quarters. There is $6.35 in her bank. How many dimes are in her bank?

6. The band members sold tickets to their concert performance. Some were $5 tickets, and some were $6 tickets. There were 16 more than twice as many $6 tickets sold as $5 tickets. The total sales were $1643. How many tickets of each price were sold?

7. A car and a bus started out from the same place, traveling in opposite directions. The average speed of the car was 15 miles per hour faster than the bus. After 5 hours, they were 535 miles apart. How fast was the car going?

8. Two cars left the same house at the same time, traveling in opposite directions. After 4 hours, they were 384 miles apart. One car was going 8 miles per hour faster than the other. How fast was each car going?

9. Three consecutive integers have a sum of 240. Find the integers.

10. Find three consecutive even numbers whose sum is negative seventy-two.

11. The sum of two numbers is 55. The larger number is 7 more than twice the smaller number. What are the numbers?

12. One number is 10 more than the other number. Twice the smaller number is 13 more than the larger number. What are the numbers?

13. The perimeter of a triangle is 43 inches. The second side is three inches longer than the first side. The third side is one inch longer than the second. Find the length of each side.

14. The perimeter of a rectangle is 80 feet. The length of the rectangle is 2 feet less than 5 times the width. What is the length and width of the rectangle?

15. Randy left town on his bicycle, traveling at an average of 18 miles per hour. Three hours later, his older brother, Jimmy, went looking for Randy on his motorcycle. Jimmy traveled at an average of 54 miles per hour. How long would it take Jimmy to catch up to Randy?

16. Keith drove west to go see his sister in California. He drove at an average of 50 miles per hour. His mom and dad left 4 hours later and drove an average of 70 miles per hour to catch up with him. How long did it take for them to catch up to Keith?

17. Priscilla bought 20 pounds of chocolate-covered peanuts for $3.50 per pound. How many pounds of chocolate-covered walnuts would she have to buy at $7.00 per pound to make boxes of nut mixtures that would cost her $4.50 per pound?

18. One solution is 15% ammonia. A second solution is 40% ammonia. How many ounces of each should be used to make 100 ounces of a 20% ammonia solution?

19. Joe, Craig, and Dylan have a combined weight of 429 pounds. Craig weighs 34 pounds more than Joe. Dylan weighs 13 pounds more than Craig. How many pounds does Craig weigh?

20. Tracie and Marcia drove to northern California to see Marcia's sister in Eureka. Tracie drove one hour more than three times as much as Marcia. The trip took a total of 17 driving hours. How many hours did Tracie drive?

21. Jesse and Larry entered a pie eating contest. Jesse ate 2 less than twice as many pies as Larry. They ate a total of 28 pies. How many pies did Larry eat?

22. Lena and Jodie are sisters and together they have 68 bottles of nail polish. Lena bought 5 more than half the bottles. How many did Jodie buy?

23. Janet and Artie wanted to play tug of war. Artie pulls with 150 pounds of force while Janet pulls with 40 pounds of force. In order to make this a fair contest, Janet enlists the help of her friends Trudi, Sherri, and Bridget who pull with 30, 25, and 40 pounds respectively. Write an inequality describing the minimum amount Janet's fourth friend, Tommy, must pull to beat Artie.

24. Jim takes great pride in decorating his float for the homecoming parade for his high school. With the $5,000 he has to spend, Jim bought 5,000 carnations at $.25 each, 4,000 tulips at $.50 each, and 300 irises at $.90 each. Write an inequality which describes how many roses, r, Jim can buy if roses cost $.80 each.

25. Mr. Chan wants to sell some or all of his shares of stock in a company. He purchased the 80 shares for $.50 last month, and the shares are now worth $4.50 each. Write an inequality which describes how much profit, p, Mr. Chan can make by selling his shares.

26. Kyle can deliver all of the newspapers on a given paper route in three hours. Jessica can deliver all the newspapers on the same route in five hours. How long would it take them to deliver newspapers if they both work together?

27. Chris can construct a robot on an erector set in 7 hours. His friend, Cristobal, can construct a robot using the same materials in 8 hours. How long would it take them to construct a robot if they both work together?

28. Anita and Reginald both enjoy rollerblading as an afternoon activity. Anita can run a particular course called the Silver Comet Trail in 7 hours. Reginald is able to rollerblade the same course in 5 hours. If Reginald and Anita start at opposite ends of the trail and rollerblade towards each other, how much time will have passed when they meet? (Hint: Think about this problem as a "working together" problem.)

Chapter 11
Introduction to Graphing

CARTESIAN COORDINATES

A **Cartesian coordinate plane** allows you to graph points with two values. A Cartesian coordinate plane is made up of two number lines. The horizontal number line is called the **x-axis**, and the vertical number line is called the **y-axis**. The point where the x and y axes intersect is called the **origin**. The x and y axes separate the Cartesian coordinate plane into four quadrants that are labeled I, II, III, and IV. The quadrants are labeled and explained on the graph below. Each point graphed on the plane is designated by an **ordered pair** of coordinates. For example, (2, −1) is an ordered pair of coordinates designated by **point B** on the plane below. The first number, 2, tells you to go over positive two on the x-axis. The −1 tells you to then go down negative one on the y-axis.

Remember: The first number always tells you how far to go right or left of 0, and the second number always tells you how far to go up or down from 0.

Quadrant II:
The x-coordinate is negative, and the y-coordinate is positive (−, +).

Quadrant III:
Both coordinates in the ordered pair are negative (−, −).

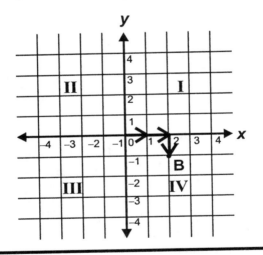

Quadrant I:
Both coordinates in the ordered pair are positive (+, +).

Quadrant IV:
The x-coordinate is positive, and the y-coordinate is negative (+, −).

Plot and label the following points on the Cartesian coordinate plane provided.

A. (2, 4) G. (−2, 5) M. (5, 5)
B. (−1, 5) H. (5, −1) N. (−2, −2)
C. (3, −4) I. (4, −4) O. (0, 0)
D. (−5, −2) J. (5, 2) P. (0, 4)
E. (5, 3) K. (−1, −1) Q. (2, 0)
F. (−3, −5) L. (3, −3) R. (−4, 0)

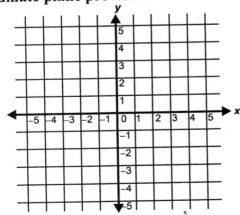

152   Copyright © American Book Company

IDENTIFYING ORDERED PAIRS

When identifying **ordered pairs**, count how far left or right of 0 to find the *x*-coordinate and then how far up or down from 0 to find the *y*-coordinate.

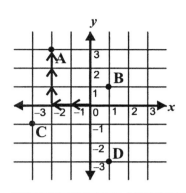

point A: Left (negative) two and up (positive) three = (−2, 3) in quadrant II
point B: Right (positive) one and up (positive) one = (1, 1) in quadrant I
point C: Left (negative) three and down (negative) one = (−3, −1) in quadrant III
point D: Right (positive) one and down (negative) three = (1, −3) in quadrant IV

Fill in the ordered pair for each point, and tell which quadrant it is in.

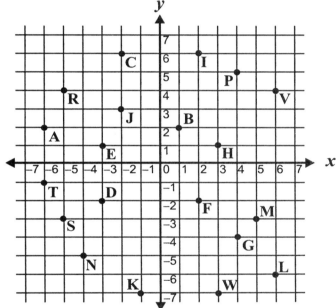

1. point A = (,) quadrant _____
2. point B = (,) quadrant _____
3. point C = (,) quadrant _____
4. point D = (,) quadrant _____
5. point E = (,) quadrant _____
6. point F = (,) quadrant _____
7. point G = (,) quadrant _____
8. point H = (,) quadrant _____
9. point I = (,) quadrant _____
10. point J = (,) quadrant _____
11. point K = (,) quadrant _____
12. point L = (,) quadrant _____
13. point M = (,) quadrant _____
14. point N = (,) quadrant _____
15. point P = (,) quadrant _____
16. point R = (,) quadrant _____
17. point S = (,) quadrant _____
18. point T = (,) quadrant _____
19. point V = (,) quadrant _____
20. point W = (,) quadrant _____

Sometimes, points on a coordinate plane fall on the *x* or *y* axis. If a point falls on the *x*-axis, then the second number of the ordered pair is 0. If a point falls on the *y*-axis, the first number of the ordered pair is 0.

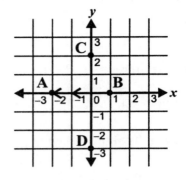

point A: Left (negative) two and up zero = (−2, 0)
point B: Right (positive) one and up zero = (1, 0)
point C: Left/right zero and up (positive) two = (0, 2)
point D: Left/right zero and down (negative) three = (0, −3)

Fill in the ordered pair for each point.

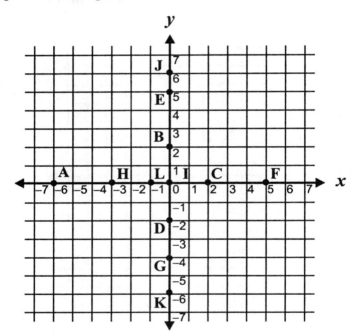

1. point A = (,   )
2. point B = (,   )
3. point C = (,   )
4. point D = (,   )
5. point E = (,   )
6. point F = (,   )
7. point G = (,   )
8. point H = (,   )
9. point I = (,   )
10. point J = (,   )
11. point K = (,   )
12. point L = (,   )

DRAWING GEOMETRIC FIGURES ON A CARTESIAN COORDINATE PLANE

You can use a **Cartesian coordinate plane** to draw geometric figures by plotting **vertices** and connecting them with line segments.

EXAMPLE 1: What are the coordinates of each vertex of quadrilateral ABCD below?

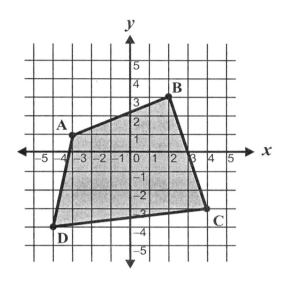

Step 1: To find the coordinates of point A, count over −3 on the *x*-axis and up 1 on the *y*-axis.
point A = (−3, 1)

Step 2: The coordinates of point B are located to the right two units on the *x*-axis and up 3 units on the *y*-axis.
point B = (2, 3)

Step 3: Point C is located 4 units to the right on the *x*-axis and down −3 on the *y*-axis.
point C = (4, −3)

Step 4: Point D is −4 units left on the *x*-axis and down −4 units on the *y*-axis.
point D = (−4, −4)

EXAMPLE 2: Plot the following points. Then construct and identify the geometric figure that you plotted.

A = (−2, −5), B = (−2, 1), C = (3, 1), D = (3, −5)

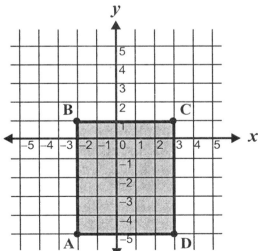

Figure ABCD is a rectangle.

Find the coordinates of the geometric figures graphed below.

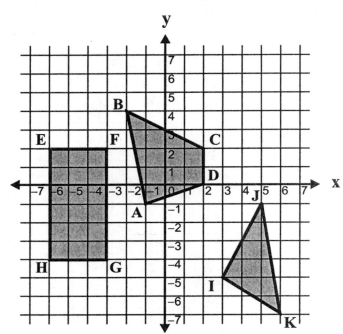

1. quadrilateral ABCD

    A = _____
    B = _____
    C = _____
    D = _____

2. rectangle EFGH

    E = _____
    F = _____
    G = _____
    H = _____

3. triangle IJK

    I = _____
    J = _____
    K = _____

---

4. parallelogram LMNO

    L = _____
    M = _____
    N = _____
    O = _____

5. right triangle PQR

    P = _____
    Q = _____
    R = _____

6. pentagon STVXY

    S = _____
    T = _____
    V = _____
    X = _____
    Y = _____

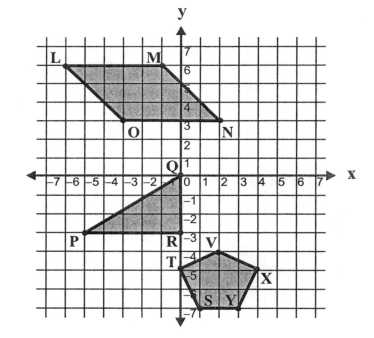

156 Copyright © American Book Company

Plot and label the following points. Then construct and identify the geometric figure you plotted. Question 1 is done for you.

Figure

1. point A = (−1, −1)
   point B = (−1, 2)
   point C = (2, 2)
   point D = (2, −1) ___square___

2. point E = (3, −2)
   point F = (5, 1)
   point G = (7, −2) _____

3. point H = (−4, 0)
   point I = (−6, 0)
   point J = (−4, 4)
   point K = (−2, 4) _____

4. point L = (−1, −3)
   point M = (4, −6)
   point N = (−1, −6) _____

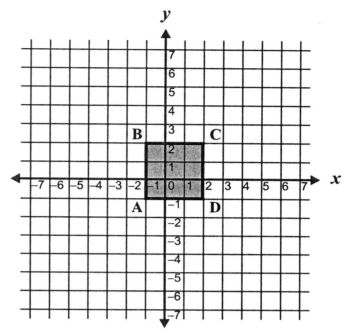

Figure

5. point A = (−2, 3)
   point B = (−3, 5)
   point C = (−1, 6)
   point D = (1, 5)
   point E = (0, 3) _____

6. point F = (−1, −3)
   point G = (−3, −5)
   point H = (−1, −7)
   point I = (1, −5) _____

7. point J = (−1, 2)
   point K = (−1, −1)
   point L = (3, −2) _____

8. point M = (6, 2)
   point N = (6, −4)
   point O = (4, −4)
   point P = (4, 2) _____

CHAPTER 11 REVIEW

Record the coordinates and quadrants of the following points.

Coordinates   Quadrant

1. A = _____   _____

2. B = _____   _____

3. C = _____   _____

4. D = _____   _____

On the same plane as above, label these additional coordinates.

5. E = (0, −3)   7. G = (4, 0)

6. F = (−3, −1)   8. H = (2, 2)

Answer the following questions.

9. In which quadrant does the point (2, 3) lie?

10. In which quadrant does the point (−5, −2) lie?

Find the coordinates of the geometric figures graphed below.

11. point A   _____   14. point D   _____

12. point B   _____   15. point E   _____

13. point C   _____   16. point F   _____

17. point G   _____

Plot and label the following points on the same graph.

18. point H = (1, 1)

19. point I = (3, 1)

20. point J = (4, −2)

21. point K = (2, −2)

22. What type figure did you plot?

Chapter 12
Graphing and Writing Equations

GRAPHING LINEAR EQUATIONS

In addition to graphing ordered pairs, the Cartesian plane can be used to graph the solution set for an equation. Any equation with two variables that are both to the first power is called a **linear equation**. The graph of a linear equation will always be a straight line.

EXAMPLE 1: Graph the solution set for $x + y = 7$.

Step 1: Make a list of some pairs of numbers that will work in the equation.

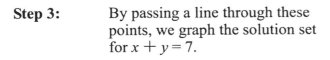

ordered pair solutions

Step 2: Plot these points on a Cartesian plane.

Step 3: By passing a line through these points, we graph the solution set for $x + y = 7$.

This means that every point on this line is a solution to the equation $x + y = 7$. For example, $(1, 6)$ is a solution; and therefore, the line passes through the point $(1, 6)$

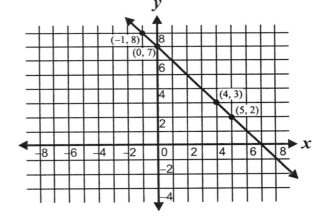

Make a table of solutions for each linear equation below. Then plot the ordered pair solutions on graph paper. Draw a line through the points. (If one of the points does not line up, you have made a mistake.)

1. $x + y = 6$
2. $y = x + 1$
3. $y = x - 2$
4. $x + 2 = y$
5. $x - 5 = y$
6. $x - y = 0$

EXAMPLE 2: Graph the equation $y = 2x - 5$.

Step 1: This equation has 2 variables, both to the first power, so we know the graph will be a straight line.

Substitute some numbers for x or y to find pairs of numbers that satisfy the equation. For the above equation, it will be easier to substitute values of x in order to find the corresponding value for y. Record the values for x and y in a table.

	x	y
If x is 0, y would be -5	0	-5
If x is 1, y would be -3	1	-3
If x is 2, y would be -1	2	-1
If x is 3, y would be 1	3	1

Step 2: Graph the ordered pairs, and draw a line through the points.

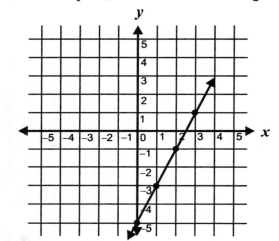

Find pairs of numbers that satisfy the equations below, and graph the line on graph paper.

1. $y = -2x + 2$
2. $2x - 2 = y$
3. $-x + 3 = y$
4. $y = x + 1$
5. $4x - 2 = y$
6. $y = 3x - 3$
7. $x = 4y - 3$
8. $2x = 3y + 1$
9. $x + 2y = 4$

GRAPHING HORIZONTAL AND VERTICAL LINES

The graph of some equations is a horizontal or a vertical line.

EXAMPLE 1: $y = 3$

Step 1: Make a list of ordered pairs that satisfy the equation $y = 3$.

x	y
0	3
1	3
2	3
3	3

No matter what value of x you choose, y is always 3.

Step 2: Plot these points on a Cartesian plane, and draw a line through the points.

The graph is a horizontal line.

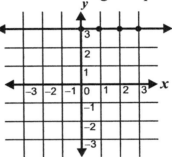

EXAMPLE 2: $2x + 3 = 0$

Step 1: For these equations with only one variable, find what x equals first.

$2x + 3 = 0$
$2x = -3$
$x = \frac{-3}{2}$

Step 2: Just like Example 1, find ordered pairs that satisfy the equation, plot the points, and graph the line.

x	y
$\frac{-3}{2}$	0
$\frac{-3}{2}$	1
$\frac{-3}{2}$	2
$\frac{-3}{2}$	3

No matter which value of y you choose, the value of x does not change.

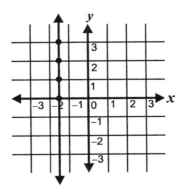

The graph is a vertical line.

Find pairs of numbers that satisfy the equations below, and graph the line on graph paper.

1. $2y + 2 = 0$
2. $x = -4$
3. $3x = 3$
4. $y = 5$
5. $4x - 2 = 0$
6. $2x - 6 = 0$
7. $4y = 1$
8. $5x + 10 = 0$
9. $3y + 12 = 0$
10. $x + 1 = 0$
11. $2y - 8 = 0$
12. $3x = -9$
13. $x = -2$
14. $6y - 2 = 0$
15. $5x - 5 = 0$
16. $2y - 4 = 0$
17. $2y - 2 = 0$
18. $3x + 1 = 0$
19. $4y = -2$
20. $-2y = 6$
21. $-4x = -8$
22. $3y = -6$
23. $x = 2$
24. $4y = 8$

FINDING THE INTERCEPTS OF A LINE

The *x*-intercept is the point where the graph of a line crosses the *x*-axis. The *y*-intercept is the point where the graph of a line crosses the *y*-axis.

> To find the *x*-intercept, set $y = 0$
> To find the *y*-intercept, set $x = 0$

EXAMPLE: Find the *x*- and *y*-intercepts of the line $6x + 2y = 18$.

Step 1: To find the *x*-intercept, set $y = 0$.

$6x + 2(0) = 18$
$\dfrac{6x}{6} = \dfrac{18}{6}$
$x = 3$ The *x*-intercept is at the point (3, 0).

Step 2: To find the *y*-intercept, set $x = 0$.

$6(0) + 2y = 18$
$\dfrac{2y}{2} = \dfrac{18}{2}$
$y = 9$ The *y*-intercept is at the point (0, 9).

Step 3: You can now use the two intercepts to graph the line.

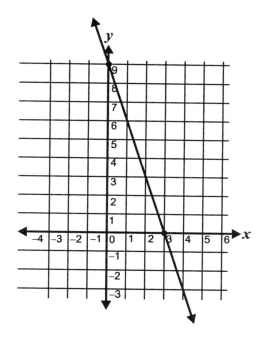

For each of the following equations, find both the *x* and the *y* intercepts of the line. For extra practice, draw each of the lines on graph paper.

1. $6x - 2y = 6$
2. $2x + 4y = 8$
3. $4x + 3y = 12$
4. $x - 3y = -4$
5. $8x + 3y = 8$
6. $5x - 4y = 10$
7. $-2x - 2y = 6$
8. $-6x + 4y = 12$
9. $6x - 2y = -6$
10. $-5x - 5y = 15$

11. $9x - 6y = -18$
12. $6x + 6y = 18$
13. $-3x - 6y = 21$
14. $8x + 3y = -8$
15. $-3x + 9y = 9$
16. $12x + 6y = 24$
17. $x - 2y = -4$
18. $-2x - 4y = 8$
19. $5x + 4y = 15$
20. $12x + 18y = 60$

21. $7x - 14y = 21$
22. $5x + 10y = 15$
23. $-12x + 16y = 48$
24. $33x - 11y = -33$
25. $-2x - 6y = -8$
26. $14x + 3y = 21$
27. $10x - 5y = 20$
28. $10x + 15y = 30$
29. $-18x + 27y = 54$
30. $21x + 42y = 63$

UNDERSTANDING SLOPE

The **slope** of a line refers to how steep a line is. When we graph a line using ordered pairs, we can easily determine the slope. Slope is often represented by the letter *m*.

$$\text{The formula for slope of a line is: } m = \frac{y_2 - y_1}{x_2 - x_1} \text{ or } \frac{\text{rise}}{\text{run}}$$

EXAMPLE 1: What is the slope of the following line that passes through the ordered pairs $(-4, -3)$ and $(1, 3)$?

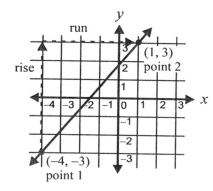

y_2 is 3, the *y*-coordinate of point 2.

y_1 is -3, the *y*-coordinate of point 1.

x_2 is 1, the *x*-coordinate of point 2.

x_1 is -4, the *x*-coordinate of point 1.

Use the formula for slope given above. $m = \dfrac{3 - (-3)}{1 - (-4)} = \dfrac{6}{5}$

The slope is $\dfrac{6}{5}$. This shows us that we can go up 6 (rise) and over 5 to the right (run) to find another point on the line.

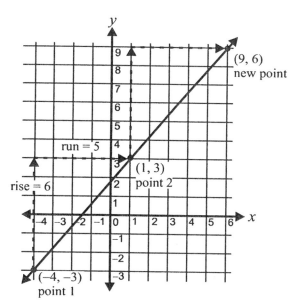

EXAMPLE 2: Find the slope of a line through the points (−2, 3) and (1, −2). It doesn't matter which pair we choose for point 1 and point 2. The answer is the same.

let point 1 be (−2, 3)
let point 2 be (1, −2)

$$\text{slope} = \frac{(y_2 - y_1)}{(x_2 - x_1)} = \frac{-2 - 3}{1 - (-2)} = \frac{-5}{3}$$

When the slope is negative, the line will slant left. For this example, the line will go **down** 5 units and then over 3 to the **right**.

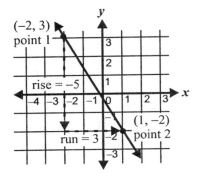

EXAMPLE 3: What is the slope of a line that passes through (1, 1) and (3, 1)?

$$\text{slope} = \frac{1-1}{3-1} = \frac{0}{2} = 0$$

When $y_2 - y_1 = 0$, the slope will equal 0, and the line will be horizontal.

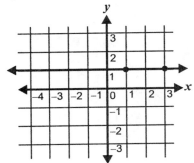

EXAMPLE 4: What is the slope of a line that passes through (2, 1) and (2, −3)?

$$\text{slope} = \frac{-3-1}{2-2} = \frac{4}{0} = \text{undefined}$$

When $x_2 - x_1 = 0$, the slope is undefined, and the line will be vertical.

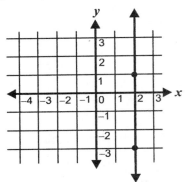

The following lines summarize what we know about slope.

slope > 0   slope < 0   slope = 0   slope is undefined

Find the slope of the line that goes through the following pairs of points. Use the formula slope = $\frac{y_2 - y_1}{x_2 - x_1}$. Then, using graph paper, graph the line through the two points, and label the rise and the run. (See Examples 1 and 2.)

1. (2, 3) (4, 5)
2. (1, 3) (2, 5)
3. (−1, 2) (4, 1)
4. (1, −2) (4, −2)
5. (3, 0) (3, 4)
6. (3, 2) (−1, 8)
7. (4, 3) (2, 4)
8. (2, 2) (1, 5)
9. (3, 4) (1, 2)
10. (3, 2) (3, 6)

11. (6, −2) (3, −2)
12. (1, 2) (3, 4)
13. (−2, 1) (−4, 3)
14. (5, 2) (4, −1)
15. (1, −3) (−2, 4)
16. (2, −1) (3, 5)
17. (2, 4) (5, 3)
18. (5, 2) (2, 5)
19. (4, 5) (6, 6)
20. (2, 1) (−1, −3)

SLOPE - INTERCEPT FORM OF A LINE

An equation that contains two variables, each to the first degree, is a **linear equation**. The graph for a linear equation is a straight line. To put a linear equation in slope-intercept form, solve the equation for y. This form of the equation shows the slope and the y-intercept. Slope-intercept form follows the pattern of $y = mx + b$. The "m" represents slope, and the "b" represents the y-intercept. The y-intercept is the point at which the line crosses the y-axis.

When the slope of a line is not 0, the graph of the equation shows a **direct variation** between y and x. When y increases, x increases in a certain proportion. The proportion stays constant. The constant is called the **slope** of the line.

EXAMPLE: Put the equation $2x + 3y = 15$ in slope-intercept form. What is the slope of the line? What is the y-intercept? Graph the line.

Step 1: Solve for y:

$$2x + 3y = 15$$
$$\underline{-2x \qquad -2x}$$
$$\frac{3y}{3} = \frac{-2x}{3} + \frac{15}{3}$$

slope-intercept form: $y = \frac{-2}{3}x + 5$

The slope is $\frac{-2}{3}$ and the y-intercept is 5

Step 2: Knowing the slope and the y-intercept, we can graph the line.

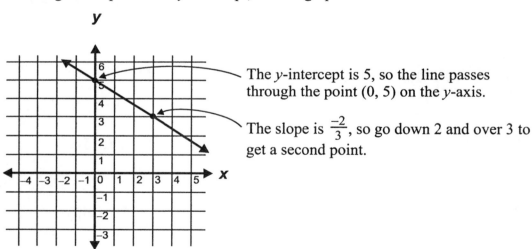

The y-intercept is 5, so the line passes through the point (0, 5) on the y-axis.

The slope is $\frac{-2}{3}$, so go down 2 and over 3 to get a second point.

Put each of the following equations in slope-intercept form by solving for *y*. On your graph paper, graph the line using the slope and *y*-intercept.

1. $4x - y = 5$
2. $2x + 4y = 16$
3. $3x - 2y = 10$
4. $x + 3y = -12$
5. $6x + 2y = 0$
6. $8x - 5y = 10$
7. $-2x + y = 4$
8. $-4x + 3y = 12$
9. $-6x + 2y = 12$
10. $x - 5y = 5$
11. $3x - 2y = -6$
12. $3x + 4y = 2$
13. $-x = 2 + 4y$
14. $2x = 4y - 2$
15. $6x - 3y = 9$
16. $4x + 2y = 8$
17. $6x - y = 4$
18. $-2x - 4y = 8$
19. $5x + 4y = 16$
20. $6 = 2y - 3x$

VERIFY THAT A POINT LIES ON A LINE

To know whether or not a point lies on a line, substitute the coordinates of the point into the formula for the line. If the point lies on the line, the equation will be true. If the point does not lie on the line, the equation will be false.

EXAMPLE 1: Does the point (5, 2) lie on the line given by the equation $x + y = 7$?

Solution: Substitute 5 for *x* and 2 for *y* in the equation. $5 + 2 = 7$. Since this is a true statement, the point (5, 2) does lie on the line $x + y = 7$.

EXAMPLE 2: Does the point (0, 1) lie on the line given by the equation $5x + 4y = 16$?

Solution: Substitute 0 for *x* and 1 for *y* in the equation. $5x + 4y = 16$. Does $5(0) + 4(1) = 16$? No, it equals 4, not 16. Therefore, the point (0, 1) is not on the line given by the equation $5x + 4y = 16$.

For each point below, state whether or not it lies on the line given by the equation that follows the point coordinates.

1. (2, 4) $6x - y = 8$
2. (1, 1) $6x - y = 5$
3. (3, 8) $-2x + y = 2$
4. (9, 6) $-2x + y = 0$
5. (3, 7) $x - 5y = -32$
6. (0, 5) $-6x - 5y = 3$
7. (2, 4) $4x + 2y = 16$
8. (9, 1) $3x - 2y = 29$
9. (6, 8) $6x - y = 28$
10. (−2, 3) $x + 2y = 4$
11. (4, −1) $-x - 3y = -1$
12. (−1, −3) $2x + y = 1$

GRAPHING A LINE KNOWING A POINT AND SLOPE

If you are given a point of a line and the slope of a line, the line can be graphed.

EXAMPLE 1: Given that line *l* has a slope of $\frac{4}{3}$ and contains the point $(2, -1)$, graph the line.

Step 1: Plot and label the point $(2, -1)$ on a Cartesian plane.

Step 2: The slope, *m*, is $\frac{4}{3}$, so the rise is 4 and the run is 3. From the point $(2, -1)$, count 4 units up and 3 units to the right.

Step 3: Draw the line through the two points.

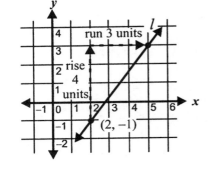

EXAMPLE 2: Given a line that has a slope of $\frac{-1}{4}$ and passes through the point $(-3, 2)$, graph the line.

Step 1: Plot the point $(-3, 2)$.

Step 2: Since the slope is negative, go **down** 1 unit and over 4 to get a second point.

Step 3: Graph the line through the two points.

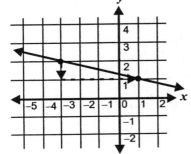

Graph a line on your own graph paper for each of the following problems. First, plot the point. Then use the slope to find a second point. Draw the line formed from the point and the slope.

1. $(2, -2)$, $m = \frac{3}{4}$
2. $(3, -4)$, $m = \frac{1}{2}$
3. $(1, 3)$, $m = \frac{-1}{3}$
4. $(2, -4)$, $m = 1$
5. $(3, 0)$, $m = \frac{-1}{2}$

6. $(-2, 1)$, $m = \frac{4}{3}$
7. $(-4, -2)$, $m = \frac{1}{2}$
8. $(1, -4)$, $m = \frac{3}{4}$
9. $(2, -1)$, $m = \frac{-1}{2}$
10. $(5, -2)$, $m = \frac{1}{4}$

11. $(-2, -3)$, $m = \frac{2}{3}$
12. $(4, -1)$, $m = \frac{-1}{3}$
13. $(-1, 5)$, $m = \frac{2}{5}$
14. $(-2, 3)$, $m = \frac{3}{4}$
15. $(4, 4)$, $m = \frac{-1}{2}$

16. $(3, -3)$, $m = \frac{-3}{4}$
17. $(-2, 5)$, $m = \frac{1}{3}$
18. $(-2, -3)$, $m = \frac{-3}{4}$
19. $(4, -3)$, $m = \frac{2}{3}$
20. $(1, 4)$, $m = \frac{-1}{2}$

FINDING THE EQUATION OF A LINE USING TWO POINTS OR A POINT AND SLOPE

If you can find the slope of a line and know the coordinates of one point, you can write the equation for the line. You know the formula for the slope of a line is:

$$m = \frac{y_2 - y_1}{x_2 - x_1} \text{ or } \frac{y_2 - y_1}{x_2 - x_1} = m$$

Using algebra, you can see that if you multiply both sides of the equation by $x_2 - x_1$, you get:

$$y - y_1 = m(x - x_1) \longleftarrow \text{ point-slope form of an equation}$$

EXAMPLE: Write the equation of the line passing through the points $(-2, 3)$ and $(1, 5)$.

Step 1: First, find the slope of the line using the two points given.

$$m = \frac{y_2 - y_1}{x_2 - x_1} = \frac{5 - 3}{1 - (-2)} = \frac{2}{3}$$

Step 2: Pick one of the points to use in the point-slope equation. For point $(-2, 3)$, we know $x_1 = -2$ and $y_1 = 3$, and we know $m = \frac{2}{3}$. Substitute these values into the point-slope form of the equation.

$$y - y_1 = m(x - x_1)$$

$$y - 3 = \frac{2}{3}[x - (-2)]$$

$$y - 3 = \frac{2}{3}x + \frac{4}{3}$$

$$y = \frac{2}{3}x + \frac{13}{3}$$

Use the point-slope formula to write an equation for each of the following lines.

1. $(1, -2)$, $m = 2$
2. $(-3, 3)$, $m = \frac{1}{3}$
3. $(4, 2)$, $m = \frac{1}{4}$
4. $(5, 0)$, $m = 1$
5. $(3, -4)$, $m = \frac{1}{2}$
6. $(-1, 4)$ $(2, -1)$
7. $(2, 1)$ $(-1, -3)$
8. $(-2, 5)$ $(-4, 3)$
9. $(-4, 3)$ $(2, -1)$
10. $(3, 1)$ $(5, 5)$
11. $(-3, 1)$, $m = 2$
12. $(-1, -2)$, $m = \frac{4}{3}$
13. $(2, -5)$, $m = -2$
14. $(-1, 3)$, $m = \frac{1}{3}$
15. $(0, -2)$, $m = -\frac{3}{2}$

WRITING AN EQUATION FROM DATA

Data is often written in a two column format. If the increases or decreases in the ordered pairs are at a constant rate, then a linear equation for the data can be found.

EXAMPLE: Write an equation for the following set of data.

Dan set his car on cruise control and noted the distance he went every 5 minutes.

Minutes in operation (x)	Distance traveled (y)
5	28,490 miles
10	28,494 miles

Step 1: Write two ordered pairs in the form (minutes, distance) for Dan's driving, (5, 28490), and (10, 28494), and find slope.

Step 2: Use the ordered pairs to write the equation in the form $y = mx + b$.
Place the slope, m, that you found and one of the pairs of points as x_1 and y_1 in the following formula:

$y - y_1 = m(x - x_1)$
$y - 28490 = \frac{4}{5}(x - 5)$
$y - 28490 + 28490 = \frac{4}{5}x - 4 + 28490$
$y = \frac{4}{5}x + 28486$

It doesn't matter which pair of points you use, the answer will be the same.

Write an equation for each of the following sets of data, assuming the relationship is linear.

1. **Doug's Doughnut Shop**

Year in Business	Total Sales
1	$55,000
4	$85,000

2. **Gwen's Green Beans**

Days Growing	Height in Inches
2	5
6	12

3. **At the Gas Pump**

Gallons Purchased	Total Cost
5	$6.00
7	$8.40

4. **Jim's Depreciation on His Jet Ski**

Years	Value
1	$4,500
6	$2,500

5. **Stepping on the Brakes**

Seconds	MPH
2	51
5	18

6. **Stepping on the Accelerator**

Seconds	MPH
4	35
7	62

GRAPHING LINEAR DATA

Many types of data are related by a constant ratio. As you learned on the previous page, this type of data is linear. The slope of the line described by linear data is the ratio between the data. Plotting linear data with a constant ratio can be helpful in finding additional values.

EXAMPLE 1: A department store prices socks per pair. Each pair of socks costs $0.75. Plot pairs of socks versus price on a Cartesian plane.

Step 1: Since the price of the socks is constant, you know that one pair of socks costs $0.75, 2 pairs of socks cost $1.50, 3 pairs of socks cost $2.25, and so on. Make a list of a few points.

Pair(s) x	Price y
1	.75
2	1.50
3	2.25

Step 2: Plot these points on a Cartesian plane, and draw a straight line through the points.

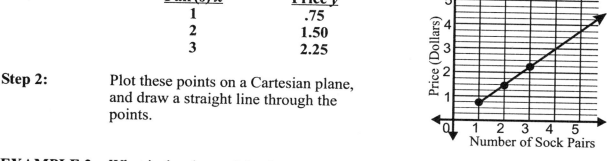

EXAMPLE 2: What is the slope of the data? What does the slope describe?

Solution: You can determine the slope either by the graph or by the data points. For this data, the slope is .75. Remember, slope is rise/run. For every $0.75 going up the y-axis, you go across one pair of socks on the x-axis. The slope describes the price per pair of socks.

EXAMPLE 3: Use the graph created in Example 1 to answer the following questions. How much would 5 pairs of socks cost? How many pairs of socks could you purchase for $3.00? Extending the line gives useful information about the price of additional pairs of socks.

Solution 1: The line that represents 5 pairs of socks intersects the data line at $3.75 on the y-axis. Therefore, 5 pairs of socks would cost $3.75.

Solution 2: The line representing the value of $3.00 on the y-axis intersects the data line at 4 on the x-axis. Therefore, $3.00 will buy exactly 4 pairs of socks.

Use the information given to make a line graph for each set of data, and answer the questions related to each graph.

1. The diameter of a circle versus the circumference of a circle is a constant ratio. Use the data given below to graph a line to fit the data. Extend the line, and use the graph to answer the next question.

   Circle

Diameter	Circumference
4	12.56
5	15.70

2. Using the graph of the data in question 1, estimate the circumference of a circle that has a diameter of 3 inches.

3. If the circumference of a circle is 3 inches, about how long is the diameter?

4. What is the slope of the line you graphed in question 1?

5. What does the slope of the line in question 4 describe?

6. The length of a side on a square and the perimeter of a square are a constant ratio. Use the data below to graph this relationship.

   Square

Length of side	Perimeter
2	8
3	12

7. Using the graph from question 6, what is the perimeter of a square with a side that measures 4 inches?

8. What is the slope of the line graphed in question 6?

9. Conversions are often constant ratios. For example, converting from pounds to ounces follows a constant ratio. Use the data below to graph a line that can be used to convert pounds to ounces.

   Measurement Conversion

Pounds	Ounces
2	32
4	64

10. Use the graph from question 9 to convert 40 ounces to pounds.

11. What does the slope of the line graphed for question 9 represent?

12. Graph the data below and create a line that shows converting weeks to days.

    Time

Weeks	Days
1	7
2	14

13. About how many days are in $2\frac{1}{2}$ weeks?

14. Graph a data line that converts feet to inches.

15. Using the graph in question 14, how many inches are in 4.5 feet?

16. What is the slope of the line converting feet to inches?

17. An electronics store sells DVD's for $25 each. Graph a data line showing total cost versus number of DVD's purchased.

18. Using the graph in question 17, how many DVD's could be purchased for $150?

IDENTIFYING GRAPHS OF LINEAR EQUATIONS

Match each equation below with the graph of the equation.

A: $x = y$
B: $x = -4$
C: $y = -x$
D: $y = 4$
E: $x = \frac{1}{2}y$
F: $y = -4$
G: $x = -2y$
H: $4x = y$
I: $-2x = y$

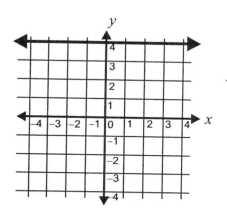
1. _____

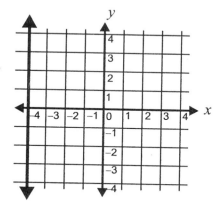
2. _____

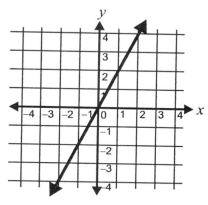
3. _____

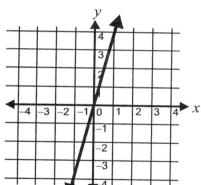
4. _____

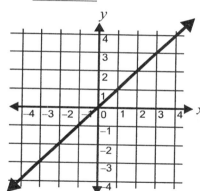
5. _____

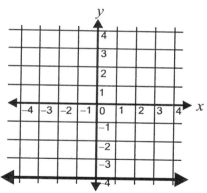
6. _____

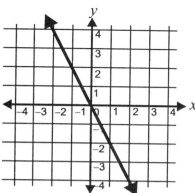
7. _____

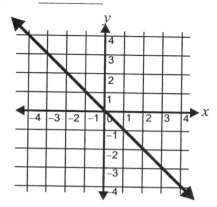
8. _____

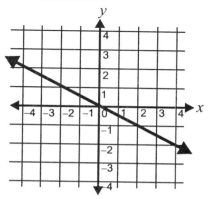
9. _____

Match each equation below with the graph of the equation.

A: $y = 4x$
B: $y = -4x$
C: $4x + y = 4$
D: $x - 2y = 6$
E: $y = 3x - 1$
F: $2x + 3y = 6$
G: $y = 3x + 2$
H: $x + 2y = 6$
I: $y = x - 3$

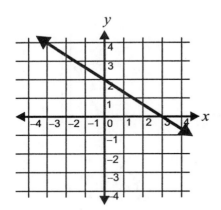
1. _____

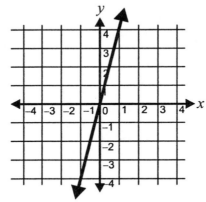
2. _____

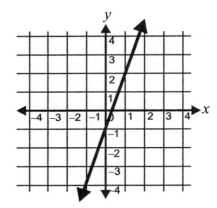
3. _____

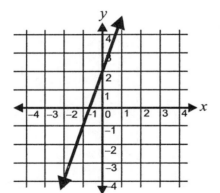
4. _____

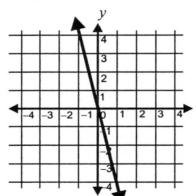
5. _____

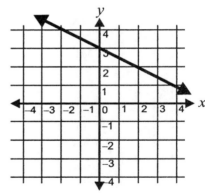
6. _____

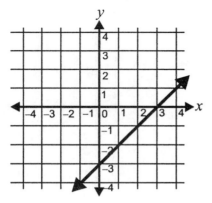
7. _____

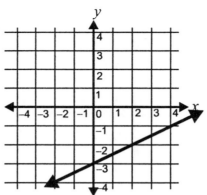
8. _____

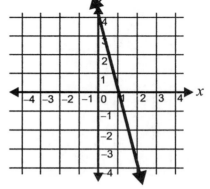
9. _____

176

GRAPHING NON-LINEAR EQUATIONS

Equations that you may encounter on the California HSEE may involve variables which are squared or cubed (raised to the second or third power). The best way to find values for the x and y variables in an equation is to plug one number into x, and then find the corresponding value for y just like you did at the beginning of this chapter. Then, plot the points and draw a line through the points.

EXAMPLE 1: $y = x^2$

Step 1: Make a table and find several values for x and y.

x	y
−2	4
−1	1
0	0
1	1
2	4

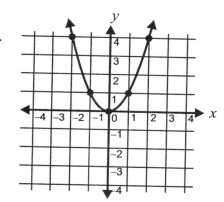

Step 2: Plot the points, and draw a curve through the points. Notice the shape of the curve. This type of curve is called a **parabola**. Equations with one squared term will be parabolas.

EXAMPLE 2: $y = x^3$

Step 1: Make a table and find several values for x and y.

x	y
−2	−8
−1	−1
0	0
1	1
2	8

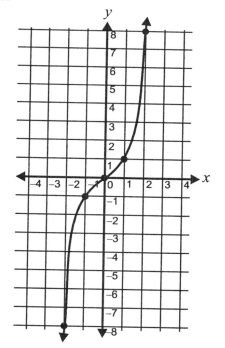

Step 2: Plot the points, and draw a curve through the points. Equations with one cubed term will always have this shape of curve.

Graph the equations below on a Cartesian plane.

1. $y = 2x^2$
2. $y = 3 - x^3$
3. $y = x^2 - 2$
4. $y = 2x^3$
5. $y = x^2 + 3$
6. $y = x^3 - 2$
7. $y = 3x^2 - 5$
8. $y = x^3 + 1$
9. $y = -x^2$
10. $y = -x^3$
11. $y = 2x^2 - 1$
12. $y = 2 - 2x^3$

CHAPTER 12 REVIEW

1. Graph the solution set for the linear equation: $x - 3 = y$.

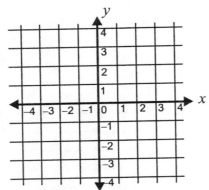

2. Which of the following is not a solution of $3x = 5y - 1$?

   A. $(3, 2)$
   B. $(7, 4)$
   C. $(-\frac{1}{3}, 0)$
   D. $(-2, -1)$

3. $(-2, 1)$ is a solution for which of the following equations?

   A. $y + 2x = 4$
   B. $-2x + 1 = 5$
   C. $x + 2y = -4$
   D. $2x - y = -5$

4. Graph the equation $2x - 4 = 0$.

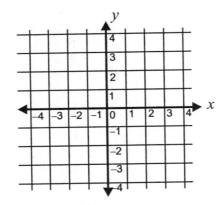

5. What is the slope of the line that passes through the points $(5, 3)$ and $(6, 1)$?

6. What is the slope of the line that passes through the points $(-1, 4)$ and $(-6, -2)$?

7. What is the x-intercept for the following equation?
$$6x - y = 30$$

8. What is the y-intercept for the following equation?
$$4x + 2y = 28$$

9. Graph the equation $3y = 9$.

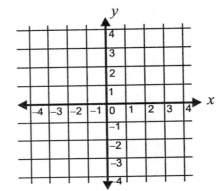

10. Write the following equation in slope-intercept form:
$$3x = -2y + 4$$

11. What is the slope of the line $y = -\frac{1}{2}x + 3$?

12. What is the x-intercept of the line $y = 5x + 6$?

13. What is the y-intercept of the line $y - \frac{2}{3}x + 2 = 0$?

14. Graph the line which has a slope of 2 and a y-intercept of −3.

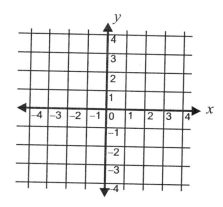

15. Graph the line which has a slope of −2 and a y-intercept of −3.

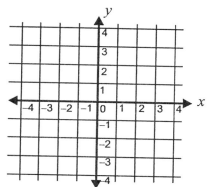

16. Which of the following points does **not** lie on the line $y = 3x - 2$?

    A. (0, −2)
    B. (1, 1)
    C. (−1, 5)
    D. (2, 4)

17. Which of the following points lies on the line $2y = -x + 1$?

    A. $(\frac{1}{2}, 0)$
    B. $(2, -\frac{1}{2})$
    C. (0, 1)
    D. (−1, −1)

18. Graph the equation $-x = 6 + 2y$.

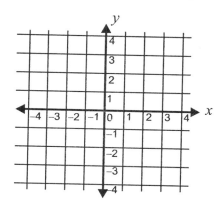

19. Find the equation of the line which contains the point (0, 2) and has a slope of $\frac{3}{4}$.

20. Which is the graph of $x - 3y = 6$?

    A.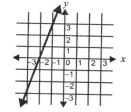

    B.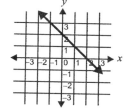

    C.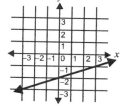

    D.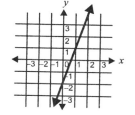

21. Which of the following is the graph of the line which has a slope of −2 and a y-intercept of (0, 3)?

A.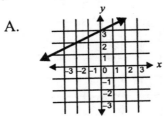

B.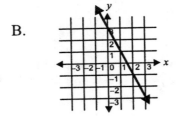

C.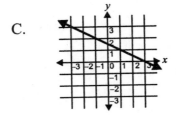

D.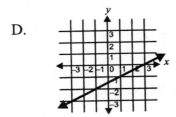

22. Given that a line contains the point (2, 3) and has a slope of $-\frac{1}{2}$, graph the line.

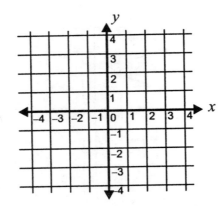

23. Paulo turned on the oven to preheat it. After one minute, the oven temperature was 200°. After 2 minutes, the oven temperature was 325°.

Oven Temperature
Minutes	Temperature
1	200°
2	325°

Assuming the oven temperature rose at a constant rate, write an equation that fits the data.

24. Write an equation that fits the data given below. Assume the data is linear.

Plumber Charges per Hour
Hour	Charge
2	$170
3	$220

25. What is the name of the curve described by the equation $y = 2x^2 - 1$?

26. Graph the following equation:
$$y = \tfrac{1}{2}x^3$$

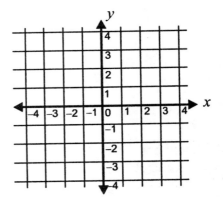

27. The data given below show conversions between miles per hour and kilometers per hour. Based on this data, graph a conversion line on the Cartesian plane below.

Speed

MPH	KPH
5	8
10	16

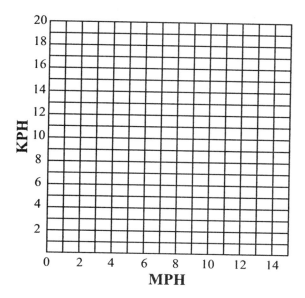

28. What would be the approximate conversion of 9 mph to kph?

29. What would be the approximate conversion of 13 kph to mph?

30. A bicyclist travels 12 mph downhill. Approximately how many kph is the bicyclist traveling?

31. Use the data given below to graph the interest versus the interest rate on $80.00 in one year.

$80.00 Principal

Interest Rate	Interest–1 year
5%	$4.00
10%	$8.00

32. About how much interest would accrue in one year at an 8% interest rate?

33. What is the slope of the line describing interest versus interest rate?

34. What information does the slope give in problem 33?

Chapter 13
Graphing Inequalities

In the previous chapter, you learned to graph linear equations. In this chapter, you will learn to graph linear inequalities.

EXAMPLE 1: In the previous chapter, you would graph the equation $x = 3$ as:

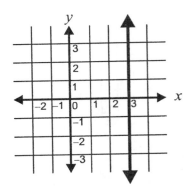

In this chapter, we graph inequalities such as $x > 3$ (read x is greater than 3). To show this, we use a broken line since the points on the line $x = 3$ are not included in the solution. We shade all points greater than 3.

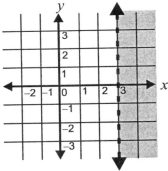

When we graph $x \geq 3$ (read x is greater than or equal to 3), we use a solid line because the points on the line $x = 3$ are included in the graph.

Graph the following inequalities on your own graph paper.

1. $y < 2$
2. $x \geq 4$
3. $y \geq 1$
4. $x < -1$
5. $y \geq -2$
6. $x \leq -4$
7. $x > -3$
8. $y \leq 3$
9. $x \leq 5$
10. $y > -5$
11. $x \geq 3$
12. $y < -1$
13. $x \leq 0$
14. $y > -1$
15. $y \leq 4$
16. $x \geq 0$
17. $y \geq 3$
18. $x < 4$
19. $x \leq -2$
20. $y < -2$
21. $y \geq -4$
22. $x \geq -1$
23. $y \leq 5$
24. $x < -3$

EXAMPLE 2: Graph $x + y \geq 3$

Step 1: First, we graph $x + y \geq 3$ by changing the inequality to an equality. Think of ordered pairs that will satisfy the equation $x + y = 3$. Then, plot the points, and draw the line.

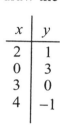
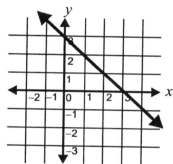

This divides the Cartesian plane into 2 half-planes, $x + y \geq 3$ and $x + y \leq 3$. One half-plane is above the line, and the other is below the line.

Step 2: To determine which side of the line to shade, first choose a test point. If the point you choose makes the inequality true, then the point is on the side you shade. If the point you choose does not make the inequality true, then shade the side that does not contain the test point.

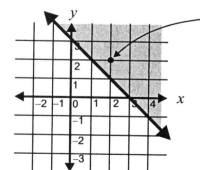

For our test point, let's choose (2, 2). Substitute (2, 2) into the inequality.

$x + y \geq 3$
$2 + 2 \geq 3$

$4 \geq 3$ is true, so shade the side that includes this point.

Use a solid line because of the $\geq$ sign.

Graph the following inequalities on your own graph paper.

1. $x + y \leq 4$
2. $x + y \geq 3$
3. $x \geq 5 - y$
4. $x \leq 1 + y$
5. $x - y \geq -2$
6. $x < y + 4$
7. $x + y < -1$
8. $x - y \leq 0$
9. $x \geq y + 2$
10. $x < -y + 1$
11. $-x + y > 1$
12. $-x - y < -2$

For more complex inequalities, it is easier to graph by first changing the inequality to an equality and second, putting the equation in slope-intercept form.

EXAMPLE: $2x + 4y \leq 8$

Step 1: Change the inequality to an equality.
$2x + 4y = 8$

Step 2: Put the equation in slope-intercept form by solving the equation for y.

$$2x + 4y = 8$$
$$\underline{-2x \qquad -2x}$$
$$\frac{4y}{4} = \frac{-2x}{4} + \frac{8}{4}$$
$$y = -\tfrac{1}{2}x + 2$$

Step 3: Graph the line. If the inequality is < or >, use a dotted line. If the inequality is ≤ or ≥, use a solid line. For this example, we should use a solid line.

Step 4: Determine which side of the line to shade. Pick a point like (0, 0) to see if it is true in the inequality.

$2x + 4y \leq 8$, so substitute (0, 0).
Is $0 + 0 \leq 8$? Yes, $0 \leq 8$, so shade the side of the line that includes the point (0, 0).

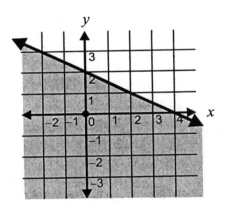

Graph the following inequalities on your own graph paper.

1. $2x + y \geq 1$
2. $3x - y \leq 3$
3. $x + 3y > 12$
4. $4x - 3y < 12$
5. $y \geq 3x + 1$
6. $x - 2y > -2$
7. $x \leq y + 4$
8. $x + y < -1$
9. $-4y \geq 2x + 1$
10. $x \leq 4y - 2$
11. $3x - y \geq 4$
12. $y \geq 2x - 5$
13. $x + 7y < 1$
14. $-2y < 4x - 1$
15. $y > 4x + 1$

CHAPTER 13 REVIEW

Graph the following inequalities on a Cartesian plane using your own graph paper.

1. $x \geq 4$
2. $x < -3$
3. $y \leq 2$
4. $y \geq -1$
5. $2y \geq 8$
6. $3x < 6$
7. $-4y \geq 12$
8. $-2x \leq 4$
9. $2x + 5y \geq 10$
10. $-2y + 4x < 8$
11. $-2x - 3y > 9$
12. $5y > -10x + 5$
13. $y \leq 2x - 6$
14. $-2x + y < 1$
15. $3x - 4y > 8$
16. $y + 5 \leq x$
17. $2x - y \geq -1$
18. $y < -5$
19. $x + y > -2$
20. $3y < 2x + 6$
21. $x \leq -2$
22. $y \geq x + 2$
23. $2x + y < 5$
24. $3 + y > x$
25. $y - 2x \leq 3$
26. $2y + 6 < x$
27. $2x - y \geq 4$
28. $y \leq 3x$
29. $y > 4x + 1$
30. $2y < 3$

Chapter 14
Systems of Equations and Systems of Inequalities

SYSTEMS OF EQUATIONS

Two linear equations considered at the same time are called a **system** of linear equations. The graph of a linear equation is a straight line. The graphs of two linear equations can show that the lines are **parallel, intersecting**, or **collinear**. Two lines that are **parallel** will never intersect and have no ordered pairs in common. If two lines are **intersecting**, they have one point in common, and in this chapter, you will learn to find the ordered pair for that one point. If the graph of two linear equations is the same line, the lines are said to be **collinear**.

If you are given a system of two linear equations, and you put both equations in slope-intercept form, you can immediately tell if the graph of the lines will be **parallel, intersecting**, or **collinear**.

If two linear equations have the same slope and the same y-intercept, then they are both equations for the same line. They are called **collinear** or **coinciding** lines. A line is made up of an infinite number of points extending infinitely far in two directions. Therefore, collinear lines have an infinite number of points in common.

EXAMPLE: $2x + 3y = -3$ In slope-intercept form: $y = -\frac{2}{3}$
$4x + 6y = -6$

If two linear equations have the same slope but different y-intercepts, they are **parallel** lines. Parallel lines never touch each other, so they have no points in common.

If two linear equations have different slopes, then they are intersecting lines and share exactly one point in common.

The chart below summarizes what we know about the graphs of two equations in slope-intercept form.

y-Intercepts	Slopes	Graphs	Number of Solutions
same	same	collinear	infinite
different	same	distinct parallel lines	none (they never touch)
same or different	different	intersecting lines	exactly one

For the pairs of equations below, put each equation in slope-intercept form, and tell whether the graphs of the lines will be collinear, parallel, or intersecting.

1. $x - y = -1$
   $-x + y = 1$ _____

2. $x - 2y = 4$
   $-x + 2y = 6$ _____

3. $y - 2 = x$
   $x + 2 = y$ _____

4. $x = y - 1$
   $-x = y - 1$ _____

5. $2x + 5y = 10$
   $4x + 10y = 20$ _____

6. $x + y = 3$
   $x - y = 1$ _____

7. $2y = 4x - 6$
   $-6x + y = 3$ _____

8. $x + y = 5$
   $2x + 2y = 10$ _____

9. $2x = 3y - 6$
   $4x = 6y - 6$ _____

10. $2x - 2y = 2$
    $3y = -x + 5$ _____

11. $x = -y$
    $x = 4 - y$ _____

12. $2x = y$
    $x + y = 3$ _____

13. $x = y + 1$
    $y = x + 1$ _____

14. $x - 2y = 4$
    $-2x + 4y = -8$ _____

15. $2x + 3y = 4$
    $-2x + 3y = 4$ _____

16. $2x - 4y = 1$
    $-6x + 12y = 3$ _____

17. $-3x + 4y = 1$
    $6x + 8y = 2$ _____

18. $x + y = 2$
    $5x + 5y = 10$ _____

19. $x + y = 4$
    $x - y = 4$ _____

20. $y = -x + 3$
    $x - y = 1$ _____

FINDING COMMON SOLUTIONS FOR INTERSECTING LINES

When two lines intersect, they share exactly one point in common.

EXAMPLE: $3x + 4y = 20$ and $4x + 2y = 12$

Put each equation in slope-intercept form.

$$3x + 4y = 20 \qquad\qquad 2y - 4x = 12$$
$$4y = -3x + 20 \qquad\qquad 2y = 4x + 12$$
$$y = -\tfrac{3}{4}x + 5 \qquad\qquad y = 2x + 6$$

slope-intercept form

Straight lines with different slopes are **intersecting lines**. Look at the graph of the lines on the same Cartesian plane.

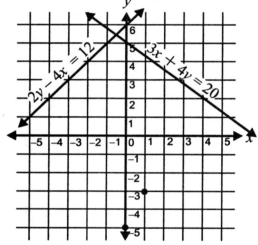

You can see from looking at the graph that the intersecting lines share one point in common. However, it is hard to tell from looking at the graph what the coordinates are for the point of intersection. To find the exact point of intersection, you can use the **substitution method** to solve the system of equations algebraically.

188   Copyright © American Book Company

SOLVING SYSTEMS OF EQUATIONS BY SUBSTITUTION

You can solve systems of equations algebraically by using the substitution method.

EXAMPLE: Find the point of intersection of the following two equations:

$$\text{Equation 1:} \quad x - y = 3$$
$$\text{Equation 2:} \quad 2x + y = 9$$

Step 1: Solve one of the equations for x or y. Let's choose to solve equation 1 for x.
Equation 1: $\quad x - y = 3$
$\quad\quad\quad\quad\quad x = y + 3$

Step 2: Substitute the value of x from equation 1 in place of x in equation 2.
Equation 2: $\quad 2x + y = 9$
$\quad\quad\quad\quad\quad 2(y + 3) + y = 9$
$\quad\quad\quad\quad\quad 2y + 6 + y = 9$
$\quad\quad\quad\quad\quad 3y + 6 = 9$
$\quad\quad\quad\quad\quad 3y = 3$
$\quad\quad\quad\quad\quad y = 1$

Step 3: Substitute the solution for y back in equation 1 and solve for x.
Equation 1: $\quad x - y = 3$
$\quad\quad\quad\quad\quad x - 1 = 3$
$\quad\quad\quad\quad\quad x = 4$

Step 4: The solution set is (4, 1). Substitute in one or both of the equations to check.

Equation 1:	$x - y = 3$	Equation 2:	$2x + y = 9$
	$4 - 1 = 3$		$2(4) + 1 = 9$
	$3 = 3$		$8 + 1 = 9$
			$9 = 9$

The point (4, 1) is common for both equations. This is the **point of intersection**.

For each of the following pairs of equations, find the point of intersection, the common solution, using the substitution method.

1. $x + 2y = 8$
   $2x - 3y = 2$

2. $x - y = -5$
   $x + y = 1$

3. $x - y = 4$
   $x + y = 2$

4. $x - y = -1$
   $x + y = 9$

5. $-x + y = 2$
   $x + y = 8$

6. $x + 4y = 10$
   $x + 5y = 12$

7. $2x + 3y = 2$
   $4x - 9y = -1$

8. $x + 3y = 5$
   $x - y = 1$

9. $-x = y - 1$
   $x = y - 1$

10. $x - 2y = 2$
    $2y + x = -2$

11. $5x + 2y = 1$
    $2x + 4y = 10$

12. $3x - y = 2$
    $5x + y = 6$

13. $2x + 3y = 3$
    $4x + 5y = 5$

14. $x - y = 1$
    $-x - y = 1$

15. $x = y + 3$
    $y = 3 - x$

GRAPHING SYSTEMS OF INEQUALITIES

Systems of inequalities are best solved graphically. Look at the following example.

EXAMPLE: Sketch the solution set of the following system of inequalities:
$y > -2x - 1$ and $y \leq 3x$

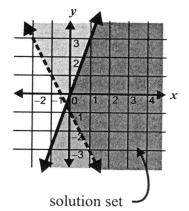

solution set

Step 1: Graph both inequalities on a Cartesian plane. Review Chapter 13 on Graphing Inequalities if you need to review.

Step 2: Shade the portion of the graph that represents the solution set to each inequality just as you did in Chapter 13.

Step 3: Any shaded region that overlaps is the solution set of both inequalities.

Graph the following systems of inequalities on your own graph paper. Shade and identify the solution set for both inequalities.

1. $2x + 2y \geq -4$
   $3y < 2x + 6$

2. $7x + 7y \leq 21$
   $8x < 6y - 24$

3. $9x + 12y < 36$
   $34x - 17y > 34$

4. $-11x - 22y \geq 44$
   $-4x + 2y \leq 8$

5. $24x < 72 + 36y$
   $11x + 22y \leq -33$

6. $15x - 60 < 30y$
   $20x + 10y < 40$

7. $-12x + 24y > -24$
   $10x < -5y + 15$

8. $y \geq 2x + 2$
   $y < -x - 3$

9. $3x + 4y \geq 12$
   $y > -3x + 2$

10. $-3x \leq 6 + 2y$
    $y \geq -x - 2$

11. $2x - 2y \leq 4$
    $3x + 3y \leq -9$

12. $-x \geq -2y - 2$
    $-2x - 2y > 4$

CHAPTER 14 REVIEW

For each pair of equations below, tell whether the graphs of the lines will be collinear, parallel, or intersecting.

1. $y = 4x + 1$
   $y = 4x - 3$

2. $y - 4 = x$
   $2x + 8 = 2y$

3. $x + y = 5$
   $x - y = -1$

4. $2y - 3x = 6$
   $4y = 6x + 8$

5. $5y = 3x - 7$
   $4x - 3y = -7$

6. $2x - 2y = 2$
   $y - x = -1$

Find the common solution for each of the following pairs of equations, using the substitution method.

7. $x - y = 2$
   $x + 4y = -3$

8. $x + y = 1$
   $x + 3y = 1$

9. $-4y = -2x + 4$
   $-x = -2y - 2$

10. $2x + 8y = 20$
    $5y = 12 - x$

11. $x = y - 3$
    $-x = y + 3$

12. $-2x + y = -3$
    $x - y = 9$

Graph the following systems of inequalities on your own graph paper. Identify the solution set to both inequalities.

13. $x + 2y \geq 2$
    $2x - y \leq 4$

14. $20x + 10y \leq 40$
    $3x + 2y \geq 6$

15. $6x + 8y \leq -24$
    $-4x + 8y \geq 16$

16. $14x - 7y \geq -28$
    $3x + 4y \leq -12$

17. $2y \geq 6x + 6$
    $2x - 4y \geq -4$

18. $9x - 6y \geq 18$
    $3y \geq 6x - 12$

Chapter 15
Polynomials

Polynomials are algebraic expressions which include **monomials** containing one term, **binomials** which contain two terms, and **trinomials**, which contain three terms. Expressions with more than three terms are all called **polynomials**. **Terms** are separated by plus and minus signs.

EXAMPLES

Monomials:	Binomials:	Trinomials:	Polynomials:
$4f$	$4t + 9$	$x^2 + 2x + 3$	$x^3 - 3x^2 + 3x - 9$
$3x^3$	$9 - 7g$	$5x^2 - 6x - 1$	$p^4 + 2p^3 + p^2 - 5p + 9$
$4g^2$	$5x^2 + 7x$	$y^4 + 15y^2 + 100$	
2	$6x^3 - 8x$		

ADDING AND SUBTRACTING MONOMIALS

Two **monomials** can be added or subtracted as long as the **variable and its exponent** are the **same**. This is called combining like terms. Use the same rules you used for adding and subtracting integers.

EXAMPLES: $4x + 5x = 9x$ $2x^2 - 9x^2 = -7x^2$ $6y^3 - 5y^3 = y^3$ $\begin{array}{r}5y\\+\ 2y\\\hline 7y\end{array}$ $\begin{array}{r}3x^4\\-\ 8x^4\\\hline -5x^4\end{array}$

> **Remember:** When the integer in front of the variable is "1", it is usually not written. $1x^2$ is the same as x^2, and $-1x$ is the same as $-x$.

Add or subtract the following monomials:

1. $2x^2 + 5x^2 =$ _____
2. $5t + 8t =$ _____
3. $9y^3 - 2y^3 =$ _____
4. $6g - 8g =$ _____
5. $7y^2 + 8y^2 =$ _____
6. $s^5 + s^5 =$ _____
7. $-2x - 4x =$ _____
8. $4w^2 - w^2 =$ _____
9. $z^4 + 9z^4 =$ _____
10. $-k + 2k =$ _____
11. $3x^2 - 5x^2 =$ _____
12. $9t + 2t =$ _____
13. $-7v^3 + 10v^3 =$ _____
14. $-2x^3 + x^3 =$ _____
15. $10y^4 - 5y^4 =$ _____

16. $\begin{array}{r}y^4\\+\ 2y^4\\\hline\end{array}$
17. $\begin{array}{r}4x^3\\-\ 9x^3\\\hline\end{array}$
18. $\begin{array}{r}8t^2\\+\ 7t^2\\\hline\end{array}$
19. $\begin{array}{r}-2y\\-\ 4y\\\hline\end{array}$
20. $\begin{array}{r}5w^2\\+\ 8w^2\\\hline\end{array}$
21. $\begin{array}{r}11t^3\\-\ 4t^3\\\hline\end{array}$
22. $\begin{array}{r}-5z\\+\ 9z\\\hline\end{array}$
23. $\begin{array}{r}4w^5\\+\ w^5\\\hline\end{array}$
24. $\begin{array}{r}7t^3\\-\ 6t^3\\\hline\end{array}$
25. $\begin{array}{r}3x\\+\ 8x\\\hline\end{array}$

ADDING POLYNOMIALS

When adding **polynomials**, make sure the exponents and variables are the same on the terms you are combining. The easiest way is to put the terms in columns with **like exponents** underneath each other. Each column is added as a separate problem. Fill in the blank spots with zeros if it helps you keep the columns straight. You never carry to the next column when adding polynomials.

EXAMPLE 1: Add $3x^2 + 14$ and $5x^2 + 2x$

$$\begin{array}{r} 3x^2 + 0x + 14 \\ (+)\ 5x^2 + 2x + 0 \\ \hline 8x^2 + 2x + 14 \end{array}$$

EXAMPLE 2: $(4x^3 - 2x) + (-x^3 - 4)$

$$\begin{array}{r} 4x^3 - 2x + 0 \\ (+)\ -x^3 + 0x - 4 \\ \hline 3x^3 - 2x - 4 \end{array}$$

Add the following polynomials.

1. $y^2 + 3y + 2$ and $2y^2 + 4$

2. $(5y^2 + 4y - 6) + (2y^2 - 5y + 8)$

3. $5x^3 - 2x^2 + 4x - 1$ and $3x^2 - x + 2$

4. $-p + 4$ and $5p^2 - 2p + 2$

5. $(w - 2) + (w^2 + 2)$

6. $4t^2 - 5t - 7$ and $8t + 2$

7. $t^4 + t + 8$ and $2t^3 + 4t - 4$

8. $(3s^3 + s^2 - 2) + (-2s^3 + 4)$

9. $(-v^2 + 7v - 8) + (4v^3 - 6v + 4)$

10. $6m^2 - 2m + 10$ and $m^2 - m - 8$

11. $-x + 4$ and $3x^2 + x - 2$

12. $(8t^2 + 3t) + (-7t^2 - t + 4)$

13. $(3p^4 + 2p^2 - 1) + (-5p^2 - p + 8)$

14. $12s^3 + 9s^2 + 2s$ and $s^3 + s^2 + s$

15. $(-9b^2 + 7b + 2) + (-b^2 + 6b + 9)$

16. $15c^2 - 11c + 5$ and $-7c^2 + 3c - 9$

17. $5c^3 + 2c^2 + 3$ and $2c^3 + 4c^2 + 1$

18. $-14x^3 + 3x^2 + 15$ and $7x^3 - 12$

19. $(-x^2 + 2x - 4) + (3x^2 - 3)$

20. $(y^2 - 11y + 10) + (-13y^2 + 5y - 4)$

21. $3d^5 - 4d^3 + 7$ and $2d^4 - 2d^3 - 2$

22. $(6t^5 - t^3 + 17) + (4t^5 + 7t^3)$

23. $4p^2 - 8p + 9$ and $-p^2 - 3p - 5$

24. $20b^3 + 15b$ and $-4b^2 - 5b + 14$

25. $(-2w + 11) + (w^3 + w - 4)$

26. $(25z^2 + 13z + 8) + (z^2 - 2z - 10)$

SUBTRACTING POLYNOMIALS

When you subtract polynomials, it is important to remember to change all the signs in the subtracted polynomial (the subtrahend) and then add.

EXAMPLE: $(4y^2 + 8y + 9) - (2y^2 + 6y - 4)$

Step 1: Copy the subtraction problem into vertical form. Make sure you line up the terms with like exponents under each other just like you did for adding polynomials.

$$\begin{array}{r} 4y^2 + 8y + 9 \\ (-)\ 2y^2 + 6y - 4 \\ \hline \end{array}$$

Step 2: Change the subtraction sign to addition and all the signs of the subtracted polynomial to the opposite sign. The bottom polynomial in the problem becomes $-2y^2 - 6y + 4$.

Step 3: Add:
$$\begin{array}{r} 4y^2 + 8y + 9 \\ (+)\ -2y^2 - 6y + 4 \\ \hline 2y^2 + 2y + 13 \end{array}$$

Subtract the following polynomials.

1. $(2x^2 + 5x + 2) - (x^2 + 3x + 1)$

2. $(8y - 4) - (4y + 3)$

3. $(11t^3 - 4t^2 + 3) - (-t^3 + 4t^2 - 5)$

4. $(-3w^2 + 9w - 5) - (-5w^2 - 5)$

5. $(6a^5 - a^3 + a) - (7a^5 + a^2 - 3a)$

6. $(14c^4 + 20c^2 + 10) - (7c^4 + 5c^2 + 12)$

7. $(5x^2 - 9x) - (-7x^2 + 4x + 8)$

8. $(12y^3 - 8y^2 - 10) - (3y^3 + y + 9)$

9. $(-3h^2 - 7h + 7) - (5h^2 + 4h + 10)$

10. $(10k^3 - 8) - (-4k^3 + k^2 + 5)$

11. $(x^2 - 5x + 9) - (6x^2 - 5x + 7)$

12. $(12p^2 + 4p) - (9p - 2)$

13. $(-2m - 8) - (6m + 2)$

14. $(13y^3 + 2y^2 - 8y) - (2y^3 + 4y^2 - 7y)$

15. $(7g + 3) - (g^2 + 4g - 5)$

16. $(-8w^3 + 4w) - (-10w^3 - 4w^2 - w)$

17. $(12x^3 + x^2 - 10) - (3x^3 + 2x^2 + 1)$

18. $(2a^2 + 2a + 2) - (-a^2 + 3a + 3)$

19. $(c + 19) - (3c^2 - 7c + 2)$

20. $(-6v^2 + 12v) - (3v^2 + 2v + 6)$

21. $(4b^3 + 3b^2 + 5) - (7b^3 - 8)$

22. $(15x^3 + 5x^2 - 4) - (4x^3 - 4x^2)$

23. $(8y^2 - 2) - (11y^2 - 2y - 3)$

24. $(-z^2 - 5z - 8) - (3z^2 - 5z + 5)$

A subtraction of polynomials problem may be stated in sentence form. Study the examples below.

EXAMPLE 1: Subtract $-5x^3 + 4x - 3$ from $3x^3 + 4x^2 - 6x$.

Step 1: Copy the problem in columns with terms with the same exponent and variable under each other. Notice the second polynomial in the sentence will be the top polynomial of the problem.

$$\begin{array}{r} 3x^3 + 4x^2 - 6x \\ (-)\; -5x^3 + 4x - 3 \\ \hline \end{array}$$

Step 2: Since this is a subtraction problem, change all the signs of the terms in the bottom polynomial. Then add.

$$\begin{array}{r} 3x^3 + 4x^2 - 6x \\ (+)\; 5x^3 - 4x + 3 \\ \hline 8x^3 + 4x^2 - 10x + 3 \end{array}$$

EXAMPLE 2: From $6y^2 + 2$ subtract $4y^2 - 3y + 8$

In a problem phrased like this one, the first polynomial will be on top, and the second will be on bottom. Change the signs on the bottom polynomial and then add.

$$\begin{array}{r} 6y^2 + 2 \\ (-)\; 4y^2 - 3y + 8 \\ \hline \end{array} \longrightarrow \begin{array}{r} 6y^2 + 2 \\ (+)\; -4y^2 + 3y - 8 \\ \hline 2y^2 + 3y - 6 \end{array}$$

Solve the following subtraction problems.

1. Subtract $3x^2 + 2x - 5$ from $5x^2 + 2$

2. From $5y^3 - 6y + 9$ subtract $8y^3 - 10$

3. From $4m^2 - 4m + 7$ subtract $2m - 3$

4. Subtract $8z^2 + 3z + 2$ from $4z^2 - 7z + 8$

5. Subtract $10t^3 + t^2 - 5$ from $-2t^3 - t^2 - 5$

6. Subtract $-7b^3 - 2b + 4$ from $-b^2 + b + 6$

7. From $10y^3 + 20$ subtract $5y^3 - 5$

8. From $14t^2 - 6t - 8$ subtract $4t^2 - 3t + 2$

9. Subtract $3p^2 + p - 2$ from $-7p^2 - 5p + 2$

10. Subtract $x^3 + 8$ from $3x^3 - 2x^2 + 9$

11. Subtract $12a^2 + 10$ from $a^3 - a^2 - 1$

12. From $6m^2 + 3m + 1$ subtract $-6m^2 - 3m$

13. From $-13z^3 - 3z^2 - 2$ subtract $-20z^3 + 20$

14. Subtract $9c^2 + 10$ from $8c^2 - 5c + 3$

15. Subtract $b^2 + b - 5$ from $5b^2 - 4b + 5$

16. Subtract $-3x - 4$ from $3x^2 + x + 9$

17. From $15y^2 + 2$ subtract $4y^2 + 3y + 7$

18. Subtract $3g^2 - 5g + 5$ from $9g^2 - 3g - 4$

19. From $-7m^2 - 8m$ subtract $3m^2 + 7$

20. Subtract $x + 1$ from $5x + 5$

21. Subtract $c^2 + c + 2$ from $-c^2 - c - 2$

22. From $8t^3 + 6t^2 - 4t + 2$ subtract $t^3 + 3t$

ADDING AND SUBTRACTING POLYNOMIALS REVIEW

Practice adding and subtracting the polynomials below.

1. Add $-3x^2 + x$ and $4x^2 - 2$

2. Subtract $(-2y^3 + 9y)$ from $(6y^3 - y + 4)$

3. $(8t^3 - 3t^2 - 9) + (-7t^3 + t - 4)$

4. $(7p^2 + 3p + 1) - (5p^2 + 4p + 6)$

5. From $4w^3 + 5w - 2$ subtract $6w^2 - 4$

6. Add $-8a^3 - 7a^2 + 10$ and $6a^3 + 4a^2$

7. $(-14b^2 + b + 2) + (6b^2 - b + 3)$

8. Subtract $(g^3 - 7g^2 - 5)$ from $(5g^3 - 10)$

9. $(4c - 6) - (2c^2 - 3c + 9)$

10. From $-m^3 + 2m^2 + m$ subtract $9m^3 + 2m$

11. $(-3v^2 + 9v - 6) + (3v^2 - 4v + 6)$

12. Add $10s^2 + 4$ and $5s - 6$

13. Subtract $-x^3 - 9x^2 - x$ from $3x^3 + 2x + 4$

14. $(-5y^2 - 4y - 1) - (5y^2 - 2y - 8)$

MULTIPLYING MONOMIALS

When two monomials have the **same variable**, they can be multiplied. The **exponents** are **added together**. If the variable has no exponent, it is understood that the exponent is 1.

EXAMPLE: $4x^4 \times 3x^2 = 12x^6$ (Add exponents, Multiply coefficients)

EXAMPLE: $2y \times 5y^2 = 10y^3$

Multiply the following monomials

1. $6a \times 9a^5 = $ ____
2. $2x^6 \times 5x^3 = $ ____
3. $4y^3 \times 3y^2 = $ ____
4. $10t^2 \times 2t^2 = $ ____
5. $2p^5 \times 4p^2 = $ ____
6. $9b^2 \times 8b = $ ____
7. $3c^3 \times 3c^3 = $ ____
8. $2d^8 \times 9d^2 = $ ____
9. $6k^3 \times 5k^2 = $ ____
10. $7m^5 \times m = $ ____
11. $11z \times 2z^7 = $ ____
12. $3w^4 \times 6w^5 = $ ____
13. $4x^4 \times 5x^3 = $ ____
14. $5n^2 \times 3n^3 = $ ____
15. $8w^7 \times w = $ ____
16. $10s^6 \times 5s^3 = $ ____
17. $4d^5 \times 4d^5 = $ ____
18. $5y^2 \times 8y^6 = $ ____
19. $7t^{10} \times 3t^5 = $ ____
20. $6p^8 \times 2p^3 = $ ____
21. $x^3 \times 2x^3 = $ ____

When problems include negative signs, follow the rules for multiplying integers.

22. $-7s^4 \times 5s^3 = -35s^7$
23. $-6a \times -9a^5 = $ ____
24. $4x \times -x = $ ____
25. $-3y^2 \times -y^3 = $ ____
26. $-5b^2 \times 3b^5 = $ ____
27. $9c^4 \times -2c = $ ____
28. $-4t^3 \times 8t^3 = $ ____
29. $10d \times -8d^7 = $ ____
30. $-3g^6 \times -2g^3 = $ ____
31. $-7s^4 \times 7s^3 = $ ____
32. $-d^3 \times -2d = $ ____
33. $11p \times -2p^5 = $ ____
34. $-5x^7 \times -3x^3 = $ ____
35. $8z^4 \times 7z^4 = $ ____
36. $-4w \times -5w^8 = $ ____
37. $-5y^4 \times 6y^2 = $ ____
38. $9x^3 \times -7x^5 = $ ____
39. $-a^4 \times -a = $ ____
40. $-7k^2 \times 3k = $ ____
41. $-15t^2 \times -t^4 = $ ____
42. $3x^8 \times 9x^2 = $ ____

MULTIPLYING MONOMIALS WITH DIFFERENT VARIABLES

Warning: You cannot add the exponents of variables that are different.

EXAMPLE: $(-4wx)(6w^3x^2)$

To work this problem, first multiply the whole numbers, $-4 \times 6 = -24$. Then multiply the w's, $w \times w^3 = w^4$. Last, multiply the x's, $x \times x^2 = x^3$. The answer is $-24\,w^4x^3$.

Multiply the following monomials.

1. $(2x^2y^2)(-4xy^3) = $ _____
2. $(9p^3q^4)(2p^2q) = $ _____
3. $(-3t^4v^2)(t^2v) = $ _____
4. $(7w^3z^2)(3wz) = $ _____
5. $(-2st^6)(-8s^2t) = $ _____
6. $(xy^3)(4x^2y^2) = $ _____
7. $(5y^2z)(3y^4z^2) = $ _____
8. $(-3a^2b^2)(-4ab^3) = $ _____
9. $(-5c^3d^2)(2c^4d^5) = $ _____
10. $(10x^4y^2)(3x^3y) = $ _____
11. $(6f^3g^5)(-f^3g) = $ _____
12. $(-4a^3v^4)(8a^4v) = $ _____
13. $(5m^8n^5)(7m^2n^4) = $ _____
14. $(7w^5y^3)(3wy) = $ _____
15. $(2x^4z^2)(-9x^2z^4) = $ _____
16. $(-4a^7c^9)(2a^2c) = $ _____
17. $(-bd^6)(-b^2d) = $ _____
18. $(3x^4y^2)(10x^3y^3) = $ _____
19. $(9p^2y)(5p^5y^3) = $ _____
20. $(-2a^7x^2)(6ax^2) = $ _____
21. $(8c^4d^3)(-2c^2d^2) = $ _____

Multiplying three monomials works the same way.

22. $(3st)(4s^3t^2)(2s^2t^4) = 24s^6t^7$
23. $(xy)(x^2y^2)(2x^3y^2) = $ _____
24. $(2a^2b^2)(a^3b^3)(2ab) = $ _____
25. $(4y^2z^4)(2y^3)(2z^2) = $ _____
26. $(5cd^3)(3c^2d^2)(d^2) = $ _____
27. $(2w^2x^3)(3x^4)(2w^3) = $ _____
28. $(a^4d^2)(ad)(a^2d^3) = $ _____
29. $(7x^3t)(2t^4)(x^2t^2) = $ _____
30. $(p^2y^2)(4py)(p^3y^3) = $ _____
31. $(4x^3y)(5xy^3)(2y^4) = $ _____
32. $(8xy^2)(x^2y^3)(2x^2y) = $ _____
33. $(6p^3t)(2t^3)(p^2) = $ _____
34. $(3bc)(b^2c)(4c^3) = $ _____
35. $(2y^4z^5)(2y^6)(y^2z^2) = $ _____
36. $(4p^3r^3)(4r^2)(p^2r) = $ _____
37. $(a^4z^6)(6a^2z^2)(3z^3) = $ _____
38. $(5c^3)(6d^2)(2c^2d) = $ _____
39. $(9s^7t^2)(3st)(s^2t^3) = $ _____
40. $(3a^3b^4)(2b^3)(3a^2) = $ _____
41. $(5wz)(w^3z^3)(3w^2z^3) = $ _____

DIVIDING MONOMIALS

When simplifying monomial fractions with exponents, all exponents need to be positive. If there are negative exponents in your answer, put the base with its negative exponent below the fraction line and remove the negative sign. Two variables that are alike should not appear in both the denominator and numerator of a reduced expression. If you have the same variable in both the denominator and numerator of the fraction, the expression has not been fully reduced.

EXAMPLE : $\dfrac{44x^4y^2}{11x^6y^6}$

Step 1: Reduce the whole numbers first. $\dfrac{44}{11} = 4$

Step 2: Simplify the x's. $\dfrac{x^4}{x^6} = x^{4-6} = x^{-2} = \dfrac{1}{x^2}$

Step 3: Simplify the y's. $\dfrac{y^2}{y^6} = y^{2-6} = y^{-4} = \dfrac{1}{y^4}$

Therefore $\dfrac{44x^4y^2}{11x^6y^6} = \dfrac{4}{x^2y^4}$

Simplify the expressions below. All answers should have only positive exponents.

1. $\dfrac{5xy^3}{x(2x^3)y^4}$

2. $\dfrac{2a^2b^5}{3a^4b^2}$

3. $\dfrac{7(2a^2)b^4}{8ab^6}$

4. $\dfrac{16(x^2y^4)^3}{20xy}$

5. $\dfrac{10a^4b^2}{5a^6b^3}$

6. $\dfrac{5(8x^2y^3)}{4(x^2y^2)^2}$

7. $\dfrac{12(3a^2)b^2}{6a^2b^2}$

8. $\dfrac{(6x^3y^4)^2}{(2x^5y)^3}$

9. $\dfrac{15a^2b^3}{3a^5b^6}$

10. $\dfrac{33x^5y^3}{22x^7y^5}$

11. $\dfrac{15(2a^6b^7)}{21a^3b^6}$

12. $\dfrac{30x^4y^2}{6(x^2y^2)^2}$

13. $\dfrac{10(8ab^5)}{20a^2b^2}$

14. $\dfrac{9x^9y^7}{45x^5y^3}$

15. $\dfrac{(a^4b^7)^4}{a^8b^7}$

16. $\dfrac{7(x^3y^5)}{8x^2y^4}$

17. $\dfrac{10(a^3b^4)}{4a^5b^2}$

18. $\dfrac{7(6x^2y^5)^2}{21x^3y^4}$

EXTRACTING MONOMIAL ROOTS

When finding the roots of monomial expressions, you must first divide the monomial expression into separate parts. Then, simplify each part of the expression.

Note: To find the square root of any variable raised to a positive exponent, simply divide the exponent by 2. For example, $\sqrt{y^8} = y^4$.

EXAMPLE: $\sqrt{16x^4y^2z^6}$

Step 1: Break each component apart.
$(\sqrt{16})(\sqrt{x^4})(\sqrt{y^2})(\sqrt{z^6})$

Step 2: Solve for each component.
$(\sqrt{16} = 4)(\sqrt{x^4} = x^2)(\sqrt{y^2} = y)(\sqrt{z^6} = z^3)$

Step 3: Recombine simplified expressions.
$(4)(x^2)(y)(z^3) = 4x^2yz^3$

Simplify the problems below.

1. $\sqrt{9a^4b^2c^8}$
2. $\sqrt{25h^{12}i^6j^8}$
3. $\sqrt{49p^{10}q^{12}r^6}$
4. $\sqrt{36a^{14}b^8c^4}$
5. $\sqrt{121t^{22}u^{18}v^4}$
6. $\sqrt{25k^6l^{16}m^{10}}$

7. $\sqrt{144s^4t^{14}u^{18}}$
8. $\sqrt{64x^6y^{18}z^{22}}$
9. $\sqrt{49a^6b^2c^4}$
10. $\sqrt{81u^8v^{12}w^{14}}$
11. $\sqrt{16x^{22}y^{14}z^6}$
12. $\sqrt{169d^{30}e^{42}f^8}$

13. $\sqrt{9f^2g^6h^{16}}$
14. $\sqrt{100l^{28}m^{16}n^2}$
15. $\sqrt{4g^{20}h^{18}i^{36}}$
16. $\sqrt{16a^{42}b^4c^{26}}$
17. $\sqrt{36j^{12}k^8l^{10}}$
18. $\sqrt{81q^2r^{10}s^{32}}$

MONOMIAL ROOTS WITH REMAINDERS

Monomial expressions which are not easily simplified under the square root symbol will also be covered under California's mathematics curriculum. Powers may be raised to odd numbers. In addition, the coefficients may not be perfect squares. Follow the example below to understand how to simplify these types of problems.

EXAMPLE: Simplify $\sqrt{20x^5 y^9 z^{13}}$

Step 1: Begin by simplifying the coefficient.
$\sqrt{20} = (\sqrt{5})(\sqrt{4})$,
$\sqrt{4} = 2$, so $\sqrt{20} = 2\sqrt{5}$

Step 2: Simplify the variable exponents.

$\sqrt{x^5} = (\sqrt{x^4})(\sqrt{x})$, $\sqrt{x^4} = x^2$, so $\sqrt{x^5} = x^2\sqrt{x}$
$\sqrt{y^9} = (\sqrt{y^8})(\sqrt{y})$, $\sqrt{y^8} = y^4$, so $\sqrt{y^9} = y^4\sqrt{y}$
$\sqrt{z^{13}} = (\sqrt{z^{12}})(\sqrt{z})$, $\sqrt{z^{12}} = z^6$, so $\sqrt{z^{13}} = z^6\sqrt{z}$

Step 3: Recombine simplified expressions.
$2x^2 y^4 z^6 \sqrt{5xyz}$

Simplify the following square root expressions.

1. $\sqrt{45 d^{14} e^{15} f^{11}}$
2. $\sqrt{50 h^{16} i^{10} j^{8}}$
3. $\sqrt{30 x^{22} y^{21} z^{7}}$
4. $\sqrt{84 p^{11} q^{7} r^{9}}$
5. $\sqrt{48 k^{25} l^{15} m^{3}}$
6. $\sqrt{54 s^{13} t^{5} u^{17}}$
7. $\sqrt{60 a^{7} b^{15} c^{21}}$
8. $\sqrt{18 p^{3} q^{22} r^{18}}$
9. $\sqrt{72 a^{19} b^{9} c^{15}}$
10. $\sqrt{68 m^{7} n^{3} p^{11}}$
11. $\sqrt{75 r^{15} s^{11} t^{13}}$
12. $\sqrt{20 g^{21} h^{14} j^{17}}$
13. $\sqrt{80 v^{3} w^{9} x^{12}}$
14. $\sqrt{44 d^{5} e^{8} f^{13}}$
15. $\sqrt{24 x^{17} y^{6} z^{11}}$
16. $\sqrt{32 a^{6} b^{13} c^{5}}$
17. $\sqrt{52 j^{12} k^{15} m^{5}}$
18. $\sqrt{96 q^{12} r^{15} s^{5}}$

MULTIPLYING MONOMIALS BY POLYNOMIALS

In Chapter 5, you learned to remove parentheses by multiplying the number outside parentheses by each term inside parentheses. $2(4x - 7) = 8x - 14$ Multiplying monomials by polynomials works the same way.

EXAMPLE: $-5t(2t^2 - 7t + 9)$

Step 1: Multiply $-5t \times 2t^2 = -10t^3$
Step 2: Multiply $-5t \times -7t = 35t^2$
Step 3: Multiply $-5t \times 9 = -45t$
Step 4: Arrange the answers horizontally in order: $-10t^3 + 35t^2 - 45t$

Remove parentheses in the following problems.

1. $3x(3x^2 + 4x - 1)$
2. $4y(y^3 - 7)$
3. $7a^2(2a^2 + 3a + 2)$
4. $-5d^3(d^2 - 5d)$
5. $2w(-4w^2 + 3w - 8)$
6. $8p(p^3 - 6p + 5)$
7. $-9b^2(-2b + 5)$
8. $2t(t^2 - 4t - 10)$
9. $10c(4c^2 + 3c - 7)$
10. $6z(2z^4 - 5z^2 - 4)$

11. $-9t^2(3t^2 + 5t + 6)$
12. $c(-3c - 5)$
13. $3p(p^3 - p^2 - 9)$
14. $-k^2(2k + 4)$
15. $-3(4m^2 - 5m + 8)$
16. $6x(-7x^3 + 10)$
17. $-w(w^2 - 4w + 7)$
18. $2y(5y^2 - y)$
19. $3d(d^5 - 7d^3 + 4)$
20. $-5t(-4t^2 - 8t + 1)$

21. $7(2w^2 - 9w + 4)$
22. $3y^2(y^2 - 11)$
23. $v^2(v^2 + 3v + 3)$
24. $8x(2x^3 + 3x + 1)$
25. $-5d(4d^2 + 7d - 2)$
26. $-k^2(-3k + 6)$
27. $3x(-x^2 - 5x + 5)$
28. $4z(4z^4 - z - 7)$
29. $-5y(9y^3 - 3)$
30. $2b^2(7b^2 + 4b + 4)$

DIVIDING POLYNOMIALS BY MONOMIALS

EXAMPLE: $\dfrac{-8wx + 6x^2 - 16wx^2}{2wx}$

Step 1: Rewrite the problem. Divide each term from the top by the denominator, $2wx$.

$$\dfrac{-8wx}{2wx} + \dfrac{6w^2}{2wx} + \dfrac{-16wx}{2wx}$$

Step 2: Simplify each term in the problem. Then combine like terms.

$$-4 + \dfrac{3w}{x} - 8 = -12 + \dfrac{3w}{x}$$

Simplify each of the following:

1. $\dfrac{bc^2 - 8bc - 2b^2c^2}{2bc}$

2. $\dfrac{3jk^2 + 12k + 9j^2k}{3jk}$

3. $\dfrac{5x^2y - 8xy^2 + 2y^3}{2xy}$

4. $\dfrac{16st^2 + st - 12s}{4st}$

5. $\dfrac{4wx^2 + 6wx - 12w^3}{2wx}$

6. $\dfrac{cd^2 + 10cd^3 + 16c^2}{2cd}$

7. $\dfrac{y^2z^3 - 2yz - 8z^2}{-2yz^2}$

8. $\dfrac{a^2b + 2ab^2 - 14ab^3}{2a^2}$

9. $\dfrac{pr^2 + 6pr + 8p^2r^2}{2pr^2}$

10. $\dfrac{6xy^2 - 3xy + 18x^2}{-3xy}$

11. $\dfrac{6x^2y + 12xy - 24y^2}{6xy}$

12. $\dfrac{5m^2n - 10mn - 25n^2}{5mn}$

13. $\dfrac{st^2 - 10st - 16s^2t^2}{2st}$

14. $\dfrac{7jk^2 - 14jk - 63j^2}{7jk}$

REMOVING PARENTHESES AND SIMPLIFYING

In the following problem, you must multiply each set of parentheses by the numbers and variables outside the parentheses, and then add the polynomials to simplify the expressions.

EXAMPLE: $8x(2x^2 - 5x + 7) - 3x(4x^2 + 3x - 8)$

Step 1: Multiply to remove the first set of parentheses.
$8x(2x^2 - 5x + 7) = 16x^3 - 40x^2 + 56x$

Step 2: Multiply to remove the second set of parentheses.
$-3x(4x^2 + 3x - 8) = -12x^3 - 9x^2 + 24x$

Step 3: Copy each polynomial in columns, making sure the terms with the same variable and exponent are under each other. Add to simplify.

$$\begin{array}{r} 16x^3 - 40x^2 + 56x \\ (+)\ \underline{-12x^3 - \ 9x^2 + 24x} \\ 4x^3 - 49x^2 + 80x \end{array}$$

Remove the parentheses, and simplify the following problems.

1. $4t(t + 7) + 5t(2t^2 - 4t + 1)$

2. $-5y(3y^2 - 5y + 3) - 6y(y^2 - 4y - 4)$

3. $-3(3x^3 + 4x) + 5x(x^2 + 3x + 2)$

4. $2b(5b^2 - 8b - 1) - 3b(4b + 3)$

5. $8d^2(3d + 4) - 7d(3d^2 + 4d + 5)$

6. $5a(3a^2 + 3a + 1) - (-2a^2 + 5a - 4)$

7. $3m(m + 7) + 8(4m^2 + m + 4)$

8. $4c^2(-6c^2 - 3c + 2) - 7c(5c^3 + 2c)$

9. $-8w(-w + 1) - 4w(3w - 5)$

10. $6p(2p^2 - 4p - 6) + 3p(p^2 + 6p + 9)$

MULTIPLYING TWO BINOMIALS

When you multiply two binomials such as $(x + 6)(x - 5)$, you must multiply each term in the first binomial by each term in the second binomial. The easiest way is to use the **FOIL** method. If you can remember the word **FOIL**, it can help you keep order when you multiply. The "F" stands for **first**, "O" stands for **outside**, "I" stands for **inside**, and "L" stands for **last**.

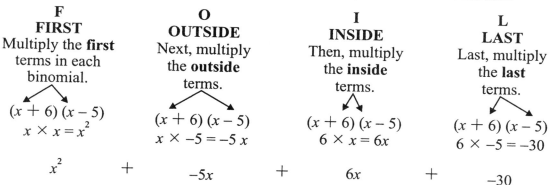

Now just combine like terms, $6x - 5x = x$, and write your answer.
$(x + 6)(x + 5) = x^2 + x - 30$

Note: It is customary for mathematicians to write polynomials in descending order. That means that the term with the highest number exponent comes first in a polynomial. The next highest exponent is second and so on. When you use the **FOIL** method, the terms will always be in the customary order. You just need to combine like terms and write your answer.

Multiply the following binomials.

1. $(y - 7)(y + 3)$
2. $(2x + 4)(x + 9)$
3. $(4b - 3)(3b - 4)$
4. $(6g + 2)(g - 9)$
5. $(7k - 5)(-4k - 3)$
6. $(8v - 2)(3v + 4)$
7. $(10p + 2)(4p + 3)$
8. $(3h - 9)(-2h - 5)$
9. $(w - 4)(w - 7)$
10. $(6x + 1)(x - 2)$
11. $(5t + 3)(2t - 1)$
12. $(4y - 9)(4y + 9)$
13. $(a + 6)(3a + 5)$
14. $(3z - 8)(z - 4)$
15. $(5c + 2)(6c + 5)$

16. $(y + 3)(y - 3)$

17. $(2w - 5)(4w + 6)$

18. $(7x + 1)(x - 4)$

19. $(6t - 9)(4t - 4)$

20. $(5b + 6)(6b + 2)$

21. $(2z + 1)(10z + 4)$

22. $(11w - 8)(w + 3)$

23. $(5d - 9)(9d + 9)$

24. $(9g + 2)(g - 2)$

25. $(4p + 7)(2p + 3)$

26. $(m + 5)(m - 5)$

27. $(8b - 8)(2b - 1)$

28. $(z + 3)(3z + 5)$

29. $(7y - 5)(y - 3)$

30. $(9x + 5)(3x - 1)$

31. $(3t + 1)(t + 10)$

32. $(2w - 9)(8w + 7)$

33. $(8s - 2)(s + 4)$

34. $(4k - 1)(8k + 9)$

35. $(h + 12)(h - 2)$

36. $(3x + 7)(7x + 3)$

37. $(2v - 6)(2v + 6)$

38. $(2x + 8)(2x - 3)$

39. $(k - 1)(6k + 12)$

40. $(3w + 11)(2w + 2)$

41. $(8y - 10)(5y - 3)$

42. $(6d + 13)(d - 1)$

43. $(7h + 3)(2h + 4)$

44. $(5n + 9)(5n - 5)$

45. $(6z + 5)(z - 8)$

46. $(4p + 5)(2p - 9)$

47. $(b + 2)(5b + 7)$

48. $(9y - 3)(8y - 7)$

SIMPLIFYING EXPRESSIONS WITH EXPONENTS

EXAMPLE 1: Simplify $(2a + 5)^2$

When you simplify an expression such as $(2a + 5)^2$, it is best to write the expression as two binomials and use FOIL to simplify.

$(2a + 5)^2 = (2a + 5)(2a + 5)$

Using FOIL we have $4a^2 + 10a + 10a + 25 = 4a^2 + 20a + 25$

EXAMPLE 2: Simplify $4(3a + 2)^2$

Using order of operations, we must simplify the exponent first.
$= 4(3a + 2)(3a + 2)$
$= 4(9a^2 + 6a + 6a + 4)$
$= 4(9a^2 + 12a + 4)$ Now multiply by 4.
$= 4(9a^2 + 12a + 4) = 36a^2 + 48a + 16$

Note: It is customary for mathematicians to write polynomials in descending order. That means that the term with the highest number exponent comes first in a polynomial. The next highest exponent is second and so on. When you use the **FOIL** method, the terms will always be in the customary order. You just need to combine like terms, and write your answer.

Multiply the following binomials.

1. $(y + 3)^2$
2. $7(2x + 4)^2$
3. $6(4b - 3)^2$
4. $5(6g + 2)^2$
5. $(-4k - 3)^2$
6. $3(-2h - 5)^2$
7. $-2(8v - 2)^2$
8. $(10p + 2)^2$
9. $6(-2h - 5)^2$
10. $6(w - 7)^2$
11. $2(6x + 1)^2$
12. $(9x + 2)^2$
13. $(5t + 3)^2$
14. $3(4y - 9)^2$
15. $8(a + 6)^2$
16. $4(3z - 8)^2$
17. $3(5c + 2)^2$
18. $4(3t + 9)^2$

CHAPTER 15 REVIEW

Simplify:

1. $3a^2 + 9a^2$

2. $(7x^2y^4)(9xy^5)$

3. $-6z^2(z+3)$

4. $(4b^2)(5b^3)$

5. $7x^2 - 9x^2$

6. $(5p-4)-(3p+2)$

7. $-5t(3t+9)^2$

8. $(3w^3y^2)(4wy^5)$

9. $3(2g+3)^2$

10. $14d^4 - 9d^4$

11. $(7w-4)(w-8)$

12. $15t^2 + 4t^2$

13. $(7c^4)(9c^2)$

14. $(9x+2)(x+5)$

15. $4y(4y^2-9y+2)$

16. $(8a^4b)(2ab^3)(ab)$

17. $(5w^6)(9w^9)$

18. $8x^3 + 12x^3$

19. $15p^5 - 11p^5$

20. $(3s^4t^2)(4st^3)$

21. $(4d+9)(2d+7)$

22. $4w(-3w^2+7w-5)$

23. $24z^6 - 10z^6$

24. $-7y^3 - 8y^3$

25. $(7x^4)(7x^5)$

26. $17p^2 + 9p^2$

27. $(a^2v)(2av)(a^3v^6)$

28. $4(6y-5)^2$

29. $(3c^2)(6c^8)$

30. $(4x^5y^3)(2xy^3)$

31. Add $2x^2 + 9x$ and $5x^2 - 8x + 2$

32. $4t(6t^2 + 4t - 6) + 8t(3t + 3)$

33. Subtract $y^2 + 4y - 6$ from $3y^2 + 7$

34. $2x(4x^2 + 6x - 3) + 4x(x + 3)$

35. $(6t - 4) - (6t^2 + t - 2)$

36. $(4x + 6) + (7x^2 - 2x + 3)$

37. Subtract $5a - 2$ from $a + 9$

38. $(-2y + 4) + (4y - 6)$

39. $2t(t + 6) - 5t(2t + 7)$

40. Add $3c - 4$ and $c^2 - 3c - 2$

41. $2b(b - 4) - (b^2 + 2b + 1)$

42. $(6k^2 + 5k) + (k^2 + k + 9)$

43. $(q^2 r^3)(3qr^2)(2q^4 r)$

44. $(5df)(d^4 f^2)(2df)$

45. $(7g^2 h^3)(g^3 h^6)(6gh^3)$

46. $(8v^2 x^3)(3v^6 x^2)(2v^4 x^4)$

47. $(3n^2 m^2)(9n^2 m)(n^3 m^7)$

48. $(11t^2 a^2)(4t^3 a^8)(2t^6 a)$

49. $\dfrac{12(2a^3)b}{3a^2 b^{-2}}$

50. $\dfrac{7(g^3 h^3)}{4(g^2 h)^{-2}}$

51. $\dfrac{16(m^2 n^3)^2}{4(m^2 n)^{-2}}$

52. $\dfrac{14p^3 q^3}{2p^2 q}$

53. $\dfrac{8(e^4 h^{-2})^{-2}}{32e^2 h^5}$

54. $\dfrac{22x^3 y^4}{5(11x^{-3} y^7)^2}$

55. $\sqrt{36r^6 s^8 t^2}$

56. $\sqrt{40g^3 h^6 j^7}$

57. $\sqrt{18m^5 n^3 p^7}$

58. $\sqrt{75a^7 b^2 c^9}$

59. $\sqrt{12x^3 y^4 z^7}$

60. $\sqrt{64f^6 g^7 h^5}$

Chapter 16
Statistics

MEAN

Statistics is a branch of mathematics. Using statistics, mathematicians organize data (numbers) into forms that are easily understood. In statistics, the **mean** is the same as the **average**. To find the **mean** of a list of numbers, first, add together all the numbers in the list, and then divide by the number of items in the list.

EXAMPLE: Find the mean of 38, 72, 110, 548.

Step 1: First add   38 + 72 + 110 + 548 = **768**

Step 2: There are 4 numbers in the list, so divide the total by 4. $4\overline{)768}$ = 192
The mean is **192**.

Practice finding the mean (average). Round to the nearest tenth if necessary.

1. Dinners served:
489 561 522 450
Mean = _____

2. Prices paid for shirts:
$4.89 $9.97 $5.90 $8.64
Mean = _____

3. Piglets born:
23 19 15 21 22
Mean = _____

4. Student absences:
6 5 13 8 9 12 7
Mean = _____

5. Paychecks received:
$89.56 $99.99 $56.54
Mean = _____

6. Choir attendance:
56 45 97 66 70
Mean = _____

7. Long distance calls:
33 14 24 21 19
Mean = _____

8. Train boxcars:
56 55 48 61 51
Mean = _____

9. Cookies eaten:
5 6 8 9 2 4 3
Mean = _____

Find the mean (average) of the following word problems.

10. Val's science grades were 95, 87, 65, 94, 78, and 97. What was her average? _____

11. Ann runs a business from her home. The number of orders for the last 7 business days were 17, 24, 13, 8, 11, 15, and 9. What was the average number of orders per day? _____

12. Melissa tracked the number of phone calls she had per day: 8, 2, 5, 4, 7, 3, 6, 1. What was the average number of calls she received? _____

FINDING DATA MISSING FROM THE MEAN

EXAMPLE: Mara knew she had an 88 average in her biology class, but she lost one of her papers. The three papers she could find had scores of 98%, 84%, and 90%. What was the score on her fourth paper?

Step 1: Figure the total score on four papers with an 88% average. $.88 \times 4 = 3.52$

Step 2: Add together the scores from the three papers you have. $.98 + .84 + .90 = 2.72$

Step 3: Subtract the scores you know from the total score. $3.52 - 2.72 = .80$ She had 80% on her fourth paper.

Find the data missing from the following problems.

1. Gabriel earned 87% on his first geography test. He wants to keep a 92% average. What does he need to get on his next test to bring his average up?

2. Rian earned $68.00 on Monday. How much money must he earn on Tuesday to have an average of $80 earned for the two days?

3. Haley, Chuck, Dana, and Chris entered a contest to see who could bake the most chocolate chip cookies in an hour. They baked an average of 75 cookies. Haley baked 55, Chuck baked 70, and Dana baked 90. How many did Chris bake?

4. Four wrestlers made a pact to lose some weight before the competition. They lost an average of 7 pounds each, over the course of 3 weeks. Carlos lost 6 pounds, Steve lost 5 pounds, and Greg lost 9 pounds. How many pounds did Wes lose?

5. Three boxes are ready for shipment. The boxes average 26 pounds each. The first box weighs 30 pounds; the second weighs 25 pounds. How much does the third box weigh?

6. The five jockeys running in the next race average 92 pounds each. Nicole weighs 89 pounds. Jon weighs 95 pounds. Jenny and Kasey weigh 90 pounds each. How much does Jordan weigh?

7. Jessica made three loaves of bread that weighed a total of 45 ounces. What was the average weight of each loaf?

8. Celeste made scented candles to give away to friends. She had 2 pounds of candle wax which she melted, scented, and poured into 8 molds. What was the average weight of each candle?

9. Each basketball player has to average a minimum of 5 points a game for the next three games to stay on the team. Ben is feeling the pressure. He scored 3 points in the first game and 2 points in the second game. How many points does he need to score in the third game to stay on the team?

MEDIAN

In a list of numbers ordered from lowest to highest, the **median** is the middle number. To find the **median,** first arrange the numbers in numerical order. If there is an odd number of items in the list, the **median** is the middle number. If there is an even number of items in the list, the median is the **average of the two middle numbers.**

EXAMPLE 1: Find the median of 42, 35, 45, 37, and 41.

Step 1: Arrange the numbers in numerical order: 35 37 (41) 42 45.

Step 2: Find the middle number. **The median is 41.**

EXAMPLE 2: Find the median of 14, 53, 42, 6, 14, and 46.

Step 1: Arrange the numbers in numerical order: 6 14 (14 42) 46 53.

Step 2: Find the average of the 2 middle numbers.
(14 + 42) ÷ 2 = 28. **The median is 28.**

Circle the median in each list of numbers.

1. 35, 55, 40, 30, and 45
2. 7, 2, 3, 6, 5, 1, and 8
3. 65, 42, 60, 46, and 90
4. 15, 16, 19, 25, and 20
5. 75, 98, 87, 65, 82, 88, and 100
6. 33, 42, 50, 22, and 19
7. 401, 758, and 254
8. 41, 23, 14, 21, and 19
9. 5, 8, 3, 10, 13, 1, and 8

10.	11.	12.	13.	14.	15.	16.
19	9	45	52	20	8	15
14	3	32	54	21	17	40
12	10	66	19	25	13	42
15	17	55	63	18	14	32
18	6	61	20	16	22	28

Find the median in each list of numbers.

17. 10, 8, 21, 14, 9, and 12 _____
18. 43, 36, 20, and 40 _____
19. 5, 24, 9, 18, 12, and 3 _____
20. 48, 13, 54, 82, 90, and 7 _____
21. 23, 21, 36, and 27 _____
22. 9, 4, 3, 1, 6, 2, 10, and 12 _____

23.	24.	25.	26.	27.	28.	29.
2	11	13	75	48	22	17
10	22	15	62	45	19	30
6	25	9	60	52	15	31
18	28	35	52	30	43	18
20	10	29	80	35	34	14
23	23	33	50	58	28	25

_____ _____ _____ _____ _____ _____ _____

MODE

In statistics, the **mode** is the number that occurs most frequently in a list of numbers.

EXAMPLE: Exam grades for a math class were as follows:
70 88 92 85 99 85 70 85 99 100 88 70 99 88 88 99 88 92 85 88.

Step 1: Count the number of times each number occurs in the list.

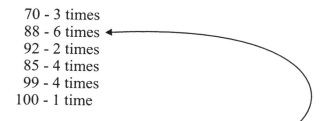

70 - 3 times
88 - 6 times
92 - 2 times
85 - 4 times
99 - 4 times
100 - 1 time

Step 2: Find the number that occurs most often.
The mode is 88 because it is listed 6 times. No other number is listed as often.

Find the mode in each of the following lists of numbers.

1. 88	2. 54	3. 21	4. 56	5. 64	6. 5	7. 12
15	42	16	67	22	4	41
88	44	15	67	22	9	45
17	56	78	19	15	8	32
18	44	21	56	14	4	16
88	44	16	67	14	7	12
17	56	21	20	22	4	12
mode ___	mode ___	mode ___	mode ___	mode ___	mode ___	mode ___

8. 48, 32, 56, 32, 56, 48, 56   **mode** _____

9. 12, 16, 54, 78, 16, 25, 20   **mode** _____

10. 5, 4, 8, 3, 4, 2, 7, 8, 4, 2   **mode** _____

11. 11, 9, 7, 11, 7, 5, 7, 7, 5   **mode** _____

12. 84, 22, 79, 22, 87, 22, 22   **mode** _____

13. 95, 87, 65, 94, 78, 95   **mode** _____

14. 8, 2, 5, 4, 7, 2, 3, 6, 1   **mode** _____

15. 89, 7, 11, 89, 17, 56   **mode** _____

16. 15, 48, 52, 41, 8, 48   **mode** _____

17. 22, 45, 48, 12, 22, 41, 22   **mode** _____

18. 62, 44, 78, 62, 54, 44, 62   **mode** _____

19. 54, 22, 54, 78, 22, 78, 22   **mode** _____

20. 14, 17, 33, 21, 33, 17, 33   **mode** _____

21. 65, 51, 8, 21, 8, 65, 70, 8   **mode** _____

22. 17, 24, 13, 8, 11, 8, 15, 9   **mode** _____

23. 51, 45, 84, 51, 65, 74, 51   **mode** _____

24. 8, 74, 65, 15, 9, 10, 74   **mode** _____

25. 62, 54, 2, 7, 89, 2, 7, 54, 2   **mode** _____

STEM-AND-LEAF PLOTS

A **stem-and-leaf plot** is a way to organize and analyze statistical data. To make a stem-and-leaf plot, first draw a vertical line.

Final Math Averages
85 92 87 62 75 84 96 52
45 77 98 75 71 79 85 82
87 74 76 68 93 77 65 84
79 65 77 82 86 84 92 60
99 75 88 74 79 80 63 84
87 90 75 81 73 69 73 75
31 86 89 65 69 75 79 76

Stem	Leaves
3	1
4	5
5	2
6	0,2,3,5,5,5,8,9,9
7	1,3,3,3,4,4,5,5,5,5,5,5,6,6,7,7,7,9,9,9
8	0,1,2,2,4,4,4,4,5,5,6,6,7,7,7,8,9
9	0,2,2,3,6,8,9

On the left side of the line, list all the numbers that are in the tens place from the set of data. Next, list each number in the ones place on the right side of the line in ascending order. It is easy to see at a glance that most of the students scored in the 70's or 80's with a majority having averages in the 70's. It is also easy to see that the maximum average is 99, and the lowest average is 31. Stem-and-leaf plots are a way to organize data making it easy to read.

Make a stem-and leaf-plot from the data below, and then answer the questions that follow.

1.   **Speeds on Turner Road**

CAR SPEED, mph
45 52 47 35 48 50 51 43
40 51 32 24 55 41 32 33
36 59 49 52 34 28 69 47
29 15 63 42 35 42 58 59
39 41 25 34 22 16 40 31
55 10 46 38 50 52 48 36
21 32 36 41 52 49 45 32
52 45 56 35 55 65 20 41

Stem	Leaves

2. What was the fastest speed recorded?

3. What was the slowest speed recorded?

4. Which speed was most often recorded?

5. If the speed limit is 45 miles per hour, how many were speeding?

6. If the speed limit is 45 miles per hour, how many were at least 20 mph over or under the speed limit?

MORE STEM-AND-LEAF PLOTS

Two sets of data can be displayed on the same stem-and-leaf plot.

EXAMPLE: The following is an example of a back-to-back stem-and-leaf plot.

Bryan's Math scores {60,65,72,78,85,90}
Bryan's English scores {78,88,89,89,92,95,100}

Math		English
5,0	6	
2,8	7	8
5	8	8,9,9
0	9	2,5
	10	0

2|7 means 72   8|9 means 89

Read the stem-and-leaf plot below and answer the questions that follow.

3rd grade Boys' Weights		3rd Grade Girls' Weights
8,7,5,3,2	4	0,2, 4, 7
6, 4, 1, 0	5	1,8,8,8, 9
5	6	0 6, 6, 8, 8
0	9	8

4|5 means 54   6|8 means 68

1. What is the median for the girls' weights?
2. What is the median for the boys' weights?
3. What is the mode for the girls' weights?
4. What is the weight of the lightest boy?
5. What is the weight of the heaviest boy?
6. What is the weight of the heaviest girl?

7. Create a stem-and-leaf plot for the data given below.

Automobile Speeds on I-85

60	65	80	75	92	81	63
65	67	75	78	79	77	69
62	57	64	65	68	71	69
71	73	56	69	69	70	74

Automobile Speeds on I-75

72	56	62	65	63	60	58
55	57	70	69	59	53	61
58	61	63	67	57	63	67
56	58	59	62	64	63	69

8. What is the median speed for I-75?
9. What is the median speed for I-85?
10. What is the mode speed for I-75?
11. What is the mode speed for I-85?
12. What was the fastest speed on either interstate?

QUARTILES AND EXTREMES

In statistics, large sets of data are separated into four equal parts. These parts are called **quartiles**. The **median** separates the data into two halves. Then, the median of the upper half is the **upper quartile**, and the median of the lower half is the **lower quartile**.

The **extremes** are the highest and lowest values in a set of data. The lowest value is called the **lower extreme**, and the highest value is called the **upper extreme**.

EXAMPLE 1: The following set of data shows the high temperatures (in degrees Fahrenheit) in cities across the United States on a particular autumn day. Find the median, the upper quartile, the lower quartile, the upper extreme, and the lower extreme of the data.

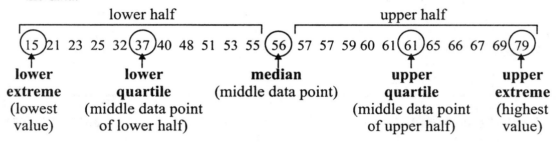

EXAMPLE 2: The following set of data shows the fastest race car qualifying speeds in miles per hour. Find the median, the upper quartile, the lower quartile, the upper extreme, and the lower extreme of the data.

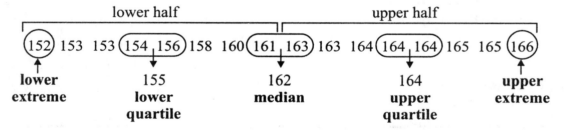

Note: When you have an even number of data points, the median is the average of the two middle points. The lower middle number is then included in the lower half of the data, and the upper middle number is included in the upper half.

Find the median, the upper quartile, the lower quartile, the upper extreme, and the lower extreme of each set of data given below.

1. 0 0 1 1 1 2 2 3 3 4 5
2. 15 16 18 20 22 22 23
3. 62 75 77 80 81 85 87 91 94
4. 74 74 76 76 77 78
5. 3 3 3 5 5 6 6 7 7 7 8 8
6. 190 191 192 192 194 195 196
7. 6 7 9 9 10 10 11 13 15
8. 21 22 24 25 27 28 32 35

220

BOX-AND-WHISKER PLOTS

Box-and-whisker plots are used to summarize data as well as to display data. A box-and-whisker plot summarizes data using the median, upper and lower quartiles, and the lower and upper extreme values. Consider the data below–a list of employees' ages at the Acme Lumber Company:

(21) 21 22 23 24 24 24 (25) 26 27 28 29 30 32 32 (33) 33 33 34 35 36 37 37 (38) 38 39 40 40 41 44 (48)

↑ lower extreme   ↑ lower quartile   ↑ median   ↑ upper quartile   ↑ upper extreme

Step 1: Find the median, upper quartile, lower quartile, upper extreme, and lower extreme just like you did on the previous page.

Step 2: Plot the 5 data points found in step 1 above on a number line as shown below.

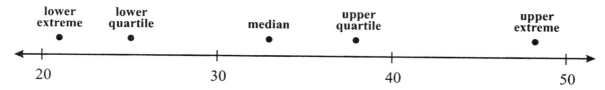

Step 3: Draw a box around the quartile values, and draw a vertical line through the median value. Draw whiskers from each quartile to the extreme value data points.

This box-and-whisker displays five types of information: lower extreme, lower quartile, median, upper quartile, and upper extreme.

Draw a box-and-whisker plot for the following sets of data.

1.
10 12 12 15 16 17 19 21 22 22 25 27 31 35 36 37 38 38 41 43 45 50 51 56 57 58 59

2.
5 5 6 7 9 9 10 11 12 15 15 16 17 18 19 19 20 22 24 26 27 27 30 31 31 35 37

Copyright © American Book Company

221

SCATTER PLOTS

A **scatter plot** is a graph of ordered pairs involving two sets of data. These plots are used to detect whether two sets of data, or variables, are truly related.

In the example to the right, two variables, income and education, are being compared to see if they are related or not. Twenty people were interviewed, ages 25 and older, and the results were recorded on the chart.

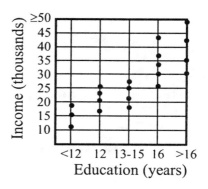

Imagine drawing a line on the scatter plot where half the points are above the line and half the points are below it. In the plot on the right, you will notice that this line slants upward and to the right. This line direction means there is a **positive** relationship between education and income. In general, for every increase in education, there is a corresponding increase in income.

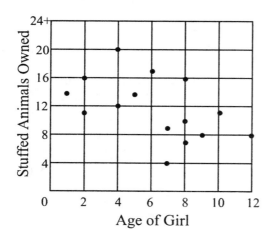

Now, examine the scatter plot on the left. In this case, 15 girls ages 2-12 were interviewed and asked, "How many stuffed animals do you currently have?" If you draw an imaginary line through the middle points, you will notice that the line slants downward and to the right. This plot demonstrates a **negative** relationship between the age of girls and their stuffed animal ownership. In general, as the girls' ages increase, the number of stuffed animals owned decreases.

Finally, look at the scatter plot shown on the right. In this plot, Rita wanted to see the relationship between the temperature in the classroom and the grades she received on tests she took at that temperature. As you look to your right, you will notice that the points are distributed all over the graph. Because this plot is not in a pattern, there is no way to draw a line through the middle of the points. This type of point pattern indicates there is **no** relationship between Rita's grades on tests and the classroom temperature.

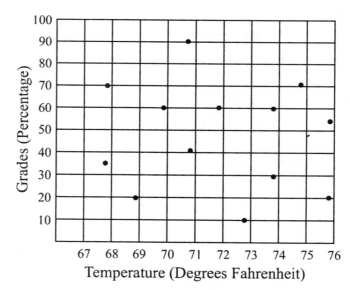

Examine each of the scatter plots below. On the line below each plot, write whether the relationship shown between the two variables is "positive", "negative", or "no relationship".

1.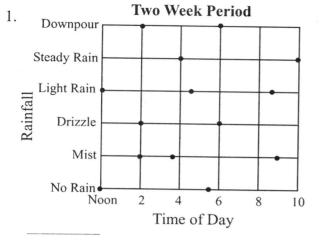

2.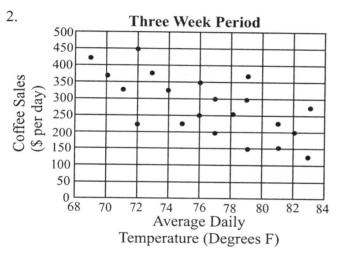

3.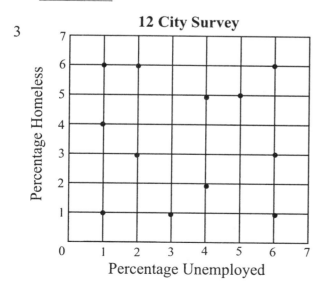

4.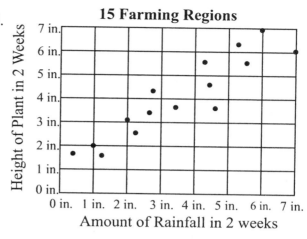

5.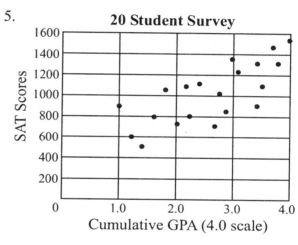

6.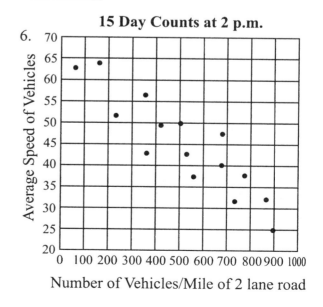

MISLEADING STATISTICS

As you read magazines and newspapers, you will see many charts and graphs which present statistical data. This data will present you with how measurements change over time or how one measurement corresponds to another measurement. However, some charts and graphs are presented to make changes in data appear greater than they actually are. The people presenting the data create these distortions to make exaggerated claims.

There is one method to arrange the data in ways which can exaggerate statistical measurements. A statistician can create a graph in which the number line does not begin with zero.

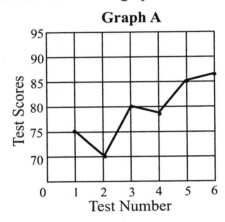
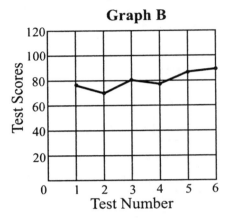

In the two graphs above, notice how each graph displays the same data. However, the way the data is displayed in graph A appears more striking than the data display in graph B. Graph A's data presentation is more striking because the test score numbers do not begin at zero.

Another form of misleading information is through the use of the wrong statistical measure to determine what is the middle. For instance, the mean, or average, of many data measurements allows **outliers** (data measurements which lie well outside the normal range) to have a large effect. Examine the measurements in the chart below.

Address	Household Income	Address	Household Income
341 Spring Drive	$19,000	346 Spring Drive	$30,000
342 Spring Drive	$17,000	347 Spring Drive	$32,000
343 Spring Drive	$26,000	348 Spring Drive	$1,870,000
344 Spring Drive	$22,000	349 Spring Drive	$31,000
345 Spring Drive	$25,000	350 Spring Drive	$28,000

Average (Mean) Household Income: $210,000
Median Household Income: $27,000

In this example, the outlier, located at 348 Spring Drive, inflates the average household income on this street to the extent that it is over eight times the median income for the area.

Read the following charts and graphs, and then answer the questions below.

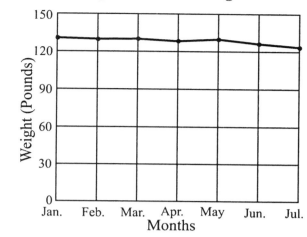

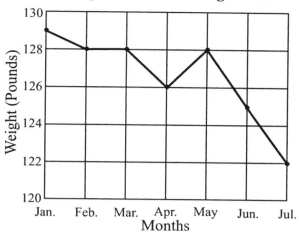

1. Which graph above presents misleading statistical information? Why is the graph misleading?

Twenty teenagers were asked how many electronic and computer games they purchased per year. The following table shows the results.

Number of Games	0	1	2	3	4	5	58
Number of Teenagers	4	2	5	3	4	1	1

2. Find the mean of the data.
3. Find the median of the data.
4. Find the mode of the data.
5. Which measurement is most misleading?
6. Which measurement would depict the data most accurately?
7. Is the *mean* of a set of data affected by outliers? Justify your answer with the example above.

Examine the two bar graphs below.

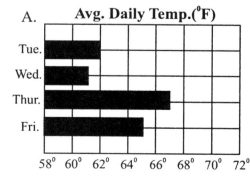

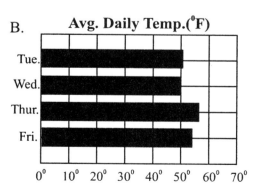

8. Which graph is misleading? Why?

CHAPTER 16 REVIEW

Find the mean, median, and mode for each of the following sets of data. Fill in the table below.

❶ Miles Run by Track Team Members

Jeff	24
Eric	20
Craig	19
Simon	20
Elijah	25
Rich	19
Marcus	20

❷ 1992 SUMMER OLYMPIC GAMES Gold Medals Won

Unified Team	45	Hungary	11
United States	37	South Korea	12
Germany	33	France	8
China	16	Australia	7
Cuba	14	Japan	3
Spain	13		

❸ Hardware Store Payroll June Week 2

Erica	$280
Dane	$206
Sam	$240
Nancy	$404
Elsie	$210
Gail	$305
David	$280

Data Set Number	Mean	Median	Mode
❶			
❷			
❸			

4. Jenica bowled three games and scored an average of 116 points per game. She scored 105 on her first game and 128 on her second game. What did she score on her third game?

5. Concession stand sales for each game in the season were $320, $540, $230, $450, $280, and $580. What was the mean sales per game?

6. Cedrick D'Amitrano works Friday and Saturday delivering pizza. He delivers 8 pizzas on Friday. How many pizzas must he deliver on Saturday to average 11 pizzas per day?

7. Long cooked three Vietnamese dinners that weighed a total of 40 ounces. What was the average weight for each dinner?

8. The Swamp Foxes scored an average of 7 points per soccer game. They scored 9 points in the first game, 4 points in the second game, and 5 points in the third game. What was their score for their fourth game?

9. Shondra is 66 inches tall, and DeWayne is 72 inches. How tall is Michael if the average height of these three students is 77 inches?

Over the past 2 years, Coach Strive has kept a record of how many points his basketball team, the Bearcats, has scored in each game:

29 32 35 35 36 38 39 40 40 41 42 43 44 44 45 47 49 50 52 53 62

10. Create a stem-and-leaf plot for the data.

Stem	Leaves

11. What is the median?
12. What is the upper quartile?
13. What is the lower quartile?
14. What is the upper extreme?
15. What is the lower extreme?

16. Create a box-and-whisker plot for the data.

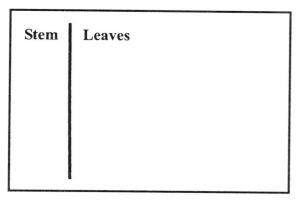

The daily high temperatures (°F) of Laughlin over the month of February are given below.

65 67 69 75 76 79 80 81 85 85 85 85 86 86
87 87 88 88 89 90 90 91 91 92 93 95 97 98

17. Create a stem-and-leaf plot for the data.

Stem	Leaves

18. Create a box-and-whisker plot for the data. Label the median, the quartiles, and the extremes.

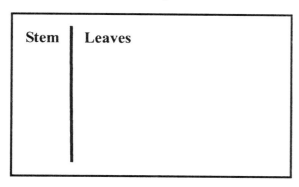

The Riveras and the Rogers families are meeting for a Fourth of July family reunion in the same park. The ages of the Rivera family members are 48, 79, 20, 2, 14, 84, 32, 61, 48, 92, 87, 54, 41, 27, 18, 21, 36, 44, 27, 66, 27, 16, 54, 48, 48, 6, and 4. The ages of the Roger's family members are 26, 84, 14, 7, 30, 50, 55, 41, 29, 33, 1, 15, 48, 16, and 20. Plot each of their ages on the stem-and-leaf plot below and answer the questions that follow.

19.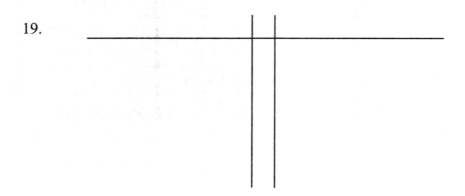

20. In the data above, which age is the mode of the data for the Rivera family?

21. Which age is the median in the Rogers family?

22. What ages are the two oldest Riveras?

23. What age are the two youngest Riveras?

24. Which family has the older median age?

On the line below each plot, write whether the relationship shown between the two variables is "positive", "negative", or "no relationship".

25.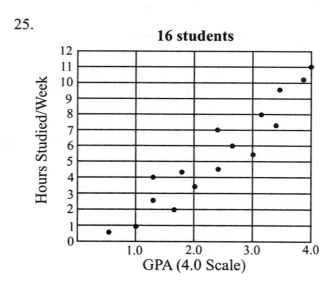

26.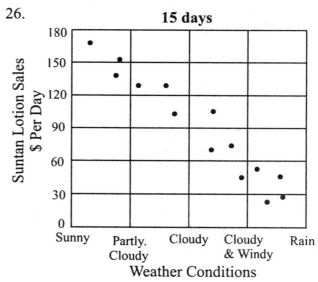

_____   _____

228   Copyright © American Book Company

27. Which of the following charts below is misleading? Why?

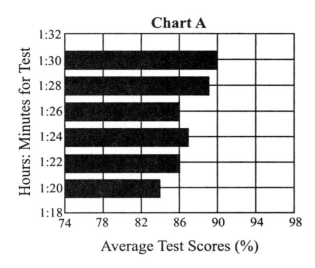

Chart A

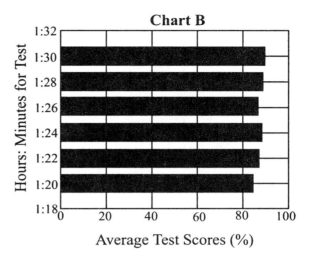

Chart B

Nine cooks were asked "What is the actual temperature inside your oven when it is set at 350°F?" The responses are in the chart below. Answer the questions that follow.

Temperature (°F)	104	347	348	349	350	351	352
# of Cooks	1	1	1	2	1	2	1

28. Find the mean of the data above.
29. Find the median of the data above.
30. Which measurement is misleading? Why?

Examine each set of information below. If you drew a scatter plot for the information in each of these tables, write whether the relationship would be "positive", "negative", or "no relationship".

Hair Color	Weight
Brown	200
Black	168
Blonde	179
Auburn	189
White	128
Blonde	100
Black	139
Red	201
Dark Blond	175

Average Monthly Temp. Degrees F	Air. Cond. Bill Dollars
65	39
73	45
59	36
88	99
79	70
68	47
48	15
39	12

31. If the hair colors were ordered from dark to light, what is the relationship?

32. If the temperatures were ordered from coldest to hottest, what is the relationship?

Chapter 17
Data Interpretation

READING TABLES

A **table** is a concise way to organize large quantities of information using rows and columns. Read each table carefully, and then answer the questions that follow.

Some employers use a tax table like the one below to figure how much Federal Income Tax should be withheld from a single person paid weekly. The number of withholding allowances claimed is also commonly referred to as the number of deductions claimed.

\multicolumn{2}{c}{Federal Income Tax Withholding Table} SINGLE Persons – WEEKLY Payroll Period					
If the wages are –		And the number of withholding allowances claimed is –			
At least	But less than	0	1	2	3
		The amount of income tax to be withheld is –			
$250	260	31	23	16	9
$260	270	32	25	17	10
$270	280	34	26	19	12
$280	290	35	28	20	13
$290	300	37	29	22	15

1. David is single, claims 2 withholding allowances, and earned $275 last week. How much Federal Income Tax was withheld? _____

2. Cecily claims 0 deductions, and she earned $297 last week. How much Federal Income Tax was withheld? _____

3. Sherri claims 3 deductions and earned $268 last week. How much Federal Income Tax was withheld from her check? _____

4. Mitch is single and claims 1 allowance. Last week, he earned $291. How much was withheld from his check for Federal Income Tax? _____

5. Ginger earned $275 this week and claims 0 deductions. How much Federal Income Tax will be withheld from her check? _____

6. Bill is single and earns $263 per week. He claims 1 withholding allowance. How much Federal Income Tax is withheld each week? _____

BAR GRAPHS

Bar graphs can be either vertical or horizontal. There may be just one bar or more than one bar for each interval. Sometimes each bar is divided into two or more parts. In this section, you will work with a variety of bar graphs. Be sure to read all titles, keys, and labels to completely understand all the data that is presented. **Answer the questions about each graph below.**

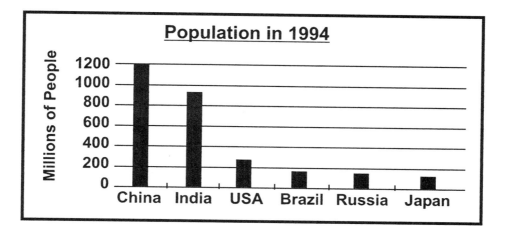

1. Which country has over 1 billion people?

2. How many countries have fewer than 200,000,000 people?

3. How many more people does India have than Japan?

4. If you added together the populations of the USA, Brazil, Russia, and Japan, would it come closer to the population of India or China?

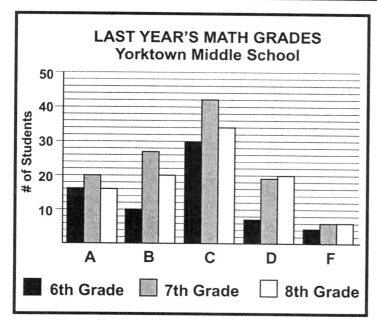

5. How many of last year's 6th graders made C's in math?

6. How many more math students made B's in the 7th grade than in the 8th grade?

7. Which letter grade is the mode of the data?

8. How many 8th graders took math last year?

9. How many students made A's in math last year?

LINE GRAPHS

Line graphs often show how data changes over time. Study the line graph below charting temperature changes for a day in Sandy, Nevada. Then answer the questions that follow.

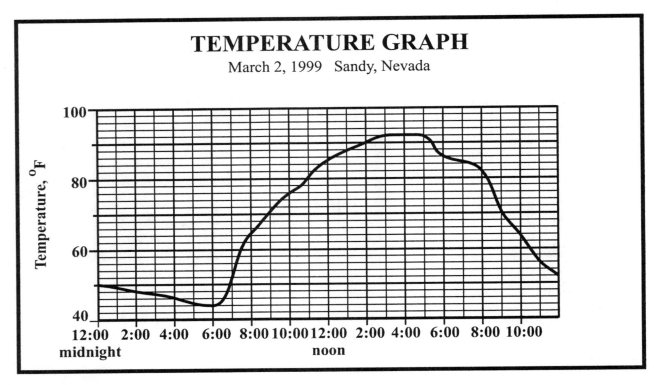

Study the graph, and then answer the questions below.

1. When was the coolest time of the day? _____
2. When was the hottest time of the day? _____
3. How much did the temperature rise between 6:00 a.m. and 2:00 p.m.? _____
4. How much did the temperature drop between 6:00 p.m. and 11:00 p.m.? _____
5. What is the difference in temperature between 8:00 a.m. and 8:00 p.m.? _____
6. Between which two hour time period was the greatest increase in temperature? _____
7. Between which hours of the day did the temperature continually increase? _____
8. Between which two hours of the day did the temperature change the least? _____
9. How much did the temperature decrease from 2:00 a.m. to 6:00 a.m.? _____
10. During which two times of day was the temperature 70°F? _____
    _____

MULTIPLE LINE GRAPHS

Multiple line graphs are a way to present a large quantity of data in a small space. It would often take several paragraphs to explain in words the same information that one graph could do.

On the graph below, there are three lines. You will need to read the **key** to understand the meaning of each.

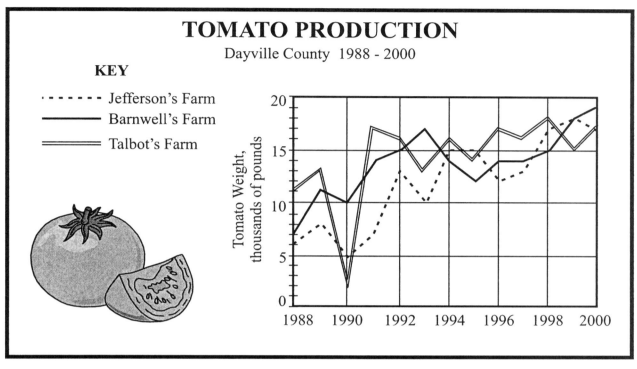

Study the graph, and then answer the questions below.

1. In what year did Barnwell's Farm produce 8,000 pounds of tomatoes more than Talbot's Farm? _____

2. In which year did Dayville County produce the most pounds of tomatoes? _____

3. In 1993, how many more pounds of tomatoes did Barnwell's Farm produce than Talbot's Farm? _____

4. How many pounds of tomatoes did Dayville County's three farms produce in 1992? _____

5. In which year did Dayville County produce the fewest pounds of tomatoes? _____

6. Which farm had the most dramatic increase in production from one year to the next? _____

7. How many more pounds of tomatoes did Jefferson's Farm produce in 1992 than in 1988? _____

8. Which farm produced the most pounds of tomatoes in 1995? _____

CIRCLE GRAPHS

Circle graphs represent data expressed in percentages of a total. The parts in a circle graph should always add up to 100%. Circle graphs are sometimes called **pie graphs** or **pie charts**.

To figure the value of a percent in a circle graph, multiply the percent by the total. Use the circle graphs below to answer the questions. The first question is worked for you as an example.

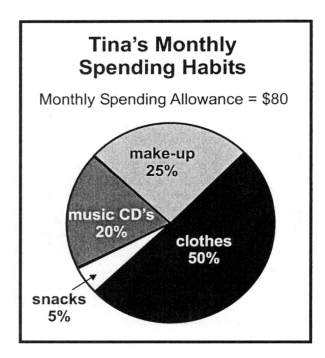

1. How much did Tina spend each month on music CD's?

    $80 × 0.20 = $16.00

    _____ $16.00

2. How much did Tina spend each month on make-up?

    _____

3. How much did Tina spend each month on clothes?

    _____

4. How much did Tina spend each month on snacks?

    _____

Fill in the following chart.

Favorite Activity	Number of Students
5. watching TV	
6. talking on the phone	
7. playing video games	
8. surfing the Internet	
9. playing sports	
10. reading	

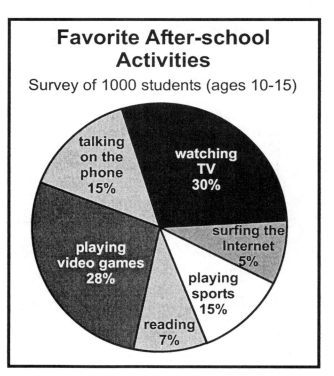

CHAPTER 17 REVIEW

KNIGHTS BASKETBALL Points Scored				
Player	game 1	game 2	game 3	game 4
Joey	5	2	4	8
Jason	10	8	10	12
Brandon	2	6	5	6
Ned	1	3	6	2
Austin	0	4	7	8
David	7	2	9	4
Zac	8	6	7	4

1. How many points did the Knights basketball team score in game 1? _____

2. How many more points did David score in game 3 than in game 1? _____

3. How many points did Jason score in the first 4 games? _____

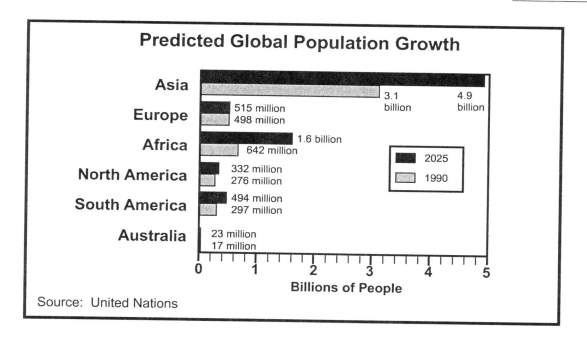

4. By how many is Asia's population predicted to increase between 1990 and 2025? _____

5. In 1990, how much larger was Africa's population than Europe's? _____

6. Where is the population expected to more than double between 1990 and 2025? _____

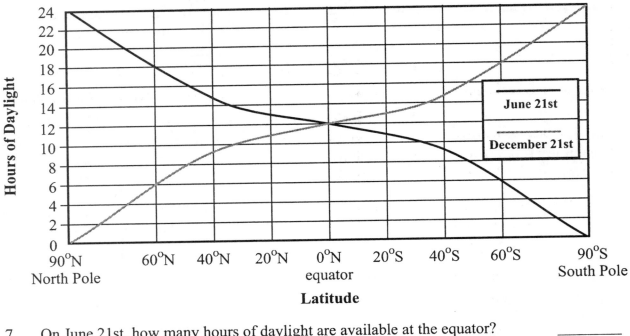

7. On June 21st, how many hours of daylight are available at the equator? _____

8. How many more hours of daylight does a person at 60°N latitude have on June 21st than a person at 60°S latitude? _____

9. Where would a person experience an entire 24 hours of daylight on December 21st? _____

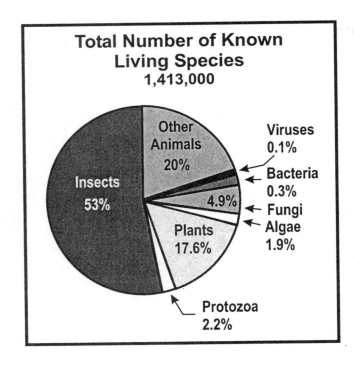

10. Which category given in the pie chart represents the greatest number of known living species?

    _____

11. Of the total 1,413,000 known living species, how many are viruses?

    _____

12. Of the total 1,413,000 known living species, how many are insects?

    _____

Chapter 18
Probability

PROBABILITY

Probability is the chance something will happen. Probability is most often expressed as a fraction, a decimal, a percent, or can also be written out in words.

EXAMPLE 1: Billy had 3 red marbles, 5 white marbles, and 4 blue marbles on the floor. His cat came along and batted one marble under the chair. What is the **probability** it was a red marble?

Step 1: The number of red marbles will be the top number of the fraction. ⟶ $\frac{3}{12}$
Step 2: The total number of marbles is the bottom number of the fraction. ⟶

The answer may be expressed in lowest terms. $\frac{3}{12} = \frac{1}{4}$

Expressed as a decimal, $\frac{1}{4} = .25$, as a percent, $\frac{1}{4} = 25\%$, and written out in words $\frac{1}{4}$ is one out of four.

EXAMPLE 2: Determine the probability that the pointer will stop on a shaded wedge or the number 1.

Step 1: Count the number of possible wedges that the spinner can stop on to satisfy the above problem. There are 5 wedges that satisfy it (4 shaded wedges and one number 1). The top number of the fraction is 5.

Step 2: Count the total number of wedges, 7. The bottom number of the fraction is 7.

The answer is $\frac{5}{7}$ or **five out of seven**.

EXAMPLE 3: Refer to the spinner above. If the pointer stops on the number 7, what is the probability that it will **not** stop on 7 on the next spin?

Step 1: Ignore the information that the pointer stopped on the number 7 on the previous spin. The probability of the next spin does not depend on the outcome of the previous spin. Simply find the probability that the spinner will **not** stop on 7. Remember, if P is the probability of an event occurring, 1−P is the probability of an event **not** occurring. In this example, the probability of the spinner landing on 7, is $\frac{1}{7}$.

Step 2: The probability is that the spinner will not stop on 7 is $1 - \frac{1}{7}$ which equals $\frac{6}{7}$.

The answer is $\frac{6}{7}$ or **six out of seven**.

Find the probability of the following problems. Express the answer as a percent.

1. A computer chose a random number between 1 and 50. What is the probability of you guessing the same number that the computer chose in 1 try?

    _____

2. There are 24 candy-coated chocolate pieces in a bag. Eight have defects in the coating that can be seen only with close inspection. What is the probability of pulling out a defective piece without looking?

    _____

3. Seven sisters have to choose which day each will wash the dishes. They put equal size pieces of paper each labeled with a day of the week in a hat. What is the probability that the first sister who draws will choose a weekend day?

    _____

4. For his garden, Clay has a mixture of 12 white corn seeds, 24 yellow corn seeds, and 16 bi-color corn seeds. If he reaches for a seed without looking, what is the probability that Clay will plant a bi-color corn seed first?

    _____

5. Mom just got a new department store credit card in the mail. What is the probability that the last digit is an odd number?

    _____

6. Alex has a paper bag of cookies that includes 8 chocolate chip, 4 peanut butter, 6 butterscotch chip, and 12 ginger. Without looking, his friend John reaches in the bag for a cookie. What is the probability that the cookie is peanut butter?

    _____

7. An umpire at a little league baseball game has 14 balls in his pockets. Five of the balls are brand A, 6 are brand B, and 3 are brand C. What is the probability that the next ball he throws to the pitcher is a brand C ball?

    _____

8. What is the probability that the spinner arrow will land on an even number?

    _____

9. The spinner in the problem above stopped on a shaded wedge on the first spin and stopped on the number 2 on the second spin. What is the probability that it will not stop on a shaded wedge or on the 2 on the third spin?

    _____

10. A company is offering 1 grand prize, 3 second place prizes, and 25 third place prizes based on a random drawing of contest entries. If you entered one of the 500 total entries, what is the probability you will win a third place prize?

    _____

11. In the contest problem above, what is the probability that you will win the grand prize or a second place prize?

    _____

12. A box of a dozen donuts has 3 lemon cream-filled, 5 chocolate cream-filled, and 4 vanilla cream-filled. If the donuts look identical, what is the probability of picking a lemon cream-filled?

    _____

INDEPENDENT AND DEPENDENT EVENTS

In mathematics, the outcome of an event may or may not influence the outcome of a second event. If the outcome of one event does not influence the outcome of the second event, these events are **independent**. However, if one event has an influence on the second event, the events are **dependent**. When someone needs to determine the probability of two events occurring, he or she will need to use an equation. These equations will change depending on whether the events are independent or dependent in relation to each other.

When finding the probability of two **independent** events, multiply the probability of each favorable outcome together.

EXAMPLE 1: One bag of marbles contains 1 white, 1 yellow, 2 blue, and 3 orange marbles. A second bag of marbles contains 2 white, 3 yellow, 1 blue, and 2 orange marbles. What is the probability of drawing a blue marble from each bag?

Solution: Probability of Favorable Outcomes

Bag 1: $\frac{2}{7}$

Bag 2: $\frac{1}{8}$

Probability of blue marble from each bag: $\frac{2}{7} \times \frac{1}{8} = \frac{2}{56} = \frac{1}{28}$

In order to find the probability of two **dependent** events, you will need to use a different set of rules. For the first event, you must divide the number of favorable outcomes by the number of possible outcomes. For the second event, you must subtract one from the number of favorable outcomes <u>only</u> if the favorable outcome is the <u>same</u>. However, you must subtract one from the number of total possible outcomes. Finally, you must multiply the probability of event one by the probability of event two.

EXAMPLE 2: One bag of marbles contains 3 red, 4 green, 7 black, and 2 yellow marbles. What is the probability of drawing a green marble, removing it from the bag, and then drawing another green marble?

	Favorable Outcomes	Total Possible Outcomes
Draw 1	4	16
Draw 2	3	15
Draw 1 × Draw 2	12	240

Answer: $\frac{12}{240}$ or $\frac{1}{20}$

EXAMPLE 3: Using the same bag of marbles, what is the probability of drawing a red marble and then drawing a black marble?

	Favorable Outcomes	Total Possible Outcomes
Draw 1	3	16
Draw 2	7	15
Draw 1 × Draw 2	21	240

Answer: $\frac{21}{240}$ or $\frac{7}{80}$

Find the probability of the following problems. Express the answer as a fraction.

1. Prithi has two boxes. Box 1 contains 3 red, 2 silver, 4 gold, and 2 blue combs. She also has a second box containing 1 black and 1 clear brush. What is the probability that Prithi selected a red brush from box 1 and a black brush from box 2?

2. Terrell cast his line into a pond containing 7 catfish, 8 bream, 3 trout, and 6 northern pike. He immediately caught a bream. What are the chances that Terrell will catch a second bream when he casts his line?

3. Gloria Quintero entered a contest in which the person who draws his or her initials out of a box containing all 26 letters of the alphabet wins the grand prize. Gloria reaches in and draws a "G", keeps it, then draws another letter. What is the probability that Gloria will next draw a "Q"?

4. Steve Marduke had two spinners in front of him. The first one was numbered 1-6, and the second was numbered 1-3. If Steve Spins each spinner once, what is the probability that the first spinner will show an odd number and the second spinner will show a "1"?

5. Carrie McCallister flipped a coin twice and got heads both times. What is the probability that Carrie will get tails the third time she flips the coin?

6. Vince Macaluso is pulling two socks out of a washing machine in the dark. The washing machine contains three tan, one white, and two black socks. If Vince reaches in and pulls the socks out one at a time, what is the probability that Vince will pull out two tan socks in his first two tries?

7. John Salome has a bag containing 2 yellow plums, 2 red plums, and 3 purple plums. What is the probability he reaches in without looking and pulls out a yellow plum and eats it, and then reaches in again without looking and pulls out a red plum to eat?

8. Artie Drake turns a spinner which is evenly divided into 11 sections numbered 1-11. On the first spin, Artie's pointer lands on "8." What is the probability that the spinner lands on an even number the second time he turns the spinner?

9. Leanne Davis played a game with a street entertainer. In this game, a ball was placed under one of three coconut halves. The vendor shifted the coconut halves so quickly that Leanne could no longer tell which coconut half contained the ball. She selected one and missed. The entertainer then shifted them around once more and asked Leanne to pick again. What is the probability that Leanne will select the coconut half containing the ball?

10. What is the probability that Jane Robelot reaches into a bag containing 1 daffodil and 2 gladiola bulbs and pulls out a daffodil bulb, and then reaches into a second bag containing 6 tulip, 3 lily, and 2 gladiola bulbs and pulls out a lily bulb?

MORE PROBABILITY

EXAMPLE: You have a cube with one number, 1, 2, 3, 4, 5 and 6 painted on each face of the cube. What is the probability that if you throw the cube 3 times, you will get the number 2 each time?

If you roll the cube once, you have a 1 in 6 chance of getting the number 2. If you roll the cube a second time, you again have a 1 in 6 chance of getting the number 2. If you roll the cube a third time, you again have a 1 in 6 chance of getting the number 2. The probability of rolling the number 2 three times in a row is:

$$\frac{1}{6} \times \frac{1}{6} \times \frac{1}{6} = \frac{1}{216}$$

Find the probability that each of the following events will occur.

There are 10 balls in a box, each with a different digit on it: 0, 1, 2, 3, 4, 5, 6, 7, 8, & 9. A ball is chosen at random and then put back in the box.

1. What is the probability that if you picked out a number ball 3 times, you would get the number 7 each time?

2. What is the probability you would pick a ball with 5, then 9, and then 3?

3. What is the probability that if you picked out a ball four times, you would always get an odd number?

---

4. A couple has 4 children ages 9, 6, 4, and 1. What is the probability that they are all girls?

There are 26 letters in the alphabet allowing a different letter to be on each of 26 cards. The cards are shuffled. After each card is chosen at random, it is put back in the stack of cards, and the cards are shuffled again.

5. What is the probability that when you pick 3 cards, one at a time, that you would draw first a "y," then an "e," and then an "s"?

6. What is the probability that you would draw 4 cards and get the letter "z" each time?

7. What is the probability that you would draw twice and get a letter that is in the word "random" both times?

---

8. If you flipped a coin 3 times, what is the probability you would get heads every time?

---

9. Marie is clueless about 4 of her multiple-choice answers. The possible answers are A, B, C, D, E, or F. What is the probability that she will guess all four answers correctly?

TREE DIAGRAMS

Drawing a **tree diagram** is another method of determining the probability of an event occuring.

EXAMPLE: If you toss two six-sided dice, what is the probability you will get two dice that add up to 9? One way to determine the probability is to make a tree diagram.

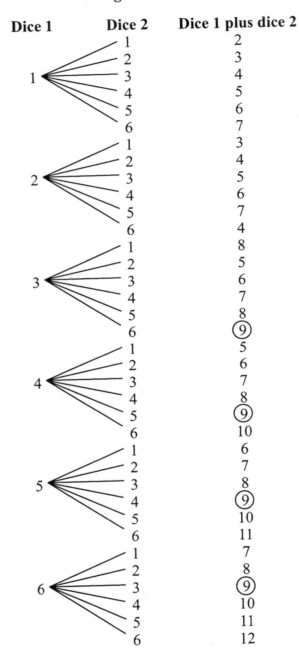

Alternative method
Write down all of the numbers in both dice which would add up to 9.

Dice 1	Dice 2
4	5
5	4
6	3
3	6

Numerator = 4 combinations

For Denominator: Multiply the number of sides on one die times the number of sides on the other die.

$6 \times 6 = 36$

Numerator: $\dfrac{4}{36} = \dfrac{1}{9}$
Denominator:

There are 36 possible ways the dice could land. Out of those 36 ways, the two dice add up to 9 only 4 times. The probability you will get two dice that add up to 9 is $\dfrac{4}{36}$ or $\dfrac{1}{9}$.

Read each of the problems below. Then answer the questions.

1. Jake has a spinner. The spinner is divided into eight equal regions numbered 1–8. In two spins, what is the probability that the numbers added together would equal 12?

2. Charlie and Libby each spin one spinner one time. The spinner is divided into 5 equal regions numbered 1–5. What is the probability that these two spins added together would equal 7?

3. Gail spins a spinner twice. The spinner is divided into 9 equal regions numbered 1–9. In two spins, what is the probability that the difference between the two numbers would equal 4?

4. Diedra throws two 10-sided dice. What is the probability that the difference between the two numbers would equal 7?

5. Cameron throws two six-sided dice. What is the probability that the difference between the two numbers would equal 3?

6. Tesla spins one spinner twice. The spinner is divided into 11 equal regions numbered 1–11. What is the probability that the two numbers added together would equal 11?

7. Samantha decides to roll two five-sided dice. What is the probability that the two numbers added together would equal 4?

8. Mary Ellen spins a spinner twice. The spinner is divided into 7 equal regions numbered 1–7. What is the probability that the product of the two numbers would equal 10?

9. Conner decides to roll two six-sided dice. What is the probability that the product of the two numbers equals 4?

10. Tabitha spins one spinner twice. The spinner is divided into 9 equal regions numbered 1–9. What is the probability that the sum of the two numbers would equal 10?

11. Darnell decides to roll two 15-sided dice. What is the probability that the difference between the two numbers would be 13?

12. Inez spins one spinner twice. The spinner is divided into 12 equal regions numbered 1–12. What is the probability that the sum of the two numbers will be equal to 10?

13. Gina spins one spinner twice. The spinner is divided into 8 equal regions numbered 1–8. What is the probability that the two numbers added together would equal 9?

14. Celia rolls two six-sided dice. What is the probability that the difference between the two numbers will be 2?

15. Brett spins one spinner twice. The spinner is divided into 4 equal regions numbered 1–4. What is the probability that the difference between the two numbers will be 3?

CHAPTER 18 REVIEW

1. There are 50 students in the school orchestra in the following sections:

    25 string section
    15 woodwind
    5 percussion
    5 brass

   One student will be chosen at random to present the orchestra director with an award. What is the probability the student will be from the woodwind section?

2. Fluffy's cat treat box contains 6 chicken-flavored treats, 5 beef-flavored treats, and 7 fish-flavored treats. If Fluffy's owner reaches in the box without looking, and chooses one treat, what is the probability that Fluffy will get a chicken-flavored treat?

3. The spinner on the right stopped on the number 5 on the first spin. What is the probability that it will not stop on the number 5 on the second spin?

4. Three cakes are sliced into 20 pieces each. Each cake contains 1 gold ring. What is the probability that one person who eats one piece of cake from each of the 3 cakes will find 3 gold rings?

5. Brianna tossed a coin 4 times. What is the probability she got all tails?

6. Sherri turned the spinner on the right 3 times. What is the probability that the pointer always landed on a shaded number?

7. A box of a dozen donuts has 3 lemon cream-filled, 5 chocolate cream-filled, and 4 vanilla cream-filled. If the donuts look identical, what is the probability that if you pick a donut at random, it will be chocolate cream-filled?

8. Erica got a new credit card in the mail. What is the probability that the last four digits are all 5's?

9. There are 26 letters in the alphabet. What is the probability that the first two letters of your new license plate will be your initials?

10. Mary has 4 green mints and 8 white mints the same size in her pocket. If she picks out one, what is the probability it will be green?

Read the following, and answer questions 11–15.

There are 9 slips of paper in a hat, each with a number from 1 to 9. The numbers correspond to a group of students who must answer a question when the number for their group is drawn. Each time a number is drawn, the number is put back in the hat.

11. What is the probability that the number 6 will be drawn twice in a row?

12. What is the probability that the first 5 numbers drawn will be odd numbers?

13. What is the probability that the second, third, and fourth numbers drawn will be even numbers?

14. What is the probability that the first five times a number is drawn it will be the number 5?

15. What is the probability that the first five numbers drawn will be 1, 2, 3, 4, 5 in that order?

Answer the following probability questions.

16. If you toss two six-sided dice, what is the probability they will add up to 7? (Make a tree diagram.)

17. Make a tree diagram to show the probability of a couple with 3 children having a boy and two girls.

Solve the following word problems. Then write whether the problem is "dependent" or "independent."

18. Felix Perez reaches into a 10 piece puzzle and pulls out one piece at random. This piece has two places where it could connect to other pieces. What is the probability that he will select another piece which fits the first one if he selects the next piece at random?

19. Barbara Stein is desperate for a piece of chocolate candy. She reaches into a bag which contains 8 peppermint, 5 butterscotch, 7 toffee, 3 mint, and 6 chocolate pieces and pulls out a toffee piece. Disappointed, she throws it back into the bag and then reaches back in and pulls out one piece of candy. What is the probability that Barbara pulled out a chocolate piece on the second try?

20. Christen Solis went to a pet shop and immediately decided to purchase a guppy she saw swimming in an aquarium. She reached into the tank containing 5 goldfish, 6 guppies, 4 miniature catfish, and 3 minnows and accidentally pulled up a goldfish. Breathing a sigh, Christen placed the goldfish back in the water. The fish were swimming so fast, it was impossible to tell what fish Christen would catch. What is the probability that Christen would catch a guppy on her second try?

Chapter 19
Patterns and Problem Solving

NUMBER PATTERNS

In each of the examples below, there is a sequence of data that follows a pattern. You must find the pattern that holds true for each number in the data. Once you determine the pattern, you can write out an equation that fits the data and figure out any other number in the sequence.

	Sequence	Pattern	Next Number in Sequence	20th Number in the Sequence
EXAMPLE 1:	3, 4, 5, 6, 7	$n + 2$	$n + 2 = 8$	$20 + 2 = 22$

In number patterns, the sequence is the output. The input can be the set of whole numbers starting with 1. However, you must determine the "rule" or pattern. Look at the table below.

input	sequence
1	→ 3
2	→ 4
3	→ 5
4	→ 6
5	→ 7

What pattern or "rule" can you come up with that gives you the first number in the sequence, 3, when you input 1? $n + 2$ will work because when $n = 1$, the first number in the sequence = 3. Does this pattern hold true for the rest of the numbers in the sequence? Yes, it does. When $n = 2$, the second number in the sequence = 4. When $n = 3$, the third number in the sequence = 5, and so on. Therefore, $n + 2$ is the pattern. Even without knowing the algebraic form of the pattern, you could figure out that 8 is the next number in the sequence. The equation describing this pattern would be $n + 2$. To find the 20th number in the pattern, use $n = 20$ to get 22.

	Sequence	Pattern	Next Number in Sequence	20th Number in the Sequence
EXAMPLE 2:	1, 4, 9, 16, 25	n^2	$n^2 = 36$	400
EXAMPLE 3:	−2, −4, −6, −8, −10	$-2n$	$-2n = -12$	−40

Find the pattern and the next number in each of the sequences below.

	Sequence	Pattern	Next Number in Sequence	20th number in the sequence
1.	−2, −1, 0, 1, 2	_____	_____	_____
2.	5, 6, 7, 8, 9	_____	_____	_____
3.	3, 7, 11, 15, 19	_____	_____	_____
4.	−3, −6, −9, −12, −15	_____	_____	_____
5.	3, 5, 7, 9, 11	_____	_____	_____
6.	2, 4, 8, 16, 32	_____	_____	_____
7.	1, 8, 27, 64, 125	_____	_____	_____
8.	0, −1, −2, −3, −4	_____	_____	_____
9.	2, 5, 10, 17, 26	_____	_____	_____
10.	4, 6, 8, 10, 12	_____	_____	_____

USING DIAGRAMS TO SOLVE PROBLEMS

Problems that require logical reasoning cannot always be solved with a set formula. Sometimes, drawing diagrams can help you see the solution.

EXAMPLE: Yvette, Barbara, Patty, and Nicole agreed to meet at the movie theater around 7:00 p.m. Nicole arrived before Yvette. Barbara arrived after Yvette. Patty arrived before Barbara but after Yvette. What is the order of their arrival?

Nicole	Yvette	Patty	Barbara
1st	2nd	3rd	4th

Arrange the names in a diagram so that the order agrees with the problem.

Use a diagram to answer each of the following questions.

1. Javy, Thomas, Pat, and Keith raced their bikes across the playground. Keith beat Thomas but lost to Pat and Javy. Pat beat Javy. Who won the race?

2. Jeff, Greg, Pedro, Lisa, Macy, and Kay eat lunch together at a round table. Kay wants to sit beside Pedro, Pedro wants to sit next to Lisa, Greg wants to sit next to Macy, and Jeff wants to sit beside Kay. Macy would rather not sit beside Lisa. Which two people should sit on each side of Jeff?

3. Three teams play a round-robin tournament where each team plays every other team. Team A beat Team C. Team B beat Team A. Team B beat Team C. Which team is the best?

4. Caleb, Thomas, Ginger, Alex, and Janice are in the lunch line. Thomas is behind Alex. Caleb is in front of Alex but behind Ginger and Janice. Janice is between Ginger and Caleb. Who is third in line?

5. Ray, Fleta, Paula, Joan, and Henry hold hands to make a circle. Joan is between Ray and Paula. Fleta is holding Ray's other hand. Paula is also holding Henry's hand. Who must be holding Henry's other hand?

6. The Bears, the Cavaliers, the Knights, and the Lions all competed in a track meet. One team from each school ran the 400 meter relay race. The Bears beat the Knights but lost to the Cavaliers. The Lions beat the Cavaliers. Who finished first, second, third, and fourth?

   1st _____
   2nd _____
   3rd _____
   4th _____

TRIAL AND ERROR PROBLEMS

Sometimes problems can only be solved by trial and error. You have to guess at a solution, and then check to see if it will satisfy the problem. If it does not, you must guess again until you get the right answer.

Solve the following problems by trial and error. Make a chart of your attempts so that you don't repeat the same attempt twice.

1. Becca had 5 coins consisting of one or more quarters, dimes, and nickels that totaled 75¢. How many quarters, dimes, and nickels did she have?

    quarters _____

    dimes _____

    nickels _____

2. Ryan needs to buy 42 cans of soda for a party at his house. He can get a six pack for $1.80, a box of 12 for $3.00, or a case of 24 for $4.90. What is the least amount of money Ryan must spend to purchase the 42 cans of soda?

    _____

3. Jana had 10 building blocks that were numbered 1 to 10. She took three of the blocks and added up the three numbers to get 27. Which three blocks did she pick?

    _____

4. Hank had 10 coins. He had 3 quarters, 3 dimes, and 4 nickels. He bought a candy bar for 75¢. How many different ways could he spend his coins to pay for the candy bar?

    _____

5. Refer to question 4. If Hank used 6 coins to pay for the candy bar, how many of his quarters did he spend?

    _____

6. The junior varsity basketball team needs to order 38 pairs of socks for the season. The coach can order 1 pair for 2.45, 6 pairs for $12.95, or 10 pairs for $20.95. What is the least amount of money he will need to spend to purchase exactly 38 pairs of socks?

    _____

7. Tyler has 5 quarters, 10 dimes, and 15 nickels in change. He wants to buy a notebook for $2.35 using the change that he has. If he wants to use as many of the coins as possible, how many quarters will he spend?

    _____

8. Kevin is packing up his room to move to another city. He has the following items left to pack.

    comic book collection ... 7 pounds
    track trophy ... 3 pounds
    coin collection ... 13 pounds
    soccer ball ... 1 pound
    model car ... 6 pounds

    If Kevin has a large box that will hold 25 pounds, what items should he pack in it to get the most weight without going over the box's weight limit?

    _____

    _____

MAKING PREDICTIONS

Use what you know about number patterns to answer the following questions.

Corn plants grow as tall as they will get in about 20 weeks. Study the chart of the rate of corn plant growth below, and answer the questions that follow.

Corn Growth	
Beginning Week	Height (inches)
2	9
7	39
11	63
14	??

1. If the growth pattern continues, how high will the corn plant be beginning week 14?

2. If the growth pattern was constant (at the same rate from week to week), how high was the corn in the beginning of the 8th week?

Peter Nichols is staining furniture for a furniture manufacturer. He stains large pieces of furniture in the beginning of the day that take longer to dry and smaller pieces of furniture as the day progresses.

Time	# Pieces Completed per Hour
Hour 1	3
Hour 3	5
Hour 6	8

3. How many pieces of furniture did Peter stain during his second hour of work?

4. How many pieces of furniture will Peter have stained by the end of an 8 hour day?

Brian Bailey is bass fishing down the Humbolt River. He has selected six locations to fish. Using his car, he drives to the first location near Golconda. His final location is near Valmy. As he travels south, he notices that the bass catches are getting larger.

Fishing Direction	Fishing Location	Number of bass caught
North	1	4
↓	2	unrecorded
	3	10
	4	unrecorded
South	5	16

5. How many bass would he likely catch in the sixth location?

6. If he fishes six locations, how many bass is he likely to catch altogether?

INDUCTIVE REASONING AND PATTERNS

Humans have always observed what happened in the past and used these observations to predict what would happen in the future. This is called **inductive reasoning**. Although mathematics is referred to as the "deductive science," it benefits from inductive reasoning. We observe patterns in the mathematical behavior of a phenomenon, and then find a rule or formula for describing and predicting its future mathematical behavior. There are lots of different kinds of predictions that may be of interest.

EXAMPLE 1: Nancy is watching her nephew, Drew, arrange his marbles in rows on the kitchen floor. The figure below shows the progression of his arrangement.

Row 1
Row 2
Row 3
Row 4

QUESTION 1: Assuming this pattern continues, how many marbles would Drew place in a fifth row?

Answer 1: It appears that Drew doubles the number of marbles in each successive row. In the 4th row he had 8 marbles, so in the 5th row we can predict 16 marbles.

QUESTION 2: How many marbles will Drew place in the *n*th row?

Answer 2: To find a rule for the number of marbles in the *n*th row, we look at the pattern suggested by the table below.

Which row	1st	2nd	3rd	4th	5th
Number of marbles	1	2	4	8	16

Observing closely, you will notice that the *n*th row contains 2^{n-1} marbles.

QUESTION 3: Suppose Nancy tells you that Drew now has 6 rows of marbles on the floor. What is the total number of marbles in his arrangement?

Answer 3: Again, organizing the data in a table could be helpful.

Number of rows	1	2	3	4	5
Total number of marbles	1	3	7	15	31

With careful observation, one will notice that the total number of marbles is always 1 less than a power of 2; indeed, for *n* rows there are $2^n - 1$ marbles total.

QUESTION 4: If Drew has 500 marbles, what is the maximum number of *complete* rows he can form?

Answer 4: With 8 complete rows, Drew will use $2^8 - 1 = 255$ marbles, and to form 9 complete rows he would need $2^9 - 1 = 511$ marbles; thus, the answer is 8 complete rows.

EXAMPLE 2: Manuel drops a golf ball from the roof of his high school while Carla videos the motion of the ball. Later, the video is analyzed, and the results are recorded concerning the height of each bounce of the ball.

QUESTION 1: What height do you predict for the fifth bounce?

Initial height	1st bounce	2nd ounce	3rd bounce	4th bounce
30 ft	18 ft	10.8 ft	6.48 ft	3.888 ft

Answer 1: To answer this question, we need to be able to relate the height of each bounce to the bounce immediately preceding it. Perhaps the best way to do this is with **ratios** as follows:

$$\frac{\text{Height of 1st bounce}}{\text{Initial bounce}} = 0.6 \quad \frac{\text{Height of 2nd bounce}}{\text{Height of 1st bounce}} = 0.6 \ldots \quad \frac{\text{Height of 4th bounce}}{\text{Height of 3rd bounce}} = 0.6$$

Since the ratio of the height of each bounce to the bounce before it appears constant, we have some basis for making predictions.

Using this, we can reason that the fifth bounce will be equal to 0.6 of the fourth bounce.

Thus, we predict the fifth bounce to have a height of **0.6 × 3.888 = 2.3328 ft.**

QUESTION 2: Which bounce will be the last one with a height of one foot or greater?

Answer 2: For this question, keep looking at predicted bounce heights until a bounce less than 1 foot is reached.

The sixth bounce is predicted to be 1.39968 ft.
The seventh bounce is predicted to be 0.839808 ft.

Thus, the last bounce with a height greater than 1 foot is predicted to be the sixth one.

Read each of the following questions carefully. Use inductive reasoning to answer each question. You may wish to make a table or a diagram to help you visualize the pattern in some of the problems. Show your work.

George is stacking his coins as shown below.

1. How many coins do you predict he will place in the fourth stack?

2. How many coins in an nth row?

3. If George has exactly 6 "complete" stacks, how many coins does he have?

4. If George has 2,000 coins, how many complete stacks can he form?

Bob and Alice have designed and created a Web site for their high school. The first week they had 5 visitors to the site; during the second week, they had 10 visitors; and during the third week, they had 20 visitors.

5. If current trends continue, how many visitors can they expect in the fifth week?

6. How many in the nth week?

7. How many weeks will it be before they get more than 500 visitors in a single week?

8. In 1979 (the first year of classes), there were 500 students at Brookstone High. In 1989, there were 1000 students. In 1999, there were 2000 students. How many students would you predict at Brookstone in 2009 if this pattern continues (and no new schools are built)?

9. The number of new drivers' licenses issued in the city of Boomtown, USA was 512 in 1992, 768 in 1994, 1,152 in 1996, and 1,728 in 1998. Estimate the number of new drivers' licenses that will be issued in 2000.

10. The average combined (math and verbal) SAT score for seniors at Brookstone High was 1,000 in 1996, 1,100 in 1997, 1,210 in 1998, and 1331 in 1999. Predict the combined SAT score for Brookstone seniors in 2000.

Juan wants to be a medical researcher, inspired in part by the story of how penicillin was discovered as a mold growing on a laboratory dish. One morning, Juan observes a mold on one of his lab dishes. Each morning thereafter, he observes and records the pattern of growth. The mold appeared to cover about 1/32 of the dish surface on the first day, 1/16 on the second day, and 1/8 on the third day.

11. If this rate of growth continues, on which day can Juan expect the entire dish to be covered with mold?

12. Suppose that whenever the original dish gets covered with mold Juan transfers half of the mold to another dish. How long will it be before *both* dishes are covered again?

13. Every year on the last day of school, the Brookstone High cafeteria serves the principal's favorite dish–Broccoli Surprise. In 1988, 1024 students chose to eat Broccoli Surprise on the last day of school, 512 students in 1992, and 256 students in 1996. Predict how many will choose Broccoli Surprise on the last day of school in 2000.

Part of testing a new drug is determining the rate at which it will break down (*decay*) in the blood. The decay results for a certain antibiotic after a 1000 milligram injection are given in the table below.

12:00 PM	1:00 PM	2:00 PM
1000 mg	800 mg	640 mg

14. Predict the number of milligrams that will be in the patient's bloodstream at 3:00 PM.

15. At which hour can the measurer expect to record a result of less than 300 mg?

16. Marie has a daylily in her mother's garden. Every Saturday morning in the spring, she measures and records its height in the table below. What height do you predict for Marie's daylily on April 29? (Hint: Look at the *change* in height each week when looking for the pattern.)

April 1	April 8	April 15	April 22
12 in	18 in	21 in	22.5 in

17. Bob puts a glass of water in the freezer and records the temperature every 15 minutes. The results are displayed in the table below. If this pattern of cooling continues, what will be the temperature at 2:15 PM? (Hint: Again, look at the *changes* in temperature in order to see the pattern.)

1:00 PM	1:15 PM	1:30 PM	1:45 PM
92°F	60°F	44°F	36°F

Suppose you cut your hand on a rusty nail that deposits 25 bacteria cells into the wound. Suppose also that each bacterium splits into two bacteria every 15 minutes.

18. How many bacteria will there be after two hours?

19. How many 15-minute intervals will pass before there are over a million bacteria?

20. Elias performed a psychology experiment at his school. He found that when someone is asked to pass information along to someone else, only about 70% of the original information is actually passed to the recipient. Suppose Elias gives the information to Brian, Brian passes it along to George, and George passes it to Montel. Using Elias's results from past experiments, what percentage of the original information does Montel actually receive?

FINDING A RULE FOR PATTERNS

EXAMPLE: Mr. Applegate wants to put desks together in his math class so that students can work in groups. The diagram below shows how he wishes to do it.

```
    2           2  3          2  3  4         2  3  4  5
1 [   ] 3    1 [    ] 4    1 [       ] 5    1 [         ] 6
    4           6  5          8  7  6         10 9  8  7
```

With 1 table he can seat 4 students, with 2 tables he can seat 6, with 3 tables 8, and with 4 tables 10.

QUESTION 1: How many students can he seat with 5 tables?

Answer 1: With 5 tables, he could seat 5 students along the sides of the tables and 1 student on each end; thus, a total of 12 students could be seated.

QUESTION 2: Write a rule that Mr. Applegate could use to tell how many students could be seated at n tables. Explain how you got the rule.

Answer 2: For n tables, there would be n students along each of 2 sides and 2 students on the ends (1 on each end); thus, a total of $2n + 2$ students could be seated at n tables.

EXAMPLE 2: When he isn't playing football for the Brookstone Bears, Tim designs Web pages. A car dealership paid Tim $500 to start a site with photos of its cars. The dealer also agreed to pay Tim $50 for each customer who buys a car first viewed on the Web site.

QUESTION 1: Write and explain a rule that tells how much the dealership will pay Tim for the sale of n cars from his Web site.

Answer 1: Tim's payment will be the initial $500 plus $50 for each sale. Translated into mathematical language, if Tim sells n cars, he will be paid a total of $500 + 50n$ dollars.

QUESTION 2: How many cars have to be sold from his site in order for Tim to get $1,000 from the dealership?

Answer 2: He earned $500 just by establishing the site, so he only needs to earn an additional $500, which at $50 per car requires the sale of only 10 cars. (Note: Another way to solve this problem is to use the rule found in the first question. In that case, you simply solve the equation $500 + 50n = 1000$ for the variable n.)

EXAMPLE 3: Eric is baking muffins to raise money for the homecoming dance. He makes 18 muffins with each batch of batter, but he must give one muffin each to his brother, his sister, and his dog, and himself (of course!) each time a batch is finished baking.

QUESTION 1: Write a rule for the number of muffins Eric produces for the fund raiser with n batches.

Answer 1: He bakes 18 with each batch, but only 14 are available for the fund raiser. Thus, with n batches he will produce $14n$ muffins for the homecoming.
The rule = $14n$

QUESTION 2: Use your rule to determine how many muffins he will contribute if he makes 7 batches.

Answer 2: The number of batches, n, equals 7. Therefore, he will produce $14 \times 7 = 98$ muffins with 7 batches.

QUESTION 3: Determine how many batches he must bake in order to contribute at least 150 muffins.

Answer 3: Ten batches will produce $10 \times 14 = 140$ muffins. Eleven batches will produce $11 \times 14 = 154$ muffins. To produce at least 150 muffins, he must bake at least 11 batches.

QUESTION 4: Determine how many muffins he would actually bake in order to contribute 150 muffins.

Answer 4: Since Eric actually bakes 18 muffins per batch, 11 batches would result in Eric baking $11 \times 18 = 198$ muffins.

Carefully read and solve the problems below. Show your work.

Tito is building a picket fence along both sides of the driveway leading up to his house. He will have to place posts at both ends and at every 10 feet along the way because the rails come in prefabricated ten-foot sections.

1. How many posts will he need for a 180 foot driveway?

2. Write and explain a rule for determining the number of posts needed for n ten-foot sections.

3. How long of a driveway can he fence along with 32 posts?

Linda is working as a bricklayer this summer. She lays the bricks for a walkway in *sections* according to the pattern depicted below.

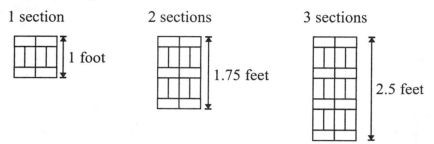

4. Write a formula for the number of bricks needed to lay *n* sections.

5. Write a formula for the number of feet covered by *n* sections.

6. How many bricks would it take to lay a walk that is 10 feet long?

Dakota's beginning pay at his new job is $300 per week. For every three months he continues to work there, he will get a $10 per week raise.

7. Write a formula for Dakota's weekly pay after *n* three-month periods.

8. After *n* years?

9. How long will he have to work before his pay gets to $400 a week?

Amanda is selling shoes this summer. In addition to her hourly wages, Amanda got a $100 bonus just for accepting the position, and she gets a $2 bonus for each pair of shoes she sells.

10. Write and explain a rule that tells how much she will make in bonuses if she sells *n* pairs of shoes.

11. How many pairs of shoes must she sell in order to make $200 in bonuses?

A certain teen telephone chat line, which sells itself as a benefit to teens but which is actually a money-making scheme, is a 900 telephone number that charges $2.00 for the first minute and $0.95 for each additional minute.

12. Write a formula for the cost of speaking *n* minutes on this line.

13. How many minutes does it take to accumulate charges of more than $50.00?

Laura's (unsharpened) pencil was initially 8 inches long. After the first sharpening, it was 7 inches long. Each sharpening thereafter, Laura noticed the pencil would be ½ inch shorter after sharpening than before.

14. Write and explain a rule that tells how long Laura's pencil will be after the nth sharpening.

15. How many sharpenings will it take to get Laura's pencil to only 3 inches long?

Ritchie's dad is tired of not being able to use his own phone whenever he wants, so he started measuring time on the phone and devised a plan for encouraging Ritchie to talk less. Ritchie will receive his ordinary allowance of $20 each week, but for each minute over two hours that Ritchie was on the phone that week, his dad deducts $0.25 from the allowance.

16. Write a rule for the allowance Ritchie receives if he talks on the phone for n minutes a week. (Hint: You actually have two rules: One for less than 120 minutes and one for 120 minutes or more.)

Every time Bob (the used car salesman) sells a car he makes a $150 commission. However, he must pay the owner of the car lot (Mike) a $32 "membership fee" for each car sold. Bob wishes to earn $1,235 for a new riding lawn mower.

17. Write a rule for the net pay that Bob earns after n sales.

18. How many cars will he have to sell in order to purchase his new mower?

Roberta works at the Oakwood movie theater. The first row in the theater has 14 seats, and each row (except the first) has four more seats than the row before it.

19. Write and explain a rule for the number of seats in the nth row at Roberta's theater.

20. Which is the first row to have more than 100 seats?

The table below displays data relating temperature in degrees Farenheit to the number of chirps per minute for a cricket.

Temp (°F)	50	52	55	58	60	64	68
Chirps/min.	40	48	60	72	80	96	112

21. Write a formula or rule that predicts the number of chirps per minute when the temperature is n degrees.

PROPORTIONAL REASONING

Proportional reasoning can be used when a selected number of individuals are tagged in a population in order to estimate the total population.

EXAMPLE: A team of scientists capture, tag, and release 50 deer in a particular national forest. One week later, they capture another 50 deer, and 2 of the deer are ones that were tagged previously. What is the approximate deer population in the national forest?

Solution: Use proportional reasoning to determine the total deer population. Preview Chapter 21 to see how to set up a proportion using the information given. You know that 50 deer out of the total deer population in the forest were tagged. You also know that 2 out of those 50 were recaptured. These two ratios should be equal because they both represent a fraction of the total deer population.

$$\frac{50 \text{ deer tagged}}{x \text{ deer total}} = \frac{2 \text{ tagged deer}}{50 \text{ deer captured}}$$

$2x = 2{,}500$ so $x = 1{,}250$ total deer

Use proportional reasoning to solve the following problems.

1. Dr. Wolf, the biologist, captures 20 fish out of a small lake behind his college. He fastens a marker onto each of these and throws them back into the lake. A week later, he again captures 20 fish. Of these, 2 have markers. How many fish could Dr. Wolf estimate are in the pond?

2. Tawanda drew 20 cards from a box. She marked each one, returned them to the box, and shook the box vigorously. She then drew 20 more cards and found that 5 of them were marked. Estimate how many cards were in the box.

3. Maureen pulls 100 pennies out of her money jar, which contains only pennies. She marks each of these, puts them back in the bank, shakes vigorously, and again pulls 100 pennies. She discovers that 2 of them are marked. Estimate how many pennies are in her money jar.

4. Mr. Kizer has a ten acre wooded lot. He catches 20 squirrels, tags them, and releases them. Several days later, he catches another 20 squirrels. One of the 20 squirrels had a tag. Estimate the number of squirrels living on Mr. Kizer's ten acres.

MATHEMATICAL REASONING/LOGIC

The California mathematics curriculum calls for skill development in mathematical **reasoning** or **logic**. The ability to use logic is an important skill for solving math problems, but it can also be helpful in real-life situations. For example, if you need to get to Park Street, and the Park Street bus always comes to the bus stop at 3 pm, then you know that you need to get to the bus stop at least by 3 pm. This is a real-life example of using logic, which many people would call "common sense."

There are many different types of statements which are commonly used to describe mathematical principles. However, using the rules of logic, the truth of any mathematical statement must be evaluated. Below are a list of tools used in logic to evaluate mathematical statements.

Logic is the discipline that studies valid reasoning. There are many forms of valid arguments, but we will just review a few here.

A **proposition** is usually a declarative sentence which may be true or false.

An **argument** is a set of two or more related propositions, called **premises**, that provide support for another proposition, called the **conclusion**.

Deductive reasoning is an argument which begins with general premises and proceeds to a more specific conclusion. Most elementary mathematical problems use deductive reasoning.

Inductive reasoning is an argument in which the truth of its premises make it likely or probable that its conclusion is true.

ARGUMENTS

Most of logic deals with the evaluation of the validity of arguments. An argument is a group of statements that includes a conclusion and at least one premise. A premise is a statement that you know is true or at least you assume to be true. Then, you draw a conclusion based on what you know or believe is true in the premise(s). Consider the following example:

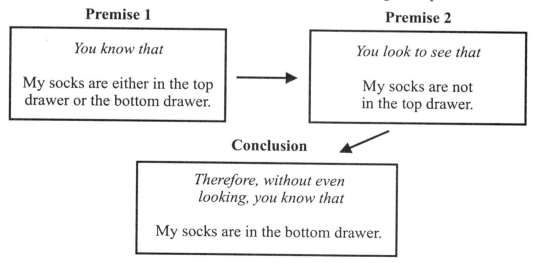

This argument is an example of deductive reasoning, where the conclusion is "deduced" from the premises and nothing else. In other words, if Premise 1 and Premise 2 are true, you don't even need to look in the bottom drawer to know that the conclusion is true.

DEDUCTIVE AND INDUCTIVE ARGUMENTS

In general, there are two types of logical arguments: **deductive** and **inductive**. Deductive arguments tend to move from general statements or theories to more specific conclusions. Inductive arguments tend to move from specific observations to general theories.

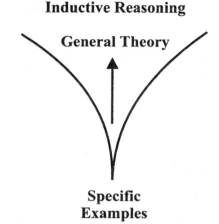

Compare the two examples below:

Deductive Argument
Premise 1 All men are mortal.
Premise 2 Socrates is a man.
Conclusion Socrates is mortal.

Inductive Argument
Premise 1 The sun rose this morning
Premise 2 The sun rose yesterday morning
Premise 3 The sun rose two days ago.
Premise 4 The sun rose three days ago.
Conclusion The sun will rise tomorrow.

An inductive argument cannot be proved beyond a shadow of a doubt. For example, it's a pretty good bet that the sun will come up tomorrow, but the sun not coming up presents no logical contradiction.

On the other hand, a deductive argument can have logical certainty, but it must be properly constructed. Consider the examples below.

True Conclusion for an Invalid Argument

All men are mortal.
Socrates is mortal.
Therefore, Socrates is a man.

Even though the above conclusion is true, the argument is based on invalid logic. Both men and women are mortal. Therefore, Socrates could be a woman.

False Conclusion from a Valid Argument

All astronauts are men.
Julia Roberts is an astronaut.
Therefore, Julia Roberts is a man.

In this case, the conclusion is false because the premises are false. However, the logic of the argument is valid because *if* the premises were true, then the conclusion would be true.

EXAMPLE 1: Which argument is valid?

If you speed on Hill Street, you will get a ticket.
If you get a ticket, you will pay a fine.

A. I paid a fine, so I was speeding on Hill Street.
B. I got a ticket, so I was speeding on Hill Street.
C. I exceeded the speed limit on Hill Street, so I paid a fine.
D. I did not speed on Hill Street, so I did not pay a fine.

Answer: C is valid.
A is incorrect. I could have paid a fine for another violation.
B is incorrect. I could have gotten a ticket for some other violation.
D is incorrect. I could have paid a fine for speeding somewhere else.

EXAMPLE 2: Assume the given proposition is true. Then determine if each statement is true or false.

Given: If a dog is thirsty, he will drink.
A. If a dog drinks, then he is thirsty.   T or F
B. If a dog is not thirsty, he will not drink.   T or F
C. If a dog will not drink, he is not thirsty.   T or F

Answer: A is false. He is not necessarily thirsty; he could just drink because other dogs are drinking or drink to show others his control of the water. This statement is the converse of the original.
B is false. The reasoning from A applies. This statement is the inverse of the original.
C is true. It is the **contrapositive** or the complete opposite of the original.

For numbers 1–5, what conclusion can be drawn from each proposition?

1. All squirrels are rodents. All rodents are mammals. Therefore,

2. All fractions are rational numbers. All rational numbers are real numbers. Therefore,

3. All squares are rectangles. All rectangles are parallelograms. All parallelograms are quadrilaterals. Therefore,

4. All Chevrolets are made by General Motors. All Luminas are Chevrolets. Therefore,

5. If a number is even and divisible by three, then it is divisible by six. Eighteen is divisible by six. Therefore,

For numbers 6–9, assume the given proposition is true. Then determine if the statements following it are true or false.

All squares are rectangles.
6. All rectangles are squares.   T or F
7. All non-squares are non-rectangles.   T or F
8. No squares are non-rectangles.   T or F
9. All non-rectangles are non-squares.   T or F

CHAPTER 19 REVIEW

Find the pattern for the following number sequences, and then find the *n*th number requested.

1. 0, 1, 2, 3, 4   pattern _____

2. 0, 1, 2, 3, 4   20th number _____

3. 1, 3, 5, 7, 9   pattern _____

4. 1, 3, 5, 7, 9   25th number _____

5. 3, 6, 9, 12, 15   pattern _____

6. 3, 6, 9, 12, 15   30th number _____

7. Andrea spent $1.24 for toothpaste. She had quarters, dimes, nickels, and pennies. How many dimes did she use if she used a total of 14 coins?

8. Cody spent 59¢ on a hotdog. He had quarters, dimes, nickels, and pennies in his pocket. He gave the cashier 9 coins for exact payment. How many quarters did he give the cashier?

9. Vince, Hal, Weng, and Carl raced on roller blades down a hill. Vince beat Carl. Hal finished before Vince but after Weng. Who won the race?

10. A veterinarian's office has a weight scale in the lobby to weigh pets. The scale will weigh up to 250 pounds. The following dogs are in the lobby:

    Pepper ... 23 pounds
    Jack ... 75 pounds
    Trooper ... 45 pounds
    Precious ... 25 pounds
    Coco ... 120 pounds

    Which dogs could you put on the scale to get as close to 250 pounds as possible, without going over? How much would they weigh?

11. Felix has set a goal to increase his running speed by 1 minute per mile every 5 weeks. He starts out running 1 mile in 11 minutes. If he can accomplish his goal, how many weeks will it take him to run a mile in 8 minutes?

Jessica Bloodsoe and Katie Turick are climbing Mt. Fuji in Japan which is 12,388 ft. high. The higher they go, the slower they climb due to lack of oxygen. The chart below shows their progress.

Days Ascending	Altitude
End of day 1	4,000 feet
End of day 2	7,200 feet
End of day 3	9,600 feet

12. If the weather holds, what will be their altitude on day 5? _____

13. If they can keep the same rate, how many days would it take them to get to the top? _____

In a large city of 200,000, there was an outbreak of tuberculosis. Immediately, health care workers began an immunization campaign. The chart below records their results.

	No. of People Immunized	No. of TB Cases
Year 1	20,000	60
Year 2	60,000	45
Year 3	100,000	30

14. About how many cases of TB would you predict for year 4? _____

Exotic goldfish are kept in different size containers of water. The larger the container, the bigger the size the goldfish can grow. The chart on the right shows how big one goldfish can grow in different size containers.

Fish Size	Container Size
1 inch	20 gallon or less
$2\frac{1}{2}$ inches	50 gallon
5 inches	100 gallon

15. Based on the chart, how large would you predict a goldfish could grow in a 140 gallon container? _____

Olivia starts, maintains, and sells ant farms as a hobby. She had 500 ants in 1996, 2,000 in 1997, and 8,000 in 1998.

16. If her hobby continues to grow as it has since 1996, how many ants will Olivia have in 2000?

17. How many in the nth year after 1996?

18. In what year would she have more than 100,000 ants?

---

19. Sean is studying bacteria and antibiotics. Using standard measurement and estimation techniques, he records a reading of about 100,000 bacteria on a lab dish. He then applies a drop of antibiotic and does another bacteria count every 30 minutes. He finds 90,000 after 30 minutes, 81,000 after 60 minutes, 72,900 after 90 minutes, and 65,610 after 120 minutes. How many do you predict he will find 150 minutes after applying the antibiotic?

Justin has just got a bill from his Internet Service Provider. The first four months of charges for his service are recorded in the table below.

	January	February	March	April
Hours	0	10	5	25
Charge	$4.95	$14.45	$9.70	$28.70

20. Write a formula for the cost of n hours of Internet service.

21. What is the greatest number of hours he can get on the Internet and still keep his bill under $20.00?

Lisa is baking cookies for the Fall Festival. She bakes 27 cookies with each batch of batter. However, she has a defective oven, which results in 5 cookies in each batch being burnt.

22. Write a formula for the number of cookies available for the festival as a result of Lisa baking n batches of cookies.

23. How many batches does she need in order to produce 300 cookies for the festival?

24. How many cookies (counting burnt ones) will she actually bake?

---

25. Jamal wonders how many ants are in his ant farm. He puts a stick in the container, and when he pulls it out, there are 15 ants on it. He gently sprays these ants with a mixture of water and food coloring, then puts them back into the container. The next day his stick draws 20 ants, 1 of which is green. Estimate how many ants Jamal has.

For numbers 26–29, assume the given proposition is true. Then determine if the statements following it are true or false.

All whales are mammals.

26. All non-whales are non-mammals.   T or F
27. If a mammal lives in the sea, it is a whale.   T or F
28. All mammals are whales.   T or F
29. All non-mammals are non-whales.   T or F

For 30-32, chose which argument is valid.

30. If I oversleep, I miss breakfast. If I miss breakfast, I cannot concentrate in class. If I do not concentrate in class, I make bad grades.

    A. I made bad grades today, so I missed breakfast.
    B. I made good grades today, so I got up on time.
    C. I could not concentrate in class today, so I overslept.
    D. I had no breakfast today, so I overslept.

31. If I do not maintain my car regularly, it will develop problems. If my car develops problems, it will not be safe to drive. If my car is not safe to drive, I cannot take a trip in it.

    A. If my car develops problems, I did not maintain it regularly.
    B. I took a trip in my car, so I maintained it regularly.
    C. If I maintain my car regularly, it will not develop problems.
    D. If my car is safe to drive, it will not develop problems.

32. If two triangles have all corresponding sides and all corresponding angles congruent, then they are congruent triangles. If two triangles are congruent, then they are similar triangles.

    A. Similar triangles have all sides and all angles congruent.
    B. If two triangles are similar, then they are congruent.
    C. If two triangles are not congruent, then they are not similar.
    D. If two triangles have all corresponding sides and angles congruent, then they are similar triangles.

Chapter 20
Measurement

When measuring length, distance, volume, mass, weight, or temperature, most of the world uses a system based on the **metric system**. The United States, however, commonly uses the **English system** of measurement. You should be familiar with both the metric system and the English system of measure.

CUSTOMARY MEASURE

Customary measure in the United States is based on the English system. The following chart gives common customary units of measure as well as the standard units for time.

English System of Measure

Measure	Abbreviations	Appropriate Instrument
Time: 1 week = 7 days 1 day = 24 hours 1 hour = 60 minutes 1 minute = 60 seconds	week = wk. hour = hr. or h. minutes = min. seconds = sec.	calendar clock clock clock
Length: 1 mile = 5,280 feet 1 yard = 3 feet 1 foot = 12 inches	mile = mi. yard = yd. foot = ft. inch = in.	odometer yard stick, tape line ruler, yard stick
Volume: 1 gallon = 4 quarts 1 quart = 2 pints 1 pint = 2 cups 1 cup = 8 ounces	gallon = gal. quart = qt. pint = pt. ounce = oz.	quart or gallon container quart container cup, pint, or quart container cup
Weight: 1 pound = 16 ounces	pound = lb. ounce = oz.	scale or balance
Temperature: Fahrenheit Celsius	°F °C	thermometer thermometer

APPROXIMATE ENGLISH MEASURE

Match the item on the left with its approximate (not exact) measure on the right. You may use some answers more than once.

_____ 1. The height of an average woman is about _____ .   A. 1 yard

_____ 2. An average candy bar weighs about _____ .   B. 2 yards

_____ 3. An average donut is about _____ across (in diameter).   C. $5\frac{1}{2}$ feet

_____ 4. A piece of notebook paper is about _____ long.   D. 4 weeks

_____ 5. A tennis ball is about _____ across (in diameter).   E. $2\frac{1}{2}$ inches

_____ 6. The average basketball is about _____ across.   F. 2 ounces

_____ 7. The average month is about _____ .   G. 1 foot

_____ 8. How long is the average lunch table?

_____ 9. About how much does a computer disk weigh?

_____ 10. What is the average height of a table?

CONVERTING UNITS USING DIMENSIONAL ANALYSIS

An easy way to solve math problems involving the conversion of units is by using **dimensional analysis**. Don't let this fancy term intimidate you. Dimensional analysis is really a simple, step-by-step way to convert units of measure by using **conversion factors**. A conversion factor is a fraction always equal to 1. The easiest way to understand dimensional analysis and conversion factors is to look at an example.

EXAMPLE 1: How many inches are in 3 feet?

Step 1: First, what is a conversion factor for feet to inches? You know that there are 12 inches to 1 foot. Therefore the conversion factors are simply the following fractions:

$$\frac{1 \text{ ft.}}{12 \text{ in.}} \quad \text{or} \quad \frac{12 \text{ in.}}{1 \text{ ft.}} \longleftarrow \textbf{conversion factors}$$

Step 2: Multiply the number with the old unit by the conversion factor to get the number with the new unit. Choose the conversion factor that will put the old unit on the bottom of the fraction so that it will cancel the other old unit.

$$3 \text{ ft.} \times \frac{12 \text{ in.}}{1 \text{ ft.}} = 36 \text{ in.}$$

Notice that the unit of "feet" cancel each other, and you are left with inches.

WRONG: If you had chosen the wrong conversion factor, this is what you would have:

$$3 \text{ ft.} \times \frac{1 \text{ ft.}}{12 \text{ in.}} = \frac{3 \text{ ft.} \times \text{ft.}}{12 \text{ in.}} \text{ or } \frac{1 \text{ ft}^2}{4 \text{ in.}}$$

Notice, feet multiplied by feet are feet squared. None of the units cancel, so you know right away that this is wrong.

EXAMPLE 2: 4 oz. equal how many pounds?

Step 1: You should know that there are 16 ounces in 1 pound, so your conversion factors are

$$\frac{16 \text{ oz.}}{1 \text{ lb.}} \quad \text{or} \quad \frac{1 \text{ lb.}}{16 \text{ oz.}}$$

Step 2: $4 \cancel{\text{oz.}} \times \frac{1 \text{ lb.}}{16 \cancel{\text{oz.}}} = \frac{4 \text{ lb.}}{16} \text{ or } \frac{1}{4} \text{ lb.}$

Again, choose the conversion factor that will cause the old unit to cancel. In this case, multiplying by the correct conversion factor means dividing 4 by 16 to get the correct units of pounds.

The two examples just given are really simple, and you may think it would be just as easy to do the math in your head. That may be true for very simple unit conversions. However, when converting more than one unit, unit conversions become more complicated. It is harder to figure out in your head whether you should divide or multiply. Get in the habit now of using dimensional analysis to solve easy problems, and more complex problems will seem very easy. Look at two more examples:

EXAMPLE 3: John drives on the interstate 60 miles per hour. How many feet per second is he traveling?

$$\frac{60 \cancel{\text{miles}}}{\cancel{\text{hour}}} \times \frac{1 \cancel{\text{hour}}}{60 \cancel{\text{min}}} \times \frac{1 \cancel{\text{min}}}{60 \text{ sec}} \times \frac{5280 \text{ ft}}{1 \cancel{\text{mile}}} = \frac{60 \times 5280}{60 \times 60} = 88 \text{ ft/sec}$$

convert hours to minutes • convert minutes to seconds • convert miles to feet • units in feet per second

EXAMPLE 4: Karen is a cashier at a grocery store. She can scan 20 items per minute. How many items can she scan in 15 seconds?

$$\frac{20 \text{ items}}{\cancel{\text{minute}}} \times \frac{1 \cancel{\text{minute}}}{60 \cancel{\text{seconds}}} \times 15 \cancel{\text{seconds}} = \frac{20 \times 15}{60} = 5 \text{ items}$$

Using dimensional analysis and the table on page 266, make the following conversions.

1. 15 yards to feet _____
2. 3 pounds to ounces _____
3. 4 feet to inches _____
4. 2 yards to inches _____
5. 2 gallons to quarts _____
6. 6 pints to cups _____
7. 1 gallon to cups _____
8. 3 inches to feet _____
9. $1\frac{1}{2}$ feet to inches _____
10. 42 inches to feet _____
11. Jared can work 54 math problems in 1 hour. How many problems can he work in 10 minutes?
12. A machine makes 360 gadgets every 24 hours. How many gadgets does it make every 20 minutes?
13. Perry's car can travel an average of 22 miles on 1 gallon of gas. How many miles can his car travel on 1 quart of gas?
14. A fast food restaurant serves an average of 40 customers per hour. On average, how many minutes does each customer wait for his or her order? (Hint: find minutes per customer, not customers per minute.)
15. Leah travels 22 feet per second on her bicycle. How many miles per hour does she travel?

THE METRIC SYSTEM

The metric system uses units based on multiples of ten. The basic units of measure in the metric system are the **meter**, the **liter**, and the **gram**. Metric prefixes tell what multiple of ten the basic unit is multiplied by. Below is a chart of metric prefixes and their values. The ones rarely used are shaded.

Prefix	kilo (k)	hecto (h)	deka (da)	unit (m, L, g)	deci (d)	centi (c)	milli (m)
Meaning	1000	100	10	1	0.1	0.01	0.001

To help you remember the order of the metric prefixes, use the following sentence:

Kings **H**ave **D**ances **U**ntil **D**ragons **C**hange **M**usic.

Or, make up your own sentence to help you memorize these prefixes.

UNDERSTANDING METERS

The basic unit of **length** in the metric system is the **meter**. Meter is abbreviated "m".

Metric Unit	Abbreviation	Memory Tip	Equivalents
1 millimeter	mm	Thickness of a dime	10 mm = 1 cm
1 centimeter	cm	Width of the tip of the little finger	100 cm = 1 m
1 meter	m	Distance from the nose to the tip of fingers (a little longer than a yard)	1000 m = 1 km
1 kilometer	km	A little more than half a mile	

UNDERSTANDING LITERS

The basic unit of **liquid volume** in the metric system is the **liter**. Liter is abbreviated "L".

The liter is the volume of a cube measuring 10 cm on each side. A milliliter is the volume of a cube measuring 1 cm on each side. A capital L is used to signify liter, so it is not confused with the number 1.

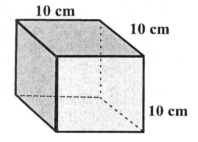

Volume = 1000 cm^3 = 1 Liter
(a little more than a quart)

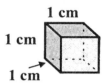

Volume = 1 cm^3 = 1 mL
(an eyedropper holds 1 mL)

UNDERSTANDING GRAMS

The basic unit of **mass** in the metric system is the **gram**. Gram is abbreviated "g".

A **gram** is the **mass** of **one cubic centimeter** of **water** at 4° C.

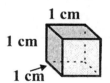

A large paper clip has a mass of about 1 gram (1g).
A nickel has a mass of 5 grams (5 g).
1000 grams = 1 kilogram (kg) = a little over 2 pounds

1 milligram (mg) = 0.001 gram. This is an extremely small amount and is used in medicine.

An aspirin tablet has a mass of 300 mg.

ESTIMATING METRIC MEASUREMENTS

Choose the best estimates.

1. The height of an average man
   A. 18 cm
   B. 1.8 m
   C. 6 km
   D. 36 mm

2. The volume of a coffee cup
   A. 300 mL
   B. 20 L
   C. 5 L
   D. 1 kL

3. The width of this book
   A. 215 mm
   B. 75 cm
   C. 2 m
   D. 1.5 km

4. The mass of an average man
   A. 5 mg
   B. 15 cg
   C. 25 g
   D. 90 kg

5. The mass of a dime
   A. 3 g
   B. 30 g
   C. 10 cg
   D. 1 kg

6. The length of a basketball court
   A. 1000 mm
   B. 250 cm
   C. 28 m
   D. 2 km

Choose the best units of measure.

7. The distance from Baton Rouge to Shreveport
   A. millimeter
   B. centimeter
   C. meter
   D. kilometer

8. The length of a house key
   A. millimeter
   B. centimeter
   C. meter
   D. kilometer

9. The thickness of a nickel
   A. millimeter
   B. centimeter
   C. meter
   D. kilometer

10. The width of a classroom
    A. millimeter
    B. centimeter
    C. meter
    D. kilometer

11. The length of a piece of chalk
    A. millimeter
    B. centimeter
    C. meter
    D. kilometer

12. The height of a pine tree
    A. millimeter
    B. centimeter
    C. meter
    D. kilometer

CONVERTING UNITS IN THE METRIC SYSTEM

In general, converting between units in the metric system is easier than converting within the English system because the units are in multiples of ten. Conversions between the same unit of measure means simply moving the decimal point.

EXAMPLE 1: Convert 34.5 meters to centimeters.

List or visualize the metric prefixes to figure out how many places to move the decimal point.

k   h   da   u . d   c   m

Start at "u" (for unit) and count over to the "c" (for centi). The decimal point needs to move to the right 2 places to convert the unit measure of meters to centimeters.

So, 3 4 . 5 0 meters is 3450 centimeters. Notice you have to add a zero.

EXAMPLE 2: Convert 250.1 millimeters to dekameters.

k   h   da   u   d   c   m .

Start at the "m" (for milli) and count over to "da" (for deka). The decimal point needs to move to the left 4 places to convert millimeters to dekameters.

So, 0 2 5 0 . 1 millimeters is .02501 dekameters. Again, add a zero.

Practice converting the following metric measurements.

1. 35 mg to g
2. 6 km to m
3. 21.5 mL to L
4. 4.9 mm to cm
5. 5.35 kL to mL
6. 32.1 mg to kg
7. 156.4 m to km
8. 25 mg to cg
9. 17.5 L to mL
10. 4.2 g to kg
11. 0.06 daL to dL
12. 0.417 kg to c
13. 18.2 cL to L
14. 81.2 dm to cm
15. 72.3 cm to mm
16. 0.003 kL to L
17. 5.06 g to mg
18. 1.058 mL to cL
19. 43 hm to km
20. 2.057 m to cm

MORE METRIC CONVERSIONS

Sometimes metric conversions can be a little more complicated. In these cases, you can use dimensional analysis.

EXAMPLE 1: How many cubic centimeters are in a liter?
First, what conversions do you know? You should know that one cubic centimeter is equal to one milliliter. You should also know that there are 1000 milliliters in one liter.

Multiply 1 liter by conversion factors until you get cubic centimeters.

$$1\cancel{L} \times \frac{1000 \cancel{mL}}{1 \cancel{L}} \times \frac{1 \text{ cm}^3}{1 \cancel{mL}} = 1000 \text{ cm}^3$$

EXAMPLE 2: How many cubic millimeters are in a cubic meter?

Again, use what you know. You know that there are 1000 millimeters in 1 meter, so the conversion would look like the following:

$$1\cancel{m^3} \times \frac{1000 \text{ mm}}{1 \cancel{m}} \times \frac{1000 \text{ mm}}{1 \cancel{m}} \times \frac{1000 \text{ mm}}{1 \cancel{m}} =$$

$$1000 \text{ mm} \times 1000 \text{ mm} \times 1000 \text{ mm} = 1,000,000,000 \text{ mm}^3$$

Remember that m^3 is $m \times m \times m$, and $mm \times mm \times mm$ is mm^3.

Practice using dimensional analysis to make the following metric or English conversions.

1. How many milliliters are in three liters?

2. How many cubic inches are in one cubic foot?

3. Holly's aspirin has a mass of 582 milligrams. What is the mass of the aspirin in grams?

4. Raymond determines that 20 drops from his eyedropper equals exactly 1 milliliter. How many drops would it take to fill a liter container?

5. A container holds 250 mL. How many liters does it hold?

6. Runners prepare to compete in the 300 meter dash. How many centimeters long is this race?

7. A hiker moves from one campsite to the next and travels 12.2 km in one day. How many millimeters did the hiker travel?

8. Sakura buys a 2 liter bottle of soda. Her drinking cup holds 400 mL. How many drinking cups can she fill with the bottle of soda? (Note: You are finding a number that does not have units, so all units must cancel.)

CHAPTER 20 REVIEW

Fill in the blanks below with the appropriate unit of measurement.

1. A box of assorted chocolates might weigh about 1 _____ (English).
2. A compact disc is about 7 _____ (English) across.
3. In Europe, gasoline is sold in _____ (metric).
4. A vitamin C tablet has a mass of 500 _____ .

Fill in the blanks below with the appropriate English or metric conversions.

5. Two gallons equals _____ cups.
6. 4.2 L equals _____ mL.
7. $3\frac{1}{2}$ yards equals _____ inches.
8. 6,800 m equals _____ kilometers.
9. 36 oz. equals _____ pounds.
10. 730 mg equals _____ kg.

Use dimensional analysis to make the following conversions.

11. Convert 30 miles per hour to feet per second.

12. An outlet pipe releases 450 gallons of water per hour. How many cups of water are released per second?

13. If Janet can type 36 words per minute, how many words can she type in 10 seconds?

14. A bicyclist travels 22 feet per second. How many miles per hour is the bicyclist traveling?

15. How many square inches are in a floor tile that measures 2 square feet?

16. A water bottle holds 500 cm^3. How many cubic inches does the water bottle hold? Use the conversion factor of 1 inch = 2.5 cm.

17. An airplane travels at 540 miles per hour. How many feet per second does it travel?

18. Jose's bathroom measures 10 feet by 10 feet. How many square centimeters does his bathroom measure? Use the conversion factor of 1 foot = 30 centimeters.

19. Paulo's faucet drips 2 cubic inches per day. How many cubic centimeters does it drip? Use the conversion factor of 1 inch = 2.5 cm.

20. Yulisa has 681 grams of candy. How many pounds of candy does she have if 1 pound = 454 grams?

Chapter 21
Ratios, Proportions, and Scale Drawings

RATIO PROBLEMS

In some word problems, you may be asked to express answers as a **ratio**. Ratios can look like fractions, they can be written with a colon, or they can be written in word form with "to" between the numbers. Numbers must be written in the order they are requested. In the following problem, 8 cups of sugar are mentioned before 6 cups of strawberries. But in the question part of the problem, you are asked for the ratio of STRAWBERRIES to SUGAR. The amount of strawberries IS THE FIRST WORD MENTIONED, so it must be the **top** number of the fraction. The amount of sugar, THE SECOND WORD MENTIONED, must be the **bottom** number of the fraction.

EXAMPLE: The recipe for jam requires 8 cups of sugar for every 6 cups of strawberries. What is the ratio of strawberries to sugar in this recipe?

First number requested $\frac{6}{8}$ cups strawberries
Second number requested   cups sugar

Answers may be reduced to lowest terms. $\frac{6}{8} = \frac{3}{4}$

This ratio is also correctly expressed as 3:4 or 3 to 4.

Practice writing ratios for the following word problems and reduce to lowest terms. DO NOT CHANGE ANSWERS TO MIXED NUMBERS. Ratios should be left in fraction form.

1. Out of the 248 seniors, 112 are boys. What is the ratio of boys to the total number of seniors?

2. It takes 7 cups of flour to make 2 loaves of bread. What is the ratio of cups of flour to loaves of bread?

3. A skyscraper that stands 620 feet tall casts a shadow that is 125 feet long. What is the ratio of the shadow to the height of the skyscraper?

4. The newborn weighs 8 pounds and is 22 inches long. What is the ratio of weight to length?

5. Jack paid $6.00 for 10 pounds of apples. What is the ratio of the price of apples to the pounds of apples?

6. Twenty boxes of paper weigh 520 pounds. What is the ratio of boxes to pounds?

SOLVING PROPORTIONS

Two **ratios (fractions)** that are **equal** to each other are called **proportions**. For example, $\frac{1}{4}=\frac{2}{8}$. Read the following example to see how to find a number missing from a proportion.

EXAMPLE: $\frac{5}{15}=\frac{8}{x}$

Step 1: To find x, you first multiply the two numbers that are diagonal to each other. $15 \times 8 = 120$

Step 2 Then divide the product (120) by the other number in the proportion (5). $120 \div 5 = 24$

Therefore, $\frac{5}{15}=\frac{8}{24}$ $x = 24$

Practice finding the number missing from the following proportions. First, multiply the two numbers that are diagonal from each other. Then divide by the other number.

1. $\frac{2}{5}=\frac{6}{x}$
2. $\frac{9}{3}=\frac{x}{5}$
3. $\frac{x}{12}=\frac{3}{4}$
4. $\frac{7}{x}=\frac{3}{9}$
5. $\frac{12}{x}=\frac{2}{5}$
6. $\frac{12}{x}=\frac{4}{3}$
7. $\frac{27}{3}=\frac{x}{2}$

8. $\frac{1}{x}=\frac{3}{12}$
9. $\frac{15}{2}=\frac{x}{4}$
10. $\frac{7}{14}=\frac{x}{6}$
11. $\frac{5}{6}=\frac{10}{x}$
12. $\frac{4}{x}=\frac{3}{6}$
13. $\frac{x}{5}=\frac{9}{15}$
14. $\frac{9}{18}=\frac{x}{2}$

15. $\frac{5}{7}=\frac{35}{x}$
16. $\frac{x}{2}=\frac{8}{4}$
17. $\frac{15}{20}=\frac{x}{8}$
18. $\frac{x}{40}=\frac{5}{100}$
19. $\frac{4}{7}=\frac{x}{28}$
20. $\frac{7}{6}=\frac{42}{x}$
21. $\frac{x}{8}=\frac{1}{4}$

RATIO AND PROPORTION WORD PROBLEMS

You can use ratios and proportions to solve problems.

EXAMPLE: A stick one meter long is held perpendicular to the ground and casts a shadow 0.4 meters long. At the same time, an electrical tower casts a shadow 112 meters long. Use ratio and proportion to find the height of the tower.

Step 1: Set up a proportion using the numbers in the problem. Put the shadow lengths on one side of the equation, and put the heights on the other side. The 1 meter height is paired with the 0.4 meter length, so let them both be top numbers. Let the unknown height be x.

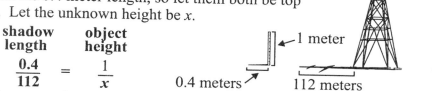

$$\begin{array}{cc} \text{shadow} & \text{object} \\ \text{length} & \text{height} \end{array}$$

$$\frac{0.4}{112} = \frac{1}{x}$$

Step 2: Solve the proportion as you did on the previous page. $112 \times 1 = 112$
$112 \div 0.4 = 280$ **Answer:** The tower height is 280 meters.

Use ratio and proportion to solve the following problems.

1. Rudolph can mow a lawn that measures 1000 square feet in 2 hours. At that rate, how long would it take him to mow a lawn 3500 square feet?

2. Faye wants to know how tall her school building is. On a sunny day, she measures the shadow of the building to be 6 feet. At the same time, she measures the shadow cast by a 5 foot statue to be 2 feet. How tall is her school building?

3. Out of every 5 students surveyed, 2 listen to country music. At that rate, how many students in a school of 800 listen to country music?

4. Bailey, a Labrador Retriever, had a litter of 8 puppies. Four of the puppies were black. At that rate, how many would be black in a litter of 10 puppies?

5. According to the instructions on a bag of fertilizer, 5 pounds of fertilizer are needed for every 100 square feet of lawn. How many square feet will a 25 pound bag cover?

6. A race car can travel 2 laps in 5 minutes. How long will it take the race car to complete 100 laps at that rate?

7. If it takes 7 cups of flour to make 4 loaves of bread, how many loaves of bread can you make from 35 cups of flour?

8. If 3 pounds of jelly beans cost $6.30, how much would 2 pounds cost?

9. For the first 4 home football games, the concession stand sold 600 hotdogs. If that ratio stays constant, how many hotdogs will sell for all 10 home games?

MAPS AND SCALE DRAWINGS

EXAMPLE 1: On a map drawn to scale, 5 cm represents 30 kilometers. A line segment connecting two cities is 7 cm long. What distance does this line segment represent?

Step 1: Set up a proportion using the numbers in the problem. Keep centimeters on one side of the equation and kilometers on the other. The 5 cm is paired with the 30 kilometers, so let them both be top numbers. Let the unknown distance be x.

$$\begin{array}{cc} \text{cm} & \text{km} \\ \dfrac{5}{7} = & \dfrac{30}{x} \end{array}$$

Step 2: Solve the proportion as you have previously. $7 \times 30 = 210$
$210 \div 5 = 42$ **Answer:** 7 cm represents 42 km.

Sometimes the answer to a scale drawing problem will be a fraction or mixed number.

EXAMPLE 2: On a scale drawing, 2 inches represents 30 feet. How many inches long is a line segment that represents 5 feet?

Step 1: Set up the proportion as you did above.

$$\begin{array}{cc} \text{inches} & \text{feet} \\ \dfrac{2}{x} = & \dfrac{30}{5} \end{array}$$

Step 2: **First, multiply the two numbers that are diagonal from each other. Then divide by the other number.**

$2 \times 5 = 10$ $10 \div 30$ is less than 1 so express the answer as a fraction and reduce.

$10 \div 30 = \dfrac{10}{30} = \dfrac{1}{3}$ inch   **Answer:** $\dfrac{1}{3}$ of an inch represents 5 feet.

Set up proportions for each of the following problems and solve.

1. If 2 inches represents 50 miles on a scale drawing, how long would a line segment be that represents 25 miles? _____

2. On a scale drawing, 2 cm represents 15 km. A line segment on the drawing is 3 cm long. What distance does this line segment represent? _____

3. On a map drawn to scale, 5 cm represents 250 km. How many kilometers are represented by a line 6 cm long? _____

4. If 2 inches represent 80 miles on a scale drawing, how long would a line segment be that represents 280 miles? _____

5. On a map drawn to scale, 5 cm represents 200 km. How long would a line segment be that represents 260 km? _____

6. On a scale drawing of a house plan, one inch represents 5 feet. How many feet wide is the bathroom if the width on the drawing is 3 inches? _____

USING A SCALE TO FIND DISTANCES

By using a **map scale**, you can determine the distance between two places in the real world. The **map scale** shows distances in both miles and kilometers. You will need your ruler to do these exercises. On the scale below, you will notice that 1 inch = 800 miles. To find the distance between Calgary and Ottawa, measure with a ruler between the two cities. You will find it measures about $2\frac{1}{2}$ inches. From the scale, you know 1 inch = 800 miles. Use multiplication to find the distance in miles. $2.5 \times 800 = 2,000$. The cities are about 2,000 miles apart.

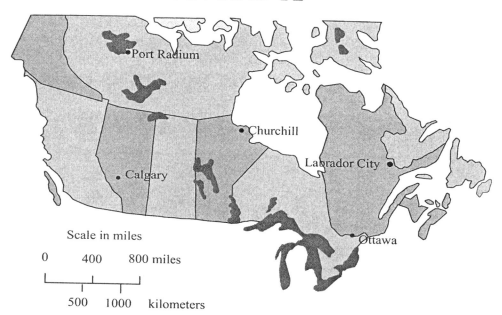

Find these distances in miles.

1. Calgary to Churchill _____
2. Churchill to Ottawa _____
3. Port Radium to Churchill _____
4. Port Radium to Ottawa _____
5. Labrador City to Ottawa _____
6. Calgary to Labrador City _____

Find these distances in kilometers.

7. Churchill to Labrador City _____
8. Ottawa to Port Radium _____
9. Port Radium to Calgary _____
10. Churchill to Ottawa _____
11. Calgary to Churchill _____
12. Calgary to Ottawa _____

USING A SCALE ON A BLUEPRINT

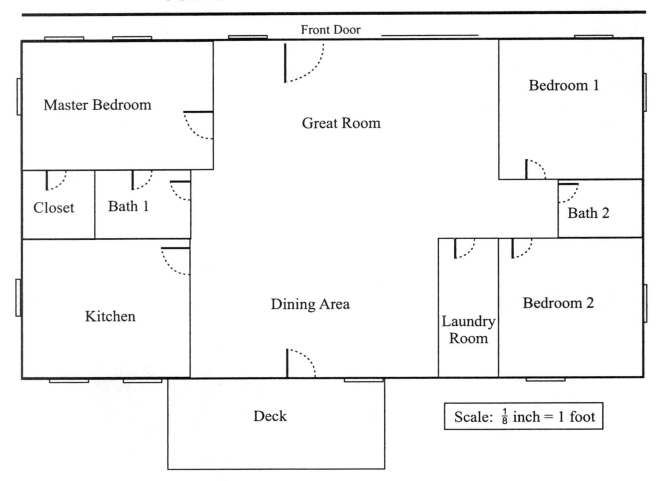

Use a ruler to find the measurements of the rooms on the blueprint above. Convert to feet using the scale. The first problem is done for you.

	long wall		short wall	
	ruler measurement	room measurement	ruler measurement	room measurement
1. Kitchen	$1\frac{3}{4}$ inch	14 feet	$1\frac{1}{2}$ inch	12 feet
2. Deck				
3. Closet				
4. Bedroom 1				
5. Bedroom 2				
6. Master Bedroom				
7. Bath 1				
8. Bath 2				

CHAPTER 21 REVIEW

1. Out of the 100 coins, 45 are in mint condition. What is the ratio of mint condition coins to the total number of coins?

   _____

2. The ratio of boys to girls in the ninth grade is 6:5. If there are 135 girls in the class, how many boys are there?

   _____

3. Twenty out of the total 235 seniors graduated with honors. What is the ratio of seniors graduating with honors to the total number of seniors?

   _____

4. Aunt Bess uses 3 cups of oatmeal to bake 6 dozen oatmeal cookies. How many cups of oatmeal would she need to bake 15 dozen cookies?

   _____

5. On a map, 2 centimeters represents 150 kilometers. If a line between two cities measures 5 centimeters, how many kilometers apart are they?

   _____

6. Shondra used six ounces of chocolate chips to make two dozen cookies. At that rate, how many ounces of chocolate chips would she need to make seven dozen cookies?

   _____

7. When Rick measures the shadow of a yard stick, it is 5 inches. At the same time, the shadow of the tree he would like to chop down is 45 inches. How tall is the tree in yards?

   _____

Solve the following proportions:

8. $\frac{8}{x} = \frac{1}{2}$

9. $\frac{2}{5} = \frac{x}{10}$

10. $\frac{x}{6} = \frac{3}{9}$

11. $\frac{4}{9} = \frac{8}{x}$

12. On a scale drawing of a house floor plan, 1 inch represents 2 feet. The length of the kitchen measures 5 inches on the floor plan. How many feet does that represent?

    _____

13. If 4 inches represents 8 feet on a scale drawing, how many feet does 6 inches represent?

    _____

14. On a scale drawing, 3 centimeters represents 100 miles. If a line segment between two points measured 5 centimeters, how many miles would it represent?

    _____

15. On a map scale, 2 centimeters represents 5 kilometers. If two towns on the map are 20 kilometers apart, how long would the line segment be between the two towns on the map?

    _____

16. If 3 inches represents 10 feet on a scale drawing, how long will a line segment be that represents 15 feet?

    _____

Chapter 22
Plane Geometry

PERIMETER

The **perimeter** is the distance around a polygon. To find the perimeter, add the lengths of the sides.

EXAMPLES:

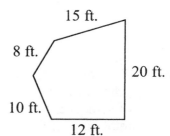

$P = 7 + 15 + 7 + 15$
$P = 44$ in.

$P = 4 + 6 + 5$
$P = 15$ cm

$P = 8 + 15 + 20 + 12 + 10$
$P = 65$ ft.

Find the perimeter of the following polygons.

1. 8 in. by 5 in. rectangle

2. Pentagon: 3 ft., 2 ft., 2 ft., 5 ft., 5 ft.

3.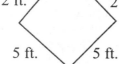
   32 cm, 29 cm, 29 cm, 33 cm, 10 cm, 35 cm

4. Triangle: 13 cm, 15 cm, 10 cm

5. Trapezoid: 12 in. (top), 8 in., 8 in., 16 in. (bottom)

6.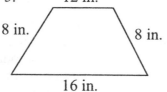
   9 ft., 7 ft., 5 ft., 6 ft.

7. Square: 6 ft. by 6 ft.

8.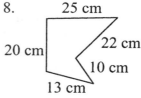
   25 cm, 22 cm, 10 cm, 13 cm, 20 cm

9. Octagon: 4 in. on each of 8 sides

10. Rectangle: 8 cm by 1 cm

11. Triangle: 7 ft., 6 ft., 8 ft.

12.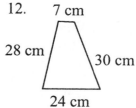
    Trapezoid: 7 cm (top), 28 cm, 30 cm, 24 cm (bottom)

282

AREA OF SQUARES AND RECTANGLES

Area - area is always expressed in square units such as in^2, cm^2, ft^2, and m^2.

The area, (A), of squares and rectangles equals length (l) times width (w). $A = l\,w$

EXAMPLE:

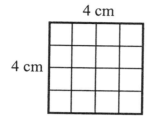

$A = l\,w$
$A = 4 \times 4$
$A = 16 \text{ cm}^2$

If a square has an area of 16 cm^2, it means that it will take 16 squares that are 1 cm on each side to cover the area of a square that is 4 cm on each side.

Find the area of the following squares and rectangles, using the formula $A = l\,w$.

1.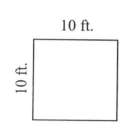

2.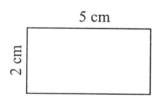

3.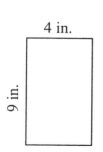

4.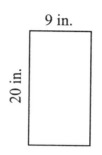

5.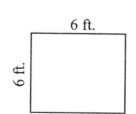

6.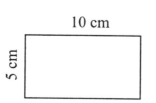

7.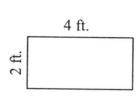

8.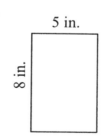

9.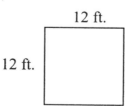

10.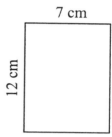

11.

12.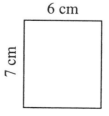

AREA OF TRIANGLES

EXAMPLE: Find the area of the following triangle.

The formula for the area of a triangle is as follows:

$A = \frac{1}{2} \times b \times h$

A = area
b = base
h = height

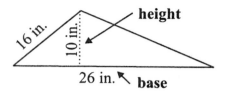

Step 1: Insert measurements from the triangle into the formula: $A = \frac{1}{2} \times 26 \times 10$

Step 2: Cancel and multiply. $A = \frac{1}{\underset{1}{2}} \times \frac{\overset{13}{\cancel{26}}}{1} \times \frac{10}{1} = 130 \text{ in}^2$

Note: Area is always expressed in square units such as in^2, ft^2, cm^2, or m^2.

Find the area of the following triangles. Remember to include units.

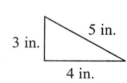

1. _____ in^2

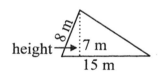

4. _____ cm^2

7. _____ m^2

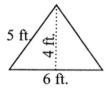

10. _____ ft^2

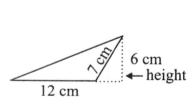

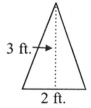

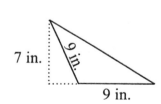

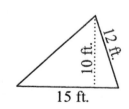

2. _____ cm^2 5. _____ ft^2 8. _____ in^2 11. _____ ft^2

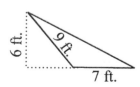

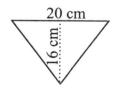

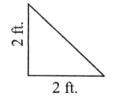

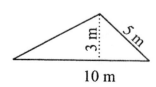

3. _____ ft^2 6. _____ cm^2 9. _____ ft^2 12. _____ m^2

AREA OF TRAPEZOIDS AND PARALLELOGRAMS

EXAMPLE 1: Find the area of the following parallelogram.

The formula for the area of a parallelogram is $A = bh$.

A = area
b = base
h = height

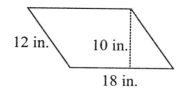

Step 1: Insert measurements from the parallelogram into the formula: $A = 18 \times 10$.

Step 2: Multiply. $18 \times 10 = 180$ in^2

EXAMPLE 2: Find the area of the following trapezoid.

The formula for the area of a trapezoid is $A = \frac{1}{2} h (b_1 + b_2)$. A trapezoid has two bases that are parallel to each other. When you add the length of the two bases together and then multiply by $\frac{1}{2}$, you find their average length.

A = area
b = base
h = height

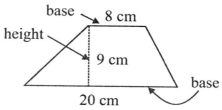

Insert measurements from the trapezoid into the formula and solve:
$\frac{1}{2} \times 9 \, (8 + 20) = 126$ cm^2.

Find the area of the following parallelograms and trapezoids.

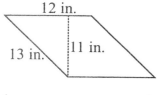

1. _____ in^2

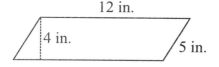

4. _____ cm^2

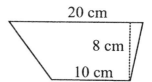

7. _____ in^2

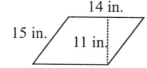

2. _____ in^2

5. _____ in^2

8. _____ cm^2

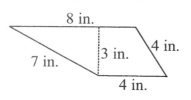

3. _____ in^2

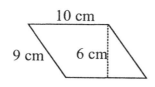

6. _____ cm^2

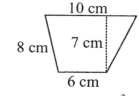

9. _____ cm^2

CIRCUMFERENCE

Circumference, C - the distance around the outside of a circle
Diameter, d - a line segment passing through the center of a circle from one side to the other
Radius, r - a line segment from the center of a circle to the edge of the circle
Pi, π - the ratio of the circumference of a circle to its diameter $\pi = 3.14$ or $\pi = \frac{22}{7}$

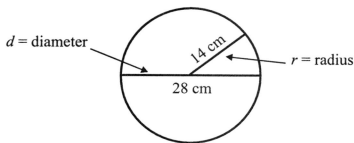

The formula for the circumference of a circle is $C = 2\pi r$ or $C = \pi d$. (The formulas are equal because the diameter is equal to twice the radius, $d = 2r$.)

EXAMPLE:

Find the circumference of the circle above.

$C = \pi d$ Use $= 3.14$
$C = 3.14 \times 28$
$C = 87.92$ cm

EXAMPLE:

Find the circumference of the circle above.

$C = 2\pi r$
$C = 2 \times 3.14 \times 14$
$C = 87.92$ cm

Use the formulas given above to find the circumference of the following circles.
Use $\pi = 3.14$.

1. 8 in.

2. 14 ft.

3. 2 cm

4. 6 m

5. 8 ft.

C = _____   C = _____   C = _____   C = _____   C = _____

Use the formulas given above to find the circumference of the following circles.
Use $\pi = \frac{22}{7}$.

6. 3 ft.

7. 12 in.

8. 6 m

9. 5 cm

10. 16 in.

C = _____   C = _____   C = _____   C = _____   C = _____

AREA OF A CIRCLE

The formula for the area of a circle is $A = \pi r^2$. The area is how many square units of measure would fit inside a circle.

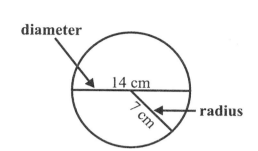

$\pi = \dfrac{22}{7}$ or $\pi = 3.14$

EXAMPLE: Find the area of the circle, using both values for π.

Let $\pi = \dfrac{22}{7}$

$A = \pi r^2$

$A = \dfrac{22}{7} \times 7^2$

$A = \dfrac{22}{7} \times \dfrac{49}{1} = 154 \text{ cm}^2$

Let $\pi = 3.14$

$A = \pi r^2$

$A = 3.14 \times 7^2$

$A = 3.14 \times 49 = 153.86 \text{ cm}^2$

Find the area of the following circles. Remember to include units.

	$\pi = 3.14$	$\pi = \dfrac{22}{7}$
1. 5 in.	$A =$ _____	$A =$ _____
2. 16 ft.	$A =$ _____	$A =$ _____
3. 8 cm	$A =$ _____	$A =$ _____
4. 3 m	$A =$ _____	$A =$ _____

Fill in the chart below. Include appropriate units.

	Radius	Diameter	Area $\pi = 3.14$	Area $\pi = \dfrac{22}{7}$
5.	9 ft.			
6.		4 in.		
7.	8 cm			
8.		20 ft.		
9.	14 m			
10.		18 cm		
11.	12 ft.			
12.		6 in.		

TWO-STEP AREA PROBLEMS

Solving the problems below will require two steps. You will need to find the area of two figures, and then either add or subtract the two areas to find the answer. **Carefully read the EXAMPLES below.**

EXAMPLE 1:

Find the area of the living room below.

Figure 1

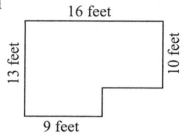

Step 1: Complete the rectangle as in Figure 2, and find the area as if it were a complete rectangle.

Figure 2

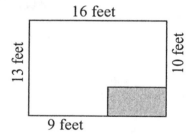

$A = \text{length} \times \text{width}$
$A = 16 \times 13$
$A = 208 \text{ ft}^2$

Step 2: Find the area of the shaded part.

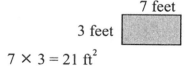

$7 \times 3 = 21 \text{ ft}^2$

Step 3: Subtract the area of the shaded part from the area of the complete rectangle.

$208 - 21 = 187 \text{ ft}^2$

EXAMPLE 2:

Find the area of the shaded sidewalk.

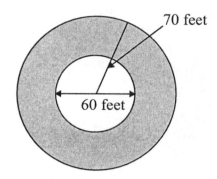

Step 1: Find the area of the outside circle.
$\pi = 3.14$
$A = 3.14 \times 70 \times 70$
$A = 15{,}386 \text{ ft}^2$

Step 2: Find the area of the inside circle.
$\pi = 3.14$
$A = 3.14 \times 30 \times 30$
$A = 2{,}826 \text{ ft}^2$

Step 3: Subtract the area of the inside circle from the area of the outside circle.

$15{,}386 - 2{,}826 = 12{,}560 \text{ ft}^2$

Find the area of the following figures.

1.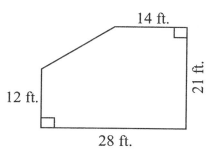

    _____

2.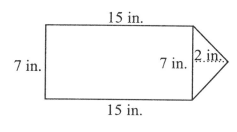

    _____

3. What is the area of the shaded circle? Use π = 3.14, and round the answer to the nearest whole number.

    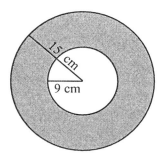

    _____

4. 1 ft.
    5 ft.
    4 ft.
    18 ft.

    _____

5. What is the area of the rectangle that is shaded? Use π = 3.14 and round to the nearest whole number.

    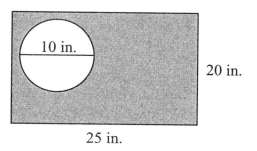

    _____

6. What is the area of the shaded part?

    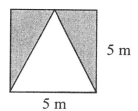

    _____

7. What is the area of the shaded part?

    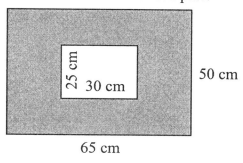

    _____

8. 24 m
    6 m
    12 m
    12 m

    _____

PERIMETER AND AREA WITH ALGEBRAIC EXPRESSIONS

You have already calculated the perimeter and area of various shapes with given measurements. Following the California curriculum for Mathematics, you must understand how to find the perimeter of shapes that are described by algebraic expressions. Study the examples below.

EXAMPLE 1: Using the equation $P = 2l + 2w$, find the perimeter of the following rectangle.

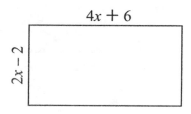

Step 1: Find $2l$ $2(4x + 6) = 8x + 12$
Step 2: Find $2w$ $2(2x - 2) = 4x - 4$
Step 3: Find $2l + 2w$ $12x + 8$

Perimeter $= 12x + 8$

EXAMPLE 2: Using the formula $A = l\,w$, find the area of the rectangle below.

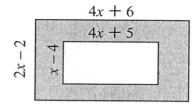

Step 1: $A = (h - 2)(h + 1)$
Step 2: $A = h^2 - 2h + h - 2$
Step 3: $A = h^2 - h - 2$

Area $= h^2 - h - 2$

EXAMPLE 3: Find the area of the shaded part in the following figure.

Step 1: Find the area of the larger rectangle.

$(4x + 6)(2x - 2) = 8x^2 - 8x + 12x - 12 = 8x^2 + 4x - 12$

Step 2: Find the area of the smaller rectangle.

$(4x + 5)(x - 4) = 4x^2 - 16x + 5x - 20 = 4x^2 - 11x - 20$

Step 3: Subtract the area of the smaller rectangle from the area of the larger rectangle.

$$\begin{array}{r} 8x^2 + 4x - 12 \\ -\,(4x^2 - 11x - 20) \end{array} \qquad \begin{array}{r} 8x^2 + 4x - 12 \\ -4x^2 + 11x + 20 \\ \hline 4x^2 + 15x + 8 \end{array}$$

← Changing signs we have
← area of shaded section

Find the perimeter of each of the following rectangles.

1. length $6x - 4$, width $2x + 8$
2. length $4x + 3$, width $4x - 5$
3. length $4x + 3$, width $3x + 2$
4. length $7x - 6$, width $5x + 7$
5. length $6x + 3$, width $4x + 2$
6. length $2x + 4$, width $x - 2$

Find the area of each of the following rectangles.

7. length $5 - 2m$, width $2 + m$
8. length $8 - 4n$, width $5 - n$
9. length $2h - 2$, width $2h - 2$
10. length $9 - h$, width $4 + 2h$
11. length n, width $n + 8$
12. length $7 + 2b$, width $2 + b$

Find the area of the shaded portion of each figure below.

13. Outer rectangle: $8x - 7$ by $6x + 3$; Inner rectangle: $7x + 2$ by $2x + 2$
14. Outer rectangle: $4x - 1$ by $3x + 2$; Inner rectangle: $3x - 4$ by $2x - 1$
15. Outer rectangle: $6x + 3$ by $5x + 4$; Inner rectangle: $3x - 4$ by $5x - 5$
16. Outer rectangle: $11x + 3$ by $10x + 8$; Inner rectangle: $9x - 4$ by $3x + 2$
17. Outer rectangle: $9x - 5$ by $3x + 4$; Inner rectangle: $5x - 7$ by $x + 5$
18. Outer rectangle: $5x - 3$ by $3x + 3$; Inner rectangle: $3x + 5$ by $2x + 7$

ESTIMATING AREA

EXAMPLE: Use the grid to estimate the area of the object.

Step 1: Count the number of whole squares that are shaded. 6

Step 2: Count the number of squares that are at least half shaded. 2

Step 3: Find the sum of the two numbers. $6 + 2 = 8$

The area is about 8 cm².

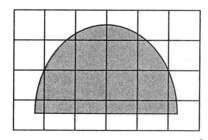

Each square is equal to 1 cm².

Estimate the area of each of the shaded figures below.

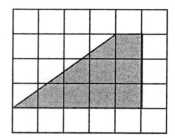 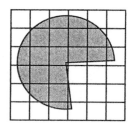 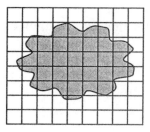

1. Each square is equal to 1 cm².

4. Each square is equal to 1 cm².

7. 1 square = 1 square yard

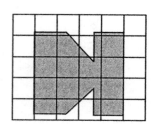 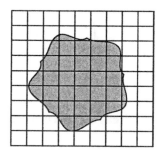 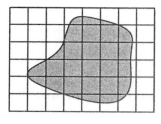

2. Each square is equal to 1 cm².

5. 1 square = 1 square mile

8. 1 square = 1 square yard

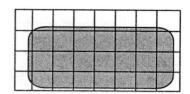 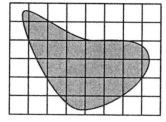 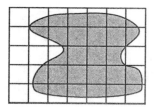

3. Each square is equal to 1 cm².

6. 1 square = 1 square mile

9. 1 square = 1 square yard

GEOMETRIC RELATIONSHIPS OF PLANE FIGURES

The California Mathematics Curriculum requires an understanding of what happens to the area of a figure when one or more of the dimensions is doubled or tripled.

EXAMPLE 1: Sam drew a square that was 2 inches on each side for art class. His teacher said the square needed to be twice as big. When Sam doubled each side to 4 inches, what happened to the area?

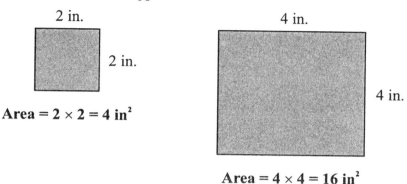

The area of the second square is 4 times larger than the first.

EXAMPLE 2: Sonya drew a circle which had a radius of 3 inches for a school project. She also needed to make a larger circle which had a radius of 9 inches. When Sonya drew the bigger circle, what was the difference in area?

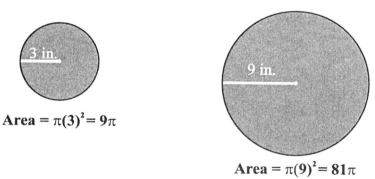

The area of the second circle is 9 times larger than the first.

From these two examples, we can determine that for every doubling or tripling of both sides or of the radius of a planar object, the total area increases by a squared value. In other words, when both sides of the square doubled, the area was 2^2 or 4 times larger. When the radius of the circle became 3 times larger, the area became 3^2 or 9 times larger.

> **Rule:** If a two-dimensional figure is increased equally in both dimensions, the multiple of the increase squared will give you the increase in the area.

Carefully read each problem below and solve.

1. Ken drew a circle with a radius of 5 cm. He then drew a circle with a radius of 10 cm. How many times larger was the area of the second circle?

2. Kobe drew a square with each side measuring 6 inches. He then drew a rectangle with a width of 6 inches and a length of 12 inches. How many times larger is the area of the rectangle than the area of the square? (**Hint:** The increase is *not* equal in both directions.)

3. Toshi drew a square 3 inches on each side. Then he drew a bigger square that was 6 inches on each side. How many times larger is the area of the second square than the area of the first square?

4. Leslie draws a triangle with a base of 5 inches and a height of 3 inches. To use her triangle pattern for a bulletin board design, it needs to be 3 times bigger. If she increases the base and the height by multiplying each by 3, how much will the area of the triangle increase?

5. Heather is using 100 square tiles that measure 1 foot by 1 foot to cover a 10 feet by 10 feet floor. If she had used tiles that measured 2 feet by 2 feet, how many tiles would she have needed?

6. The area of circle B is 9 times larger than the area of circle A. If the radius of circle A is represented by x, how would you represent the radius of circle B?

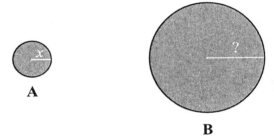

7. How many squares will it take to fill the rectangle below?

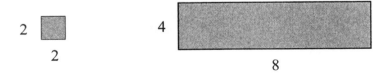

8. If the area of diamond B is one-fourth the area of diamond A, what are the dimensions of diamond B?

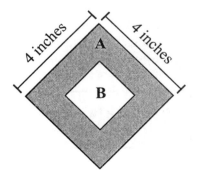

CONGRUENT FIGURES

Two figures are **congruent** when they are exactly the same size and shape. If the corresponding sides and angles of two figures are congruent, then the figures themselves are congruent. For example, look at the two triangles below.

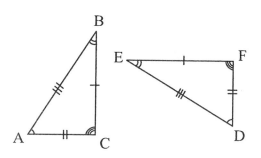

Compare the lengths of the sides of the triangles. The slash marks indicate that $\overline{AB}$ and $\overline{ED}$ have the same length. Therefore, they are congruent, which can be expressed as AB ≅ ED. We can also see that BC ≅ EF and AC ≅ FD. In other words, the corresponding sides are congruent. Now, compare the corresponding angles. The arc markings show that the corresponding angles have the same measure and are, therefore, congruent: ∠A ≅ ∠D, ∠B ≅ ∠E, and ∠C ≅ ∠F. Because the corresponding sides and angles of the triangles are congruent, we say that the triangles are congruent: △ABC ≅ △DEF.

EXAMPLE 1: Decide whether the figures in each pair below are congruent or not.

PAIR 1

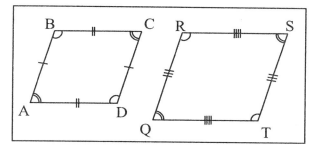

In Pair 1, the two parallelograms have congruent corresponding angles. However, because the corresponding sides of the parallelogram are not the same size, the figures are not congruent.

PAIR 2

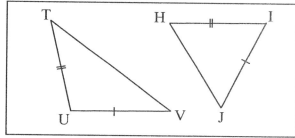

In Pair 2, the two triangles have two corresponding sides which are congruent. However, the hypotenuses of these triangles are not congruent (indicated by the lack of a triple hash mark).

PAIR 3

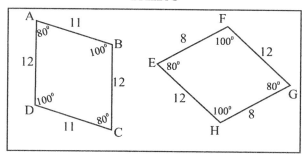

In Pair 3, all of the corresponding angles of these parallelograms are congruent; however, the corresponding sides are not congruent. Therefore, these figures are not congruent.

PAIR 4

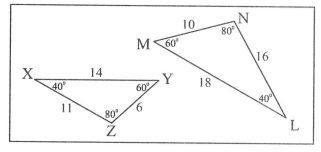

In Pair 4, the triangles share congruent corresponding angles, but the measures for all three corresponding sides of the triangles are not congruent. Therefore, the triangles are not congruent.

Examine the pairs of corresponding figures below. On the first line below the figures, write whether the figures are congruent or not congruent. On the second line, write a brief explanation of how you chose your answer.

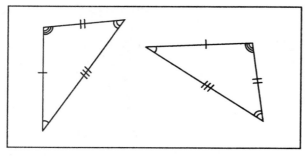

1. _____

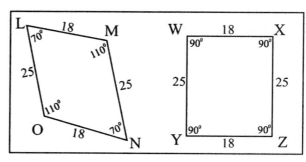

4. _____

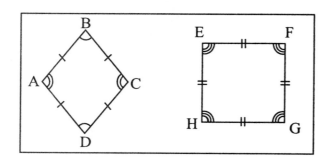

2. _____

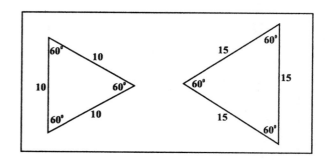

5. _____

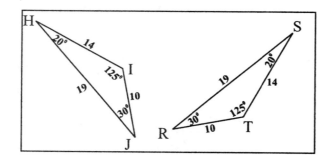

3. _____

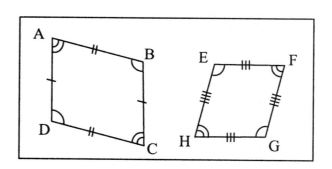

6. _____

SIMILAR TRIANGLES

Two triangles are similar if the measurements of the three angles in both triangles are the same. If the three angles are the same, then their corresponding sides are proportional.

CORRESPONDING SIDES - The triangles below are similar. Therefore, the two shortest sides from each triangle, *c* and *f*, are corresponding. The two longest sides from each triangle, *a* and *d*, are corresponding. The two medium length sides, *b* and *e*, are corresponding.

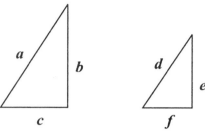

PROPORTIONAL - The corresponding sides of similar triangles are proportional to each other. This means if we know all the measurements of one triangle, and we only know one measurement of the other triangle, we can figure out the measurements of the other two sides with proportion problems. The two triangles below are similar.

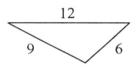 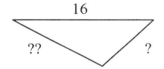

Note: **To set up the proportion correctly, it is important to keep the measurements of each triangle on opposite sides of the equal sign.**

To find the short side:	To find the medium length side:
Step 1: Set up the proportion.	**Step 1:** Set up the proportion.
$\dfrac{long\ side}{short\ side}\quad \dfrac{12}{6}=\dfrac{16}{?}$	$\dfrac{long\ side}{medium}\quad \dfrac{12}{9}=\dfrac{16}{??}$
Step 2: Solve the proportion as you did on the previous page.	**Step 2:** Solve the proportion as you did on the previous page.
$16 \times 6 = 96$ $96 \div 12 = 8$	$16 \times 9 = 144$ $144 \div 12 = 12$

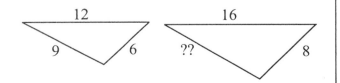 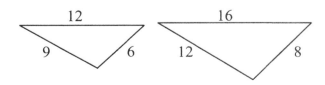

SIMILAR TRIANGLES

Find the missing side from the following similar triangles.

1.
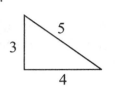 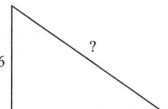

5.
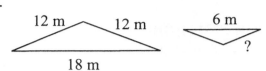

2.
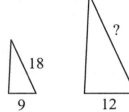

6.
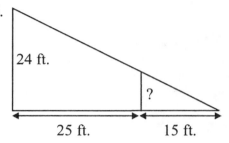

3.
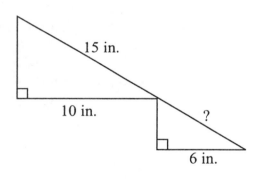

7.
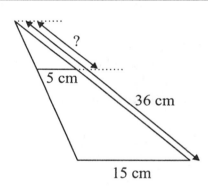

4.
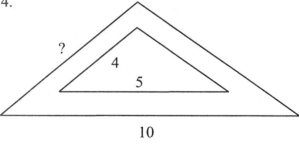

8.
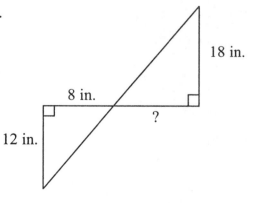

PYTHAGOREAN THEOREM

Pythagoras was a Greek mathematician and philosopher who lived around 600 B.C. He started a math club among Greek aristocrats called the Pythagoreans. Pythagoras formulated the **Pythagorean Theorem** which states that in a **right triangle**, the sum of the squares of the legs of the triangle are equal to the square of the hypotenuse. Most often you will see this formula written as $a^2 + b^2 = c^2$. **This relationship is only true for right triangles.**

EXAMPLE: Find the length of side c.

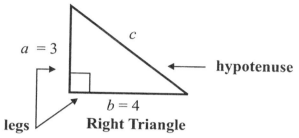

Formula: $a^2 + b^2 = c^2$
$3^2 + 4^2 = c^2$
$9 + 16 = c^2$
$25 = c^2$
$25 = c^2$
$\sqrt{25} = \sqrt{c^2}$
$5 = c$

Find the hypotenuse of the following triangles. Round the answers to two decimal places.

1.

$c = $ _____

4.
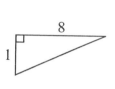
$c = $ _____

7.
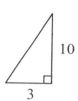
$c = $ _____

2.
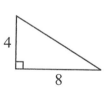
$c = $ _____

5.
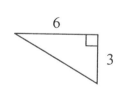
$c = $ _____

8.
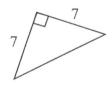
$c = $ _____

3.
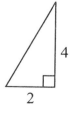
$c = $ _____

6.
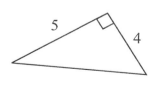
$c = $ _____

9.
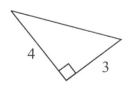
$c = $ _____

FINDING THE MISSING LEG OF A RIGHT TRIANGLE

In the right triangle below, the measurement of the hypotenuse is known as well as one of the legs. To find the measurement of the other leg, use the Pythagorean Theorem by filling in the known measurements, and then solve for the unknown side.

In the formula, $a^2 + b^2 = c^2$, a and b are the legs, and c is always the hypotenuse.
$9^2 + b^2 = 41^2$. Now solve for b algebraically.
$$81 + b^2 = 1681$$
$$b^2 = 1681 - 81$$
$$b^2 = 1600$$
$$\sqrt{b^2} = \sqrt{1600}$$
$$b = 40$$

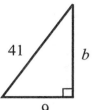

Practice finding the measure of the missing leg in each right triangle below.

1.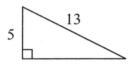

2.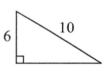

3.

4.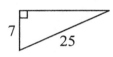

5.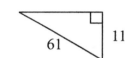

6.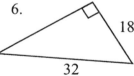

7.

8.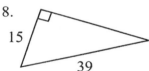

9.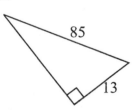

300 Copyright © American Book Company

CHAPTER 22 REVIEW

1. What is the length of line segment $\overline{WY}$?

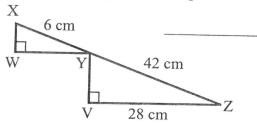

_____

2. Find the missing side.

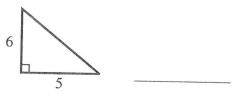

_____

3. Find the area of the shaded region of the figure below.

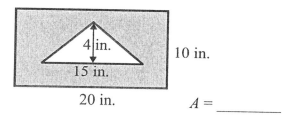

$A =$ _____

4. Calculate the perimeter of the following figure.

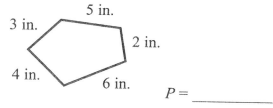

$P =$ _____

Calculate the perimeter and area of the following figures.

5.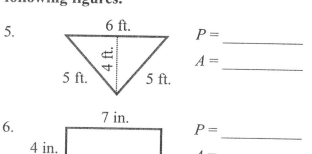

$P =$ _____
$A =$ _____

6.

$P =$ _____
$A =$ _____

Calculate the circumference and the area of the following circles.

7.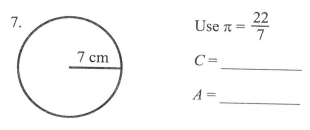

Use $\pi = \frac{22}{7}$

$C =$ _____

$A =$ _____

8.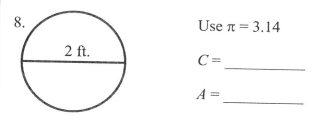

Use $\pi = 3.14$

$C =$ _____

$A =$ _____

9. Use $\pi = 3.14$ to find the area of the shaded part. Round your answer to the nearest whole number.

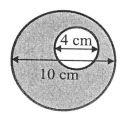

$A =$ _____

10. Tara is using 6-inch squares to make a quilt. The quilt dimensions are 5 feet by 12 feet. How many 6-inch squares will she need to complete the quilt?

_____

11. John has a small frisbee with a diameter of 6 inches. He has a larger frisbee with a diameter of 18 inches. How much larger is the area of the 18 inch frisbee than the 6 inch frisbee?

_____

12. Find the measure of the missing leg of the right triangle below.

    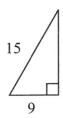

    _____

13. The shaded area below represents Grimes National Park. On the grid below, each square represents 10 square miles. Estimate the area of Grimes National Park.

    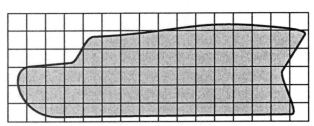

    Area is about _____

14. The following two triangles are similar. Find the length of the missing side.

    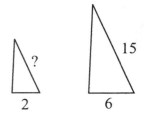

    _____

Find the area of the following figures.

15.

    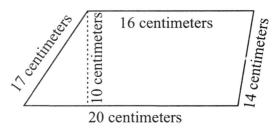

    $A =$ _____

16. How many smaller squares will fit in the larger rectangle? (Figures are **not** drawn to scale.)

    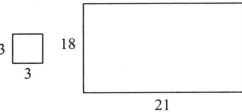

    _____

17.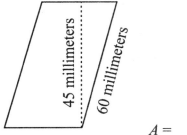

    $A =$ _____

18. Which of the following statements are true about the figures below.

    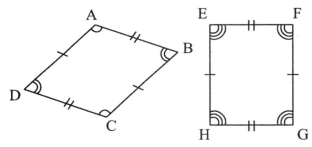

    A. The figures are congruent because the sides are equal.
    B. The figures are congruent because the angles and sides are equal.
    C. The figures are not congruent because the angles are not equal.
    D. The figures are not congruent because the sides are not equal.

19. What is the area of a square which measures 8 inches on each side?

    $A =$ _____

20. If the radius of a circle is doubled, how is the area of the circle affected?

    _____

Chapter 23
Solid Geometry

In this chapter, you will learn about the following three-dimensional shapes.

SOLIDS

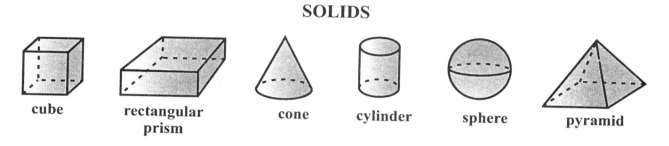

cube   rectangular prism   cone   cylinder   sphere   pyramid

UNDERSTANDING VOLUME

Volume - Measurement of volume is expressed in cubic units such as in^3, ft^3, m^3, cm^3, or mm^3. The volume of a solid is the number of cubic units that can be contained in the solid.

First, let's look at rectangular solids.

EXAMPLE: How many 1 cubic centimeter cubes will it take to fill up the figure below?

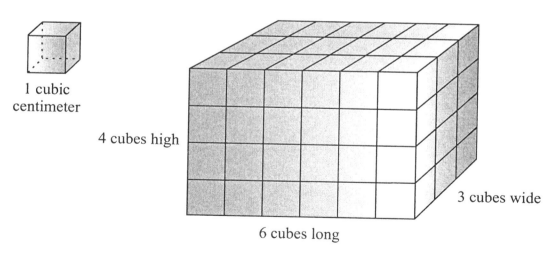

1 cubic centimeter

4 cubes high

3 cubes wide

6 cubes long

To find the volume, you need to multiply the length, times the width, times the height.

Volume of a rectangular solid = length × width × height $(V = l\,w\,h)$.

$V = 6 \times 3 \times 4 = 72\ cm^3$

VOLUME OF RECTANGULAR PRISMS

You can calculate the volume (*V*) of a rectangular prism (box) by multiplying the length (*l*) by the width (w) by the height (*h*), as expressed in the formula $V = (l\ w\ h)$.

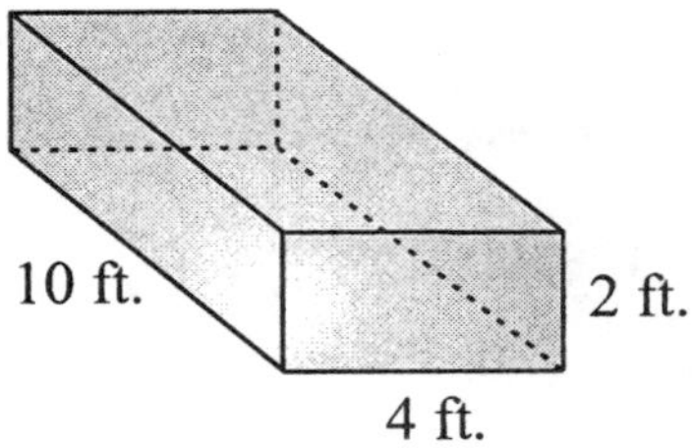
10 ft.
2 ft.
4 ft.

EXAMPLE: Find the volume of the box pictured on the right.

Step 1: Insert measurements from the figure into the formula.

Step 2: Multiply to solve. $10 \times 4 \times 2 = 80\text{ ft}^3$

Note: **Volume is always expressed in cubic units such as in^3, ft^3, m^3, cm^3, or mm^3.**

Find the volume of the following rectangular prisms (boxes).

1.

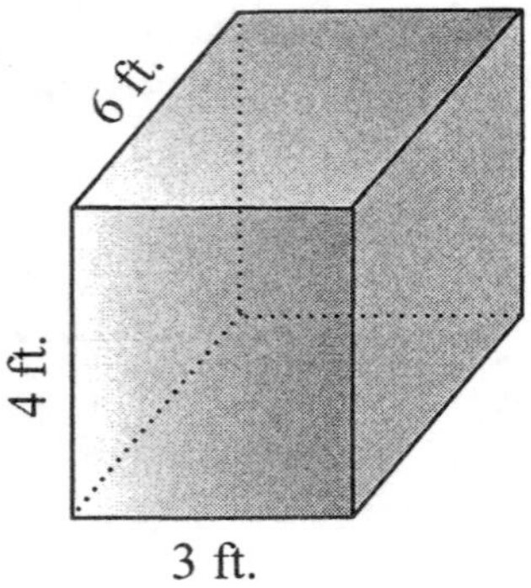
6 ft.
4 ft.
3 ft.

V = ____________

4.

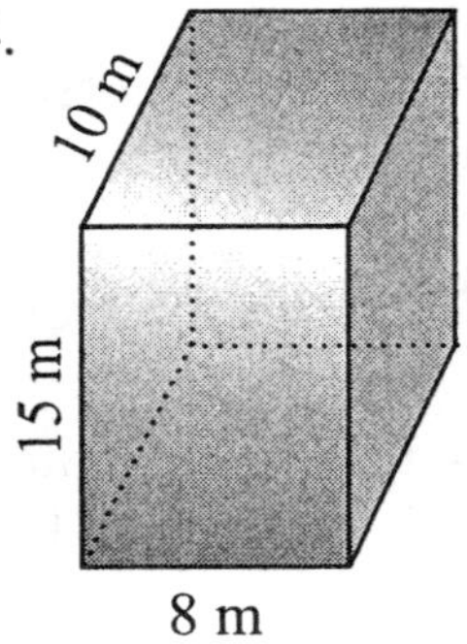
10 m
15 m
8 m

V = ____________

7.

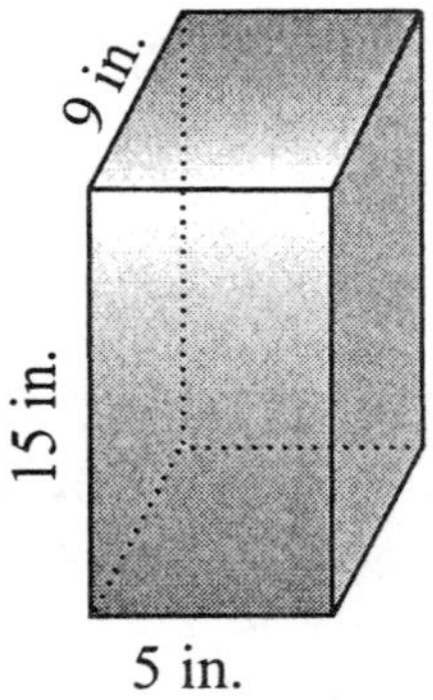
9 in.
15 in.
5 in.

V = ____________

2.

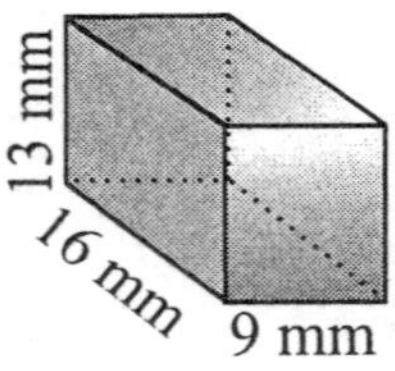
13 mm
16 mm
9 mm

V = ____________

5.

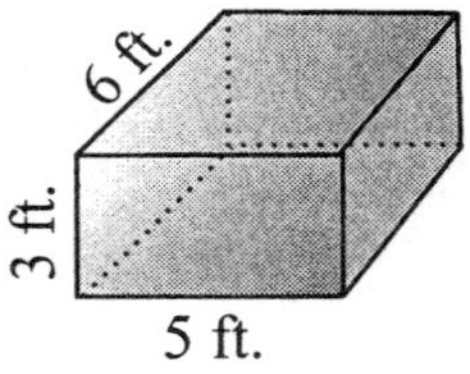
6 ft.
3 ft.
5 ft.

V = ____________

8.

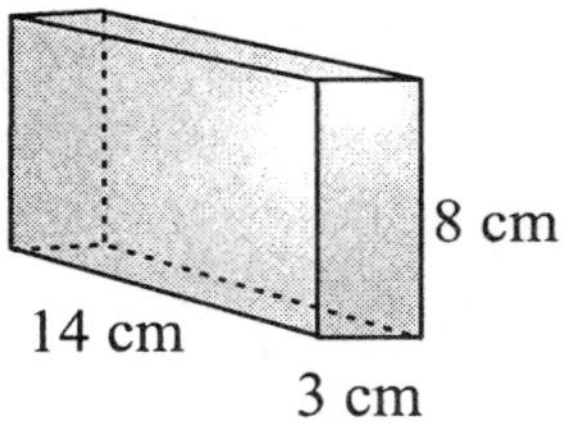
8 cm
14 cm
3 cm

V = ____________

3.

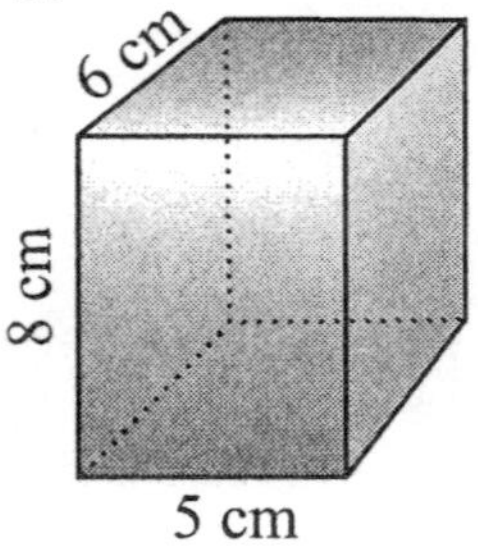
6 cm
8 cm
5 cm

V = ____________

6.

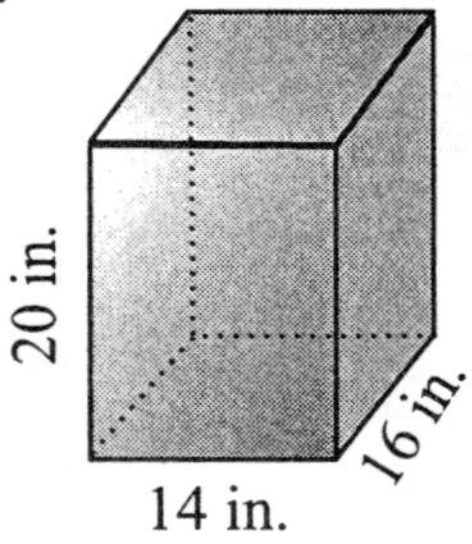
20 in.
16 in.
14 in.

V = ____________

9.

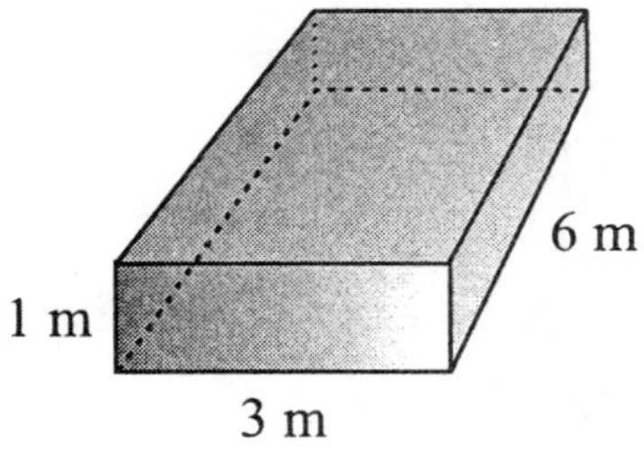
6 m
1 m
3 m

V = ____________

VOLUME OF CUBES

A **cube** is a special kind of rectangular prism (box). Each side of a cube has the same measure. So, the formula for the volume of a cube is $V = s^3$ ($s \times s \times s$).

EXAMPLE: Find the volume of the cube pictured at the right.

Step 1: Insert measurements from the figure into the formula.

Step 2: Multiply to solve. $5 \times 5 \times 5 = 125$ cm^3

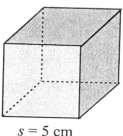

$s = 5$ cm

Note: Volume is always expressed in cubic units such as in^3, ft^3, m^3, cm^3, or mm^3.

Answer each of the following questions about cubes.

1. If a cube is 3 centimeters on each edge, what is the volume of the cube?

    _____

2. If the measure of the edge is doubled to 6 centimeters on each edge, what is the volume of the cube?

    _____

3. What if the edge of a 3 centimeter cube is tripled to become 9 centimeters on each edge? What will the volume be?

    _____

4. How many cubes with edges measuring 3 centimeters would you need to stack together to make a solid 12 centimeter cube?

    _____

5. What is the volume of a 2 centimeter cube?

    _____

6. Jerry built a 2 inch cube to hold his marble collection. He wants to build a cube with a volume 8 times larger. How much will each edge measure?

    _____

Find the volume of the following cubes.

7.

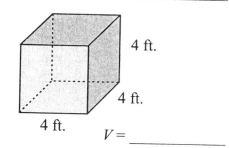

$s = 7$ in. $V =$ _____

8.

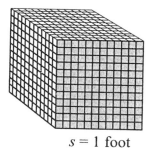

4 ft.   4 ft.   4 ft. $V =$ _____

9.   12 inches = 1 foot

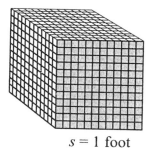

$s = 1$ foot

How many cubic inches are in a cubic foot? _____

VOLUME OF SPHERES, CONES, CYLINDERS, AND PYRAMIDS

To find the volume of a solid, insert the measurements given for the solid into the correct formula and solve. Remember, volumes are expressed in cubic units such as in^3, ft^3, m^3, cm^3, or mm^3.

Sphere
$V = \frac{4}{3}\pi r^3$

Cone
$V = \frac{1}{3}\pi r^2 h$

Cylinder
$V = \pi r^2 h$

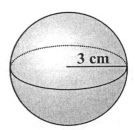
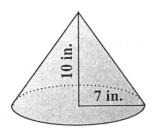
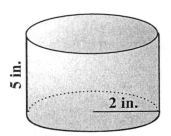

$V = \frac{4}{3}\pi r^3 \quad \pi = 3.14$
$V = \frac{4}{3} \times 3.14 \times 27$
$V = 113.04 \text{ cm}^3$

$V = \frac{1}{3}\pi r^2 h \quad \pi = 3.14$
$V = \frac{1}{3} \times 3.14 \times 49 \times 10$
$V = 512.87 \text{ in}^3$

$V = \pi r^2 h \quad \pi = \frac{22}{7}$
$V = \frac{22}{7} \times 4 \times 5$
$V = 62\frac{6}{7} \text{ in}^3$

Pyramids

$V = \frac{1}{3}Bh \quad B = $ area of rectangular base

$V = \frac{1}{3}Bh \quad B = $ area of triangular base

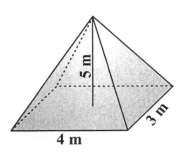
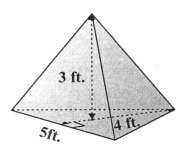

$V = \frac{1}{3}Bh \quad B = l \times w$
$V = \frac{1}{3} \times 4 \times 3 \times 5$
$V = 20 \text{ m}^3$

$B = \frac{1}{2} \times 5 \times 4 = 10 \text{ ft}^2$
$V = \frac{1}{3} \times 10 \times 3$
$V = 10 \text{ ft}^3$

Find the volume of the following shapes. Use π = 3.14.

1. 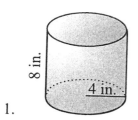 V = _____

2. 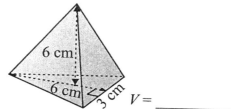 V = _____

3. 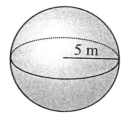 V = _____

4.  V = _____

5. 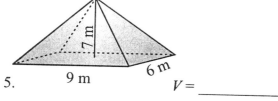 V = _____

6. 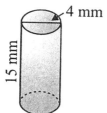 V = _____

7. 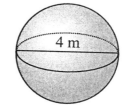 V = _____

8. 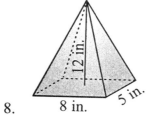 V = _____

9. 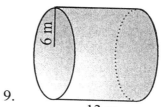 V = _____

10. 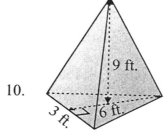 V = _____

11. 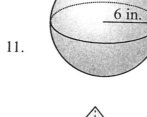 V = _____

12. 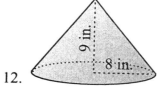 V = _____

Copyright © American Book Company

TWO-STEP VOLUME PROBLEMS

Some objects are made from two geometric figures, for example the tower below.

EXAMPLE: Find the maximum volume of the funnel pictured at the right.

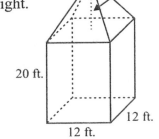

Step 1: Determine which formulas you will need. The tower is made from a pyramid and a prism, so you will need the formulas for the volume of these two figures.

Step 2: Find the volume of each part of the tower.
The bottom of the tower is a rectangular prism. $V = lwh$
$V = 12 \times 12 \times 20 = 2{,}880 \text{ ft}^3$

The top of the tower is a rectangular pyramid. $V = \frac{1}{3}Bh$
$V = \frac{1}{3} \times 12 \times 12 \times 10 = 480 \text{ ft}^3$

Step 3: Add the two volumes together. $2880 \text{ ft}^3 + 480 \text{ ft}^3 = 3{,}360 \text{ ft}^3$

Find the volume of the geometric figures below. *Hint:* **If part of a solid has been removed, find the volume of the hole, and subtract it from the volume of the total object.**

1.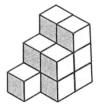

2. Each side of the cubes in the figure below measures 3 inches.

3. A rectangular hole passes through the middle of the figure below. The hole measures 1 cm on each side.

    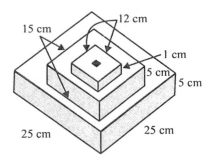

4. In the figure below, 3 cylinders are stacked on top of one another. The radii of the cylinders are 2 inches, 4 inches, and 6 inches. The height of each cylinder is 1 inch.

5.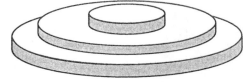

6. A hole, 1 meter in diameter, has been cut through the cylinder below.

    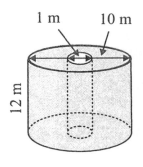

ESTIMATING VOLUME

Use the dimensions given to estimate the volume of the figures below. Use the formula for volume that is most similar to the figure.

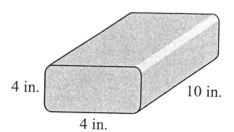

1. Volume is about _____ in³

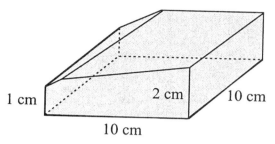

4. Volume is about _____ cm³

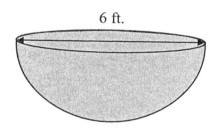

2. Volume is about _____ ft³

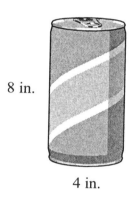

5. Volume is about _____ in³.

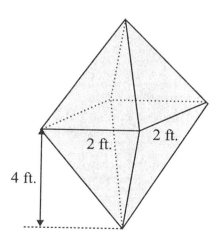

3. Volume is about _____ ft³

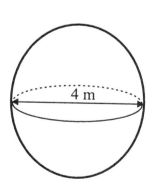

6. Volume is about _____ m³

GEOMETRIC RELATIONSHIPS OF SOLIDS

In the previous chapter, you looked at geometric relationships between 2-dimensional figures. Now you will learn about the relationships between 3-dimensional figures. The formulas for finding the volumes of geometric solids are given below.

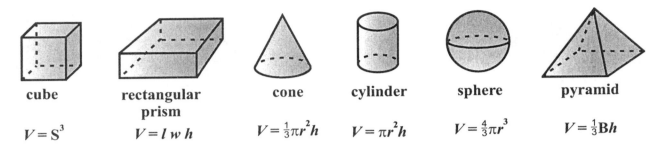

cube	rectangular prism	cone	cylinder	sphere	pyramid
$V = s^3$	$V = lwh$	$V = \frac{1}{3}\pi r^2 h$	$V = \pi r^2 h$	$V = \frac{4}{3}\pi r^3$	$V = \frac{1}{3}Bh$

By studying each formula and by comparing formulas between different solids, you can determine general relationships.

EXAMPLE 1: How would doubling the radius of a sphere affect the volume?

The volume of a sphere is $V = \frac{4}{3}\pi r^3$. Just by looking at the formula, can you see that by doubling the radius, the volume would increase by 2^3 or 8? So, a sphere with a radius of 2 would have a volume 8 times greater than a sphere with a radius of 1.

EXAMPLE 2: A cylinder and a cone have the same radius and the same height. What is the difference between their volumes?

Compare the formulas for the volume of a cone and the volume of a cylinder. They are identical except that the cone is multiplied by $\frac{1}{3}$. Therefore, the volume of a cone with the same height and radius as a cylinder would be one-third less. Or, the volume of a cylinder with the same height and radius as a cone would be three times greater.

EXAMPLE 3: If you double one dimension of a rectangular prism, how will the volume be affected? How about doubling two dimensions? How about doubling all three dimensions?

Do you see that doubling just one of the dimensions of a rectangular prism will also double the volume? Doubling two of the dimensions will cause the volume to increase by 2^2 or 4. Doubling all three dimensions will cause the volume to increase by 2^3 or 8.

EXAMPLE 4: A cylinder holds 100 cubic centimeters of water. If you triple the radius of the cylinder but keep the height the same, how much water would you need to fill the new cylinder?

Tripling the radius of a cylinder causes the volume to increase by 3^2 or 27. The volume of the new cylinder would hold 27×100 or 2,700 cubic centimeters of water.

Answer the following questions by comparing the volumes of two solids that share some of the same dimensions.

1. If you have a cylinder with a height of 8 inches and a radius of 4 inches, and you have a cone with the same height and radius, how many times greater is the volume of the cylinder than the volume of the cone?

2.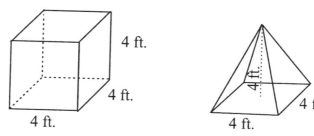

   In the two figures above, how many times larger is the volume of the cube than the volume of the pyramid?

3. How many times greater is the volume of a cylinder if you double the radius?

4. How many times greater is the volume of a cylinder if you double the height?

5. In a rectangular solid, how many times greater is the volume if you double the length?

6. In a rectangular solid, how many times greater is the volume if you double the length and the width?

7. In a rectangular solid, how many times greater is the volume if you double the length and the width and the height?

8. In the following two figures, how many cubes like Figure 1 will fit inside Figure 2?

   Figure 1 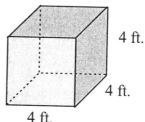 Figure 2

9. A sphere has a radius of 1. If the radius is increased to 3, how many times greater will the volume be?

10. It takes 2 liters of water to fill cone A below. If the cone is stretched so the radius is doubled, but the height stays the same, how much water is needed to fill the new cone, B?

    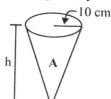 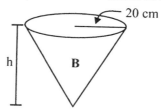

SURFACE AREA

The **surface area of a solid** is the total area of all the sides of a solid.

CUBE

There are six sides on a cube. To find the surface area of a cube, find the area of one side and multiply by 6.

Area of each side of the cube:
$3 \times 3 = 9$ cm^2

Total surface area: $9 \times 6 = 54$ cm^2

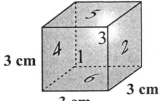

RECTANGULAR PRISM

There are 6 sides on a rectangular prism. To find the surface area, add the areas of the six rectangular sides.

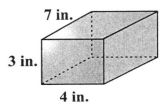

Top and Bottom

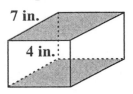

Front and Back

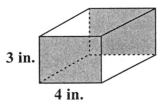

Left and Right

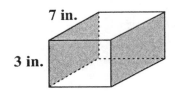

Area of top side:
7 in. $\times$ 4 in. = 28 in^2
Area of top and bottom:
28 in. $\times$ 2 in. = 56 in^2

Area of front:
3 in. $\times$ 4 in. = 12 in^2
Area of front and back:
12 in. $\times$ 2 in. = 24 in^2

Area of left side:
3 in. $\times$ 7 in. = 21 in^2
Area of left and right:
21 in. $\times$ 2 in. = 42 in^2

Total surface area: 56 in^2 + 24 in^2 + 42 in^2 = 122 in^2

Find the surface area of the following cubes and prisms.

1.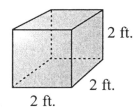

   SA = _____

2.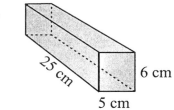

   SA = _____

3.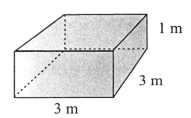

   SA = _____

4.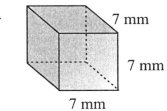

   SA = _____

5.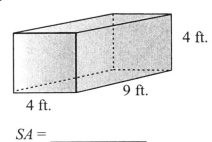

   SA = _____

6.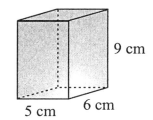

   SA = _____

7.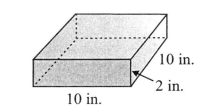

   SA = _____

8.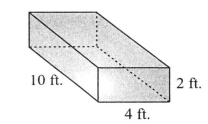

   SA = _____

9.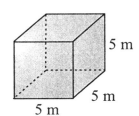

   SA = _____

10.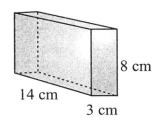

    SA = _____

PYRAMID

The pyramid below is made of a square base with 4 triangles on the sides.

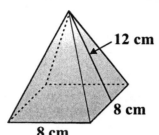

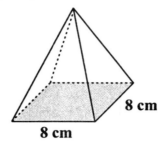

Area of square base:
$A = l \times w$
$A = 8 \times 8 = 64$ cm^2

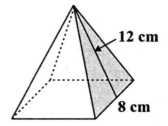

Area of sides:
Area of 1 side = $\frac{1}{2}bh$
$A = \frac{1}{2} \times 8 \times 12 = 48$ cm^2
Area of 4 sides = $48 \times 4 = 192$ cm^2

Total surface area: $64 + 192 = 256$ cm^2

Find the surface area of the following pyramids.

1. ![pyramid] 3 ft., 2 ft., 2 ft.

SA = _____

2. ![pyramid] 12 mm, 6 mm, 6 mm

SA = _____

3. ![pyramid] 15 m, 10 m, 10 m

SA = _____

4. 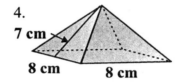 7 cm, 8 cm, 8 cm

SA =

5. 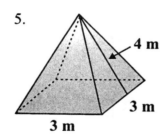 4 m, 3 m, 3 m

SA = _____

6. 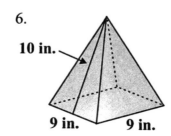 10 in., 9 in., 9 in.

SA = _____

7. ![pyramid] 9 m, 4 m, 4 m

SA = _____

8. 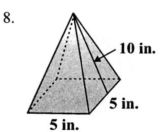 10 in., 5 in., 5 in.

SA = _____

9. 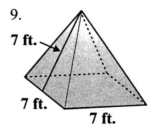 7 ft., 7 ft., 7 ft.

SA = _____

CYLINDER

If the side of a cylinder were slit from top to bottom and laid flat, its shape would be a rectangle. The length of the rectangle is the same as the circumference of the circle that is the base of the cylinder. The width of the rectangle is the height of the cylinder.

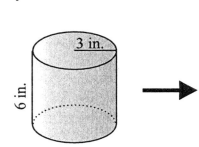 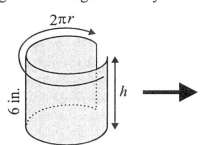 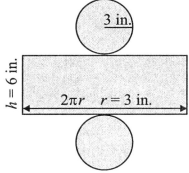

Total Surface Area of a Cylinder = $2\pi r^2 + 2\pi rh$

Area of top and bottom:
Area of a circle = πr^2
Area of top = $3.14 \times 3^2 = 28.26$ in.^{2}
Area of top and bottom = $2 \times 28.26 = 56.52$ in.^{2}

Area of side:
Area of rectangle = $l \times h$
$l = 2\pi r = 2 \times 3.14 \times 3 = 18.84$ in.
Area of rectangle = $18.84 \times 6 = 113.04$ in.^{2}

Total surface area = $56.52 + 113.04 = 169.56$ in^2

Find the surface area of the following cylinders. Use $\pi = 3.14$

1.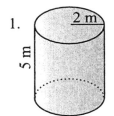

SA = _____

4.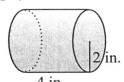

SA = _____

7.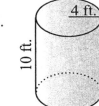

SA = _____

2.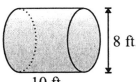

SA = _____

5.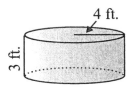

SA = _____

8.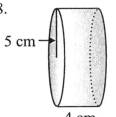

SA = _____

3.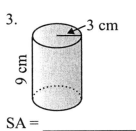

SA = _____

6.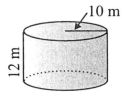

SA = _____

9.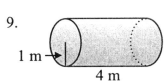

SA = _____

SOLID GEOMETRY WORD PROBLEMS

1. If an Egyptian pyramid has a square base that measures 500 yards by 500 yards, and the pyramid stands 300 yards tall, what would be the volume of the pyramid? Use the formula for volume of a pyramid, $V=\frac{1}{3}Bh$ where B is the area of the base.

   V = _____

2. Robert is using a cylindrical barrel filled with water to flatten the sod in his yard. The circular ends have a radius of 1 foot. The barrel is 3 feet wide. How much water will the barrel hold? The formula for volume of a cylinder is $V = \pi r^2 h$. Use $\pi = 3.14$.

   V = _____

3. If a basketball measures 24 centimeters in diameter, what volume of air will it hold? The formula for volume of a sphere is $V = \frac{4}{3}\pi r^3$. Use $\pi = 3.14$.

   V = _____

4. What is the volume of a cone that is 2 inches in diameter and 5 inches tall? The formula for volume of a cone is $V = \frac{1}{3}\pi r^2 h$. Use $\pi = 3.14$.

   V = _____

5. Kelly has a rectangular fish aquarium that measures 24 inches wide, 12 inches deep, and 18 inches tall. What is the maximum amount of water that the aquarium will hold?

   V = _____

6. Jenny has a rectangular box that she wants to cover in decorative contact paper. The box is 10 cm long, 5 cm wide, and 5 cm high. How much paper will she need to cover all 6 sides?

   SA = _____

7. Gasco needs to construct a cylindrical, steel gas tank that measures 6 feet in diameter and is 8 feet long. How many square feet of steel will be needed to construct the tank? Use the following formulas as needed: $A = l \times w$, $A = \pi r^2$, $C = 2\pi r$. Use $\pi = 3.14$.

   SA = _____

8. Craig wants to build a miniature replica of San Francisco's Transamerica Pyramid out of glass. His replica will have a square base that measures 6 cm by 6 cm. The 4 triangular sides will be 6 cm wide and 60 cm tall. How many square centimeters of glass will he need to build his replica? Use the following formulas as needed: $A = l \times w$ and $A = \frac{1}{2}bh$.

   SA = _____

9. Jeff built a wooden, cubic toy box for his son. Each side of the box measures 2 feet. How many square feet of wood did he use to build the toy box? How many cubic feet of toys will the box hold?

   SA = _____

   V = _____

CHAPTER 23 REVIEW

Find the volume and/or the surface area of the following solids.

1.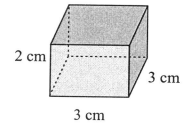

   $V =$ _____

   $SA =$ _____

2.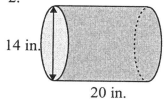

   $V = \pi r^2 h$
   $SA = 2\pi r^2 + 2\pi rh$

   Use $\pi = \frac{22}{7}$

   $V =$ _____

   $SA =$ _____

3.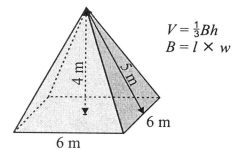

   $V = \frac{1}{3}Bh$
   $B = l \times w$

   $V =$ _____

   $SA =$ _____

4.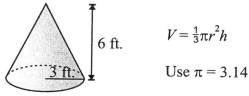

   $V = \frac{1}{3}\pi r^2 h$

   Use $\pi = 3.14$

   $V =$ _____

5.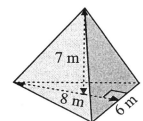

   $V = \frac{1}{3}Bh$
   $B =$ area of the triangular base

   $V =$ _____

6.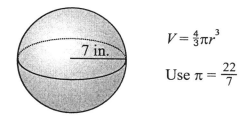

   $V = \frac{4}{3}\pi r^3$

   Use $\pi = \frac{22}{7}$

   $V =$ _____

7. The sandbox at the local elementary school is 60 inches wide and 100 inches long. The sand in the box is 6 inches deep. How many cubic inches of sand are in the sandbox?

   _____

8. If you have cubes that are two inches on each edge, how many would fit in a cube that was 16 inches on each edge?

9. If you double each edge of a cube, how many times larger is the volume?

10. It takes 8 cubic inches of water to fill the cube below. If each side of the cube is doubled, how much water is needed to fill the new cube?

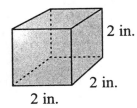

11. If a ball is 4 inches in diameter, what is its volume? Use $\pi = 3.14$

12. A grain silo is in the shape of a cylinder. If a silo has an inside diameter of 10 feet and a height of 35 feet, what is the maximum volume inside the silo?

Use $\pi = \frac{22}{7}$

13. A closed cardboard box is 30 centimeters long, 10 centimeters wide, and 20 centimeters high. What is the total surface area of the box?

14. Siena wants to build a wooden toy box with a lid. The dimensions of the toy box are 3 feet long, 4 feet wide, and 2 feet tall. How many square feet of wood will she need to construct the box?

15. How many 1-inch cubes will fit inside a larger 1 foot cube? (Figures are not drawn to scale.)

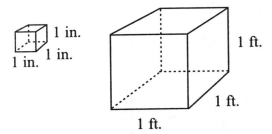

16. The cylinder below has a volume of 240 cubic inches. The cone below has the same radius and the same height as the cylinder. What is the volume of the cone?

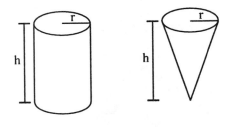

17. Estimate the volume of the figure below.

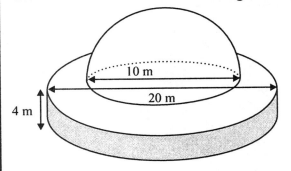

18. Estimate the volume of the figure below.

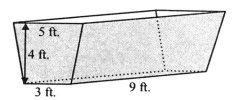

_____

19. Find the volume of the figure below.

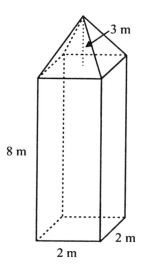

_____

20. Find the volume of the figure below. Each side of each cube measures 4 feet.

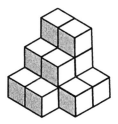

_____

Chapter 24
Reflections, Translations, and Plotted Shapes

REFLECTIONS

A **reflection** of a geometric figure is a mirror image of the object. Placing a mirror on the **line of reflection** will give you the position of the reflected image. On paper, folding an image across the line of reflection will give you the position of the reflected image.

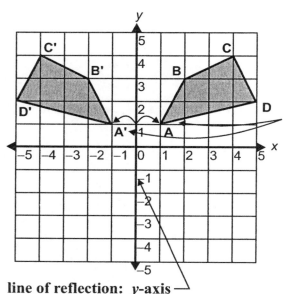

line of reflection: y-axis

Quadrilateral ABCD is reflected across the y-axis to form quadrilateral A'B'C'D'. The y-axis is the line of reflection. Point A' (read as A prime) is the reflection of point A, point B' corresponds to point B, C' to C, and D' to D.

Point A is +1 space from the y-axis. Point A's mirror image, point A', is −1 space from the y-axis.

Point B is +2 spaces from the y-axis. Point B' is −2 spaces from the y-axis.

Point C is +4 spaces from the y-axis and point C' is −4 spaces from the y-axis.

Point D is +5 spaces from the y-axis and point D' is −5 spaces from the y-axis.

Triangle FGH is reflected across the x-axis to form triangle F'G'H'. The x-axis is the line of reflection. Point F' reflects point F. Point G' corresponds to point G, and H' mirrors H.

Point F is +3 spaces from the x-axis. Likewise, point F' is −3 spaces from the x-axis.

Point G is +1 space from the x-axis, and point G' is −1 space from the x-axis.

Point H is 0 spaces from the x-axis, so point H' is also 0 spaces from the x-axis.

line of reflection: x-axis

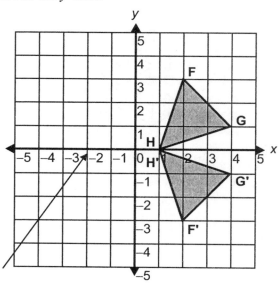

Reflecting Across a 45° Line

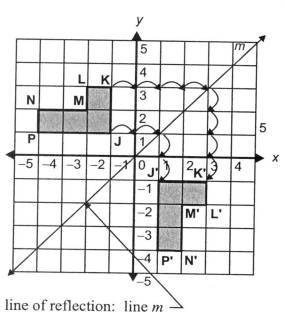

line of reflection: line *m*

Figure JKLMNP is reflected across line *m* to form figure J'K'L'M'N'P'. Line *m* is at a 45° angle. Point J corresponds to J', K to K', L to L', M to M', N to N' and P to P'. Line *m* is the line of reflection. **Pay close attention to how to determine the mirror image of figure JKLMNP across line *m* described below. This method only works when the line of reflection is at a 45° angle.**

Point J is 2 spaces over from line *m*, so J' must be 2 spaces down from line *m*.

Point K is 4 spaces over from line *m*, so K' is 4 spaces down from line *m*, and so on.

Draw the following reflections, and record the new coordinates of the reflection. The first problem is done for you.

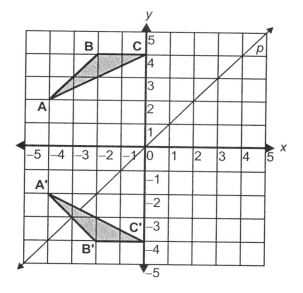

1. Reflect figure ABC across the *x*-axis. Label vertices A'B'C' so that point A' is the reflection of point A, B' is the reflection of B, and C' is the reflection of C.

    A' = **(–4, –2)**   B' = **(–2, –4)**   C' = **(0, –4)**

2. Reflect figure ABC across the *y*-axis. Label vertices A"B"C" so that point A" is the reflection of point A, B" is the reflection of B, and C" is the reflection of C.

    A" = _____   B" = _____   C" = _____

3. Reflect figure ABC across line *p*. Label vertices A'''B'''C''' so that point A''' is the reflection of point A, B''' is the reflection of B, and C''' is the reflection of C.

    A''' = _____   B''' = _____   C''' = _____

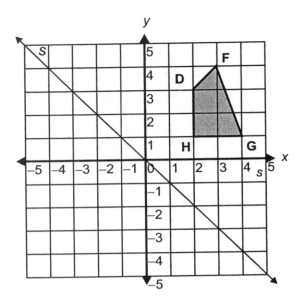

4. Reflect figure DFGH across the y-axis. Label vertices D'F'G'H' so that point D' is the reflection of point D, F' is the reflection of F, G' is the reflection of G, and H' is the reflection of H.

   D' = _____   G' = _____

   F' = _____   H' = _____

5. Reflect figure DFGH across the x-axis. Label vertices D", F", G", H" so that point D" is the reflection of D, F" is the reflection of F, G" is the reflection of G, and H" is the reflection of H.

   D" = _____   G" = _____

   F" = _____   H" = _____

6. Reflect figure DFGH across line s. Label vertices D'''F'''G'''H''' so that point D''' is the reflection of D, F''' corresponds to F, G''' to G, and H''' to H.

   D''' = _____   G''' = _____

   F''' = _____   H''' = _____

---

7. Reflect quadrilateral MNOP across the y-axis. Label vertices M'N'O'P' so that point M' is the reflection of point M, N' is the reflection of N, O' is the reflection of O, and P' is the reflection of P.

   M' = _____   O' = _____

   N' = _____   P' = _____

8. Reflect figure MNOP across the x-axis. Label vertices M", N", O", P" so that point M" is the reflection of M, N" is the reflection of N, O" is the reflection of O and P" is the reflection of P.

   M" = _____   O" = _____

   N" = _____   P" = _____

9. Reflect figure MNOP across line w. Label vertices M'''N'''O'''P''' so that point M''' is the reflection of M, N''' corresponds to N, O''' to O, and P''' to P.

   M''' = _____   O''' = _____

   N''' = _____   P''' = _____

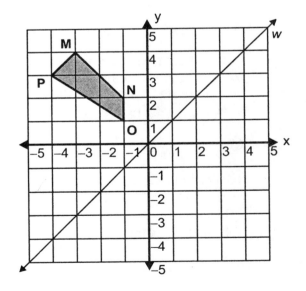

TRANSLATIONS

To make a **translation** of a geometric figure, first duplicate the figure. Then slide it along a path.

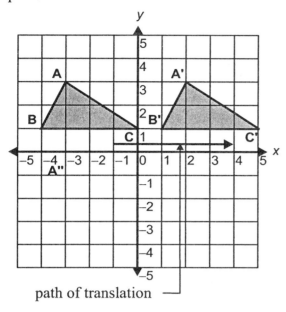

path of translation

Triangle A'B'C' is a translation of triangle ABC. Each point is translated 5 spaces to the right. In other words, the triangle slid 5 spaces to the right. Look at the path of translation. It gives the same information as above. Count the number of spaces across given by the path of translation, and you will see it represents a move 5 spaces to the right. Each new point is found at $(x + 5, y)$.

Point A is at $(-3, 3)$. Therefore, A' is found at $(-3 + 5, 3)$ or $(2, 3)$.

B is at $(-4, 1)$, so B' is at $(-4 + 5, 1)$ or $(1, 1)$.

C is at $(0, 1)$, so C' is at $(0 + 5, 1)$ or $(5, 1)$.

Quadrilateral FGHI is translated 5 spaces to the right and 3 spaces down. The path of translation shows the same information. It points right 5 spaces and down 3 spaces. Each new point is found at $(x + 5, y - 3)$.

Point F is located at $(-4, 3)$. Point F' is located at $(-4 + 5, 3 - 3)$ or $(1, 0)$.

Point G is at $(-2, 5)$. Point G' is at $(-2 + 5, 5 - 3)$ or $(3, 2)$.

Point H is at $(-1, 4)$. Point H' is at $(-1 + 5, 4 - 3)$ or $(4, 1)$.

Point I is at $(-1, 2)$. Point I' is at $(-1 + 5, 2 - 3)$ or $(4, -1)$.

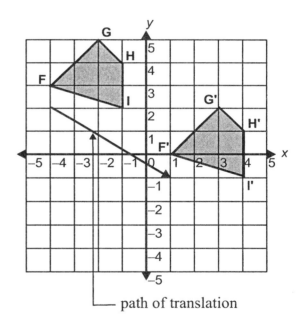

path of translation

Draw the following translations, and record the new coordinates of the translation. The figure for the first problem is drawn for you.

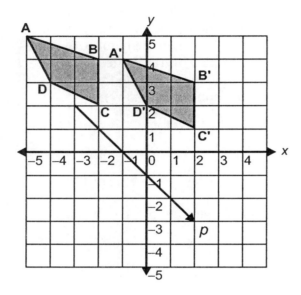

1. Translate figure ABCD 4 spaces to the right and 1 space down. Label the vertices of the translated figure A', B', C', and D' so that point A' corresponds to the translation of point A, B' corresponds to B, C' to C, and D' to D.

   A' = _____ C' = _____

   B' = _____ D' = _____

2. Translate figure ABCD 5 spaces down. Label the vertices of the translated figure A", B", C", and D" so that point A" corresponds to the translation of point A, B" corresponds to B, C" to C, and D" to D.

   A" = _____ C" = _____

   B" = _____ D" = _____

3. Translate figure ABCD along the path of translation, p. Label the vertices of the translated figure A''', B''', C''', and D''' so that point A''' corresponds to the translation of point A, B''' corresponds to B, C''' to C, and D''' to D.

   A''' = _____ C''' = _____

   B''' = _____ D''' = _____

---

4. Translate triangle FGH 6 spaces to the left and 3 spaces up. Label the vertices of the translated figure F', G', and H' so that point F' corresponds to the translation of point F, G' corresponds to G, and H' to H.

   F' = _____ G' = _____ H' = _____

5. Translate triangle FGH 4 spaces up and 1 space to the left. Label the vertices of the translated triangle F"G"H" so that point F" corresponds to the translation of point F, G" corresponds to G, and H" to H.

   F" = _____ G" = _____ H" = _____

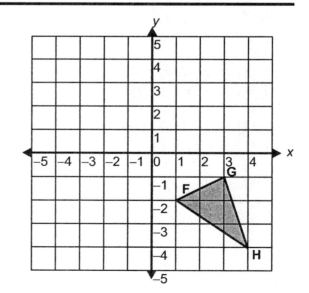

FINDING LENGTHS OF PLOTTED SHAPES ON CARTESIAN PLANES

Once pairs of values are labeled as points and lines drawn to connect these points, shapes are drawn. Use the Cartesian number lines to find the lengths of lines, perimeters of shapes, and area. On some problems, you may be given only the data points and asked to draw lines connecting the points. In this case, you must first plot the points. Then draw the lines on a separate sheet of paper.

EXAMPLE 1: What are the lengths of the rectangle ABCD?
Data points: (−4, 2), (3, 2) (3, −2) (−4,−2)
Examine the plotted rectangle to your right. You can find the lengths of each side by counting the number of blocks used. For instance, line segment $\overline{AB}$ on the Cartesian plane is 7 blocks in length. Line segment $\overline{BC}$ is 4 blocks in length, and so on.

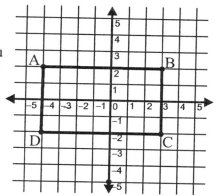

EXAMPLE 2: What is the length of the hypotenuse of triangle ABC?
Data points: (−4, 4) (−4,−2) (4,−2)
Examine the plotted triangle to your right. First, find the length of the two legs. By counting the blocks, you will find side $\overline{AB}$ has a length of 6 and side $\overline{BC}$ has a length of 8. Next, use the Pythagorean Theorem to find the length of the diagonal side, $\overline{AC}$. $6^2 + 8^2 = 100$.
$\sqrt{100} = 10$ Therefore, the length of the hypotenuse, side $\overline{AC}$, = 10.

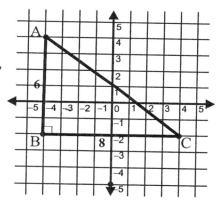

EXAMPLE 3: What is the radius of the plotted circle?
Finding the radius and the diameter of a circle on a Cartesian plane is simple. First, examine the circle. Next, draw an imaginary line through the center of the circle. Third, count the number of blocks which span the length of the line segment to find the diameter. To find the radius, simply divide this number by 2. In this example, the diameter = 5, so the radius = 2.5.

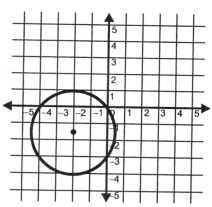

Read each of the following graphs carefully. Using the methods taught on the previous page, find the length of each side and the area of each shape. (You may want to refer back to Chapter 22 for methods of finding area.)

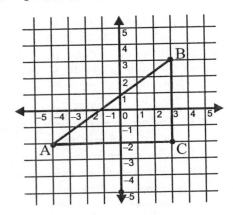
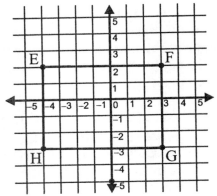

1. $\overline{AB}$ = _____
2. $\overline{BC}$ = _____
3. $\overline{CA}$ = _____
4. Area = _____

7. $\overline{EF}$ = _____
8. $\overline{FG}$ = _____
9. Area = _____

10. $\overline{GH}$ = _____
11. $\overline{HE}$ = _____

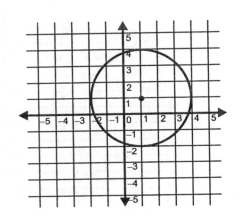
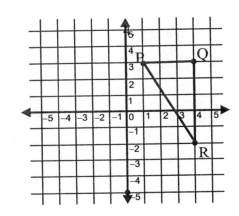

5. radius = _____
6. Area = _____

12. $\overline{PQ}$ = _____
13. $\overline{QR}$ = _____
14. $\overline{RP}$ = _____
15. Area = _____

For questions 16-21, plot the points given. Next, calculate the perimeter and the area of each plotted figure.

16. (2,1) (-3,1) (-3,4)
17. (-3,4) (2,4) (-3,-2) (2,-2)
18. (3,3) (3,-1) (-2,-1)

19. (0,0) (0, -3) (0,4) (-3,4)
20. (4,4) (-2,4) (-2, -3) (4,-3)
21. Circle: center at point (2,1). Point (2,-1) on circle. Find radius and area.

CHAPTER 24 REVIEW

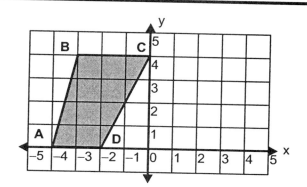

1. Draw the reflection of image ABCD over the y-axis. Label the points A', B', C', and D'. List the coordinates of these points below.

2. A' _____   4. C' _____

3. B' _____   5. D' _____

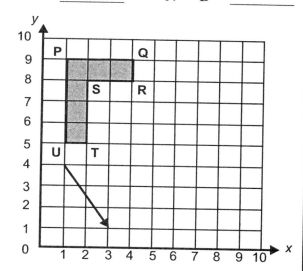

6. Use the translation described by the arrow to translate the polygon above. Label the points P', Q', R', S', T', and U'. List the coordinates of each.

7. P' _____   10. S' _____

8. Q' _____   11. T' _____

9. R' _____   12. U' _____

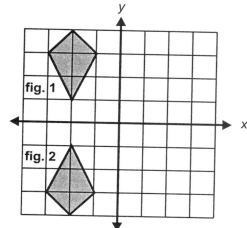

13. Figure 1 goes through a transformation to form figure 2. Which of the following descriptions fits the transformation shown?

    A. reflection across the x-axis
    B. reflection across the y-axis
    C. translation down 5 units
    D. translation down 2 units

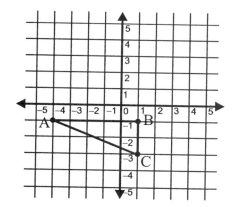

14. What is the length of $\overline{AC}$?

15. What is the perimeter of $\triangle ABC$?

16. What is the area of $\triangle ABC$?

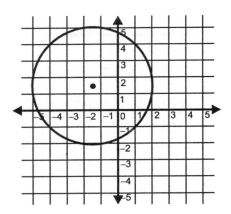

17. What is the length of the radius of the circle above?

18. What is the area of the circle?

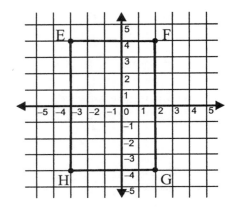

19. What is the length of $\overline{EF}$?
20. What is the length of $\overline{FG}$?
21. What is the perimeter of the above figure?
22. What is the area of the above figure?

Given the following points, find the perimeter and area of the following figures.

23. (−3, 4) (−1,0) (−1,4) (−3,0)
24. (4,1) (0,4) (0,1)
25. (2,3) (−2,−2) (2,−2) (−2,3)
26. (4,−3) (−2,−3) (−2,1)
27. (0,3) (−3,−4) (−3,3) (0,−4)
28. (−4,0) (4,4) (4,0)

California Math Review Progress Test 1

1. Lisa's exam scores for history are listed below. What is her average score for the tests?

Test 1	95
Test 2	105
Test 3	80

   A. 91   C. 100
   B. 92   D. 93

2. 11 quarts are equal to:

   A. 22 pints
   B. 2.5 gallons
   C. 10.1 gallons
   D. 44 pints

3. A box contains spools of thread: 3 spools of red, 4 spools of blue, 2 spools of green, and 3 spools of yellow. What is the probability of reaching in the box without looking and picking a red spool?

   A. $\frac{1}{4}$ C. $\frac{2}{7}$
   B. $\frac{1}{5}$ D. $\frac{3}{4}$

4. Scientists believe that the sun is about 4,600,000,000 miles away from earth. How do you write this distance in scientific notation?

   A. 4.6×10^8
   B. 46×10^7
   C. $4,600 \times 10^6$
   D. 4.6×10^9

5. Casey, our pet iguana, whacked his tail into a pile of marbles. One of the marbles went under the couch, never to be seen again. There were 6 red marbles, 11 orange marbles, 4 blue marbles, and 7 multicolored marbles. What is the probability the one missing is **not** orange?

   A. .39   C. .61
   B. .036   D. .96

6. John must run at least 35 miles every week. At this rate, write an inequality that best expresses the miles (*m*) John runs in 10 days.

   A. $m \geq 24\frac{1}{2}$ C. $m \geq 75$
   B. $m \geq 50$ D. $m \geq 10$

7. Brent sold a $6,000 used car. His commission was $480. What percent commission did he receive?

   A. 4.8%   C. 8.6%
   B. 8%   D. 12.5%

8. A rectangular box and a rectangular pyramid have the same dimensions for their bases and heights. How does the volume of the box compare to the volume of the pyramid?

   A. The volumes are the same.
   B. The volume of the box is twice the volume of the pyramid.
   C. The volume of the box is three times as large as the pyramid.
   D. The volume of the box is four times as large as the pyramid.

9. Examine the charts and answer the question below.

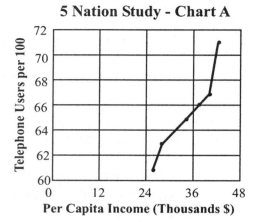

Which of the following is an accurate statement?

A. Chart B is misleading because it minimizes the differences between income and telephone ownership.
B. Chart A is misleading because it shows a dramatic link between income and telephone ownership.
C. Chart B is misleading because the data points indicate a strong link between income and telephone usage.
D. Chart A is misleading because the data describing telephone usage does not begin with zero.

10. Use the tree diagram below to predict the probability of flipping 3 coins and getting one head and two tails.

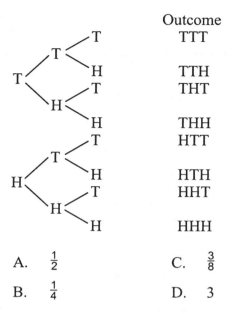

A. $\frac{1}{2}$ C. $\frac{3}{8}$
B. $\frac{1}{4}$ D. 3

11. In a basketball shooting contest, which of the following players had the lowest percentage of shots made?

A. Erica made 2 out of 7 shots.
B. Greg made 60% of his shots.
C. Bob made $\frac{3}{8}$ of his shots.
D. Kent made 5 out of 8 shots.

12. It takes $1\frac{2}{3}$ yards of fabric to cover one chair. If Shung Kim bought 12 yards of fabric, how many yards would she have left if she covered 6 chairs?

A. 2 yards
B. $4\frac{1}{3}$ yards
C. $6\frac{2}{3}$ yards
D. 10 yards

13. Ian worked 37 hours at Burger Inn and earned $5.50 per hour. What were Ian's total earnings?

A. $20.35   C. $55.00
B. $42.50   D. $203.50

14. $\left(\frac{1}{4}\right)^2 =$

    A. $\frac{1}{4}$ C. $\frac{1}{8}$
    B. $\frac{1}{16}$ D. $\frac{2}{8}$

15. Jeff was making $6.25 per hour. His boss gave him a $.75 per hour raise. What percent raise did Jeff get?

    A. 12% C. 40%
    B. 25% D. 70%

16. What is the volume of the following box?

    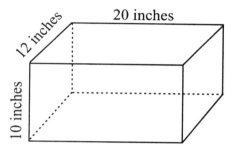

    A. 120 cubic inches
    B. 240 cubic inches
    C. 1,200 cubic inches
    D. 2,400 cubic inches

17. Joy can sell 4 magazine subscriptions in one hour. Tito can sell 3 magazine subscriptions in the same amount of time. How long would it take them to sell 15 magazine subscriptions if they work together?

    A. $1\frac{7}{10}$ hours C. $2\frac{1}{7}$ hours
    B. 2 hours D. $\frac{7}{15}$ hours

18. $(2x)^{-4} =$

    A. $\frac{1}{2x^4}$ C. $\frac{2}{x^4}$
    B. $\frac{1}{16x^4}$ D. $2x^{\frac{1}{4}}$

19. A school store buys packages of graph paper to sell to students for a small profit. The school buys 50 packages of graph paper for $8.00. They sell the packs of graph paper for 25¢ each. How many packages of graph paper do they have to sell to generate $18.00 worth of profit?

    A. 200 C. 176
    B. 225 D. 113

20. Cecil noticed that his pet mice became very upset and ate very little when the music in his room was too loud. Based on this observation, we can assume that

    A. plants grow faster when someone talks to them.
    B. a chipmunk living near an airport runway will be skinny.
    C. elephants eat more when they are played with.
    D. better tasting food will entice the mice to eat more.

21. Which of the following can be used to compute $\frac{1}{8} + \frac{3}{4}$?

    A. $\frac{1+3}{8+4}$
    B. $\frac{1 \cdot 1}{8 \cdot 1} + \frac{3 \cdot 2}{4 \cdot 2}$
    C. $\frac{1}{8 \cdot 1} + \frac{3}{4 \cdot 2}$
    D. $\frac{1}{8 \cdot 4} + \frac{3}{4 \cdot 4}$

22. Simplify: $\frac{(3a^2)^3}{a^3}$

    A. $27a^3$ C. $9a^3$
    B. $\frac{9a^6}{a^3}$ D. $\frac{3a^6}{a^3}$

23. $\sqrt{6}$ is between

    A. 5 and 6 C. 4 and 5
    B. 2 and 3 D. 3 and 4

24. $-|6| =$

    A. -6 C. $\frac{1}{6}$
    B. 6 D. 6^2

25. Jack is four years older than half his brother's age, **b**. Which algebraic expression below represents Jack's age?

    A. $b + 4$ C. $4 - 2b$
    B. $\frac{1}{2}b + 4$ D. $b - 4$

26. What is the solution to

    $2(5-2)^2 - 15 \div 5$?

    A. $-\frac{3}{5}$

    B. $\frac{3}{5}$

    C. 15

    D. $4\frac{1}{5}$

27. According to the bar graph below, what year had the greatest increase in the kangaroo population at **Zoo Down Under**?

    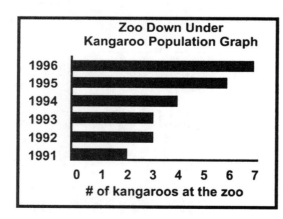

    A. 1996   C. 1995
    B. 1993   D. 1994

28.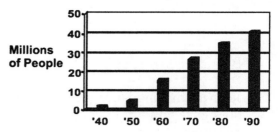

    How many people received Social Security checks in 1980?

    A.   35 people
    B.   35,000 people
    C.   350,000 people
    D.   35,000,000 people

29.

    How many more billion dollars did the USA spend on the military than Russia?

    A. $191 billion
    B. $200 billion
    C. $254 billion
    D. $317 billion

30. What is the slope of the equation graphed below?

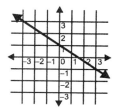

   A. $\frac{2}{3}$ C. $-\frac{2}{3}$
   B. $\frac{3}{2}$ D. $-\frac{3}{2}$

31. Which of the following is a graph of the equation with a slope of $\frac{1}{2}$ and a y-intercept of 1?

   A.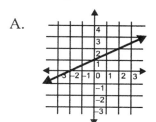

   B.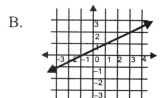

   C.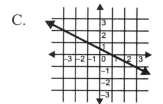

   D.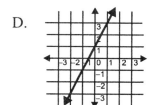

32. Which of the following is an accurate graph of the point values below?

Equilateral Triangle	
Length	Perimeter
1	3
2	6

   A.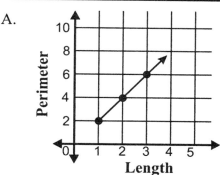

   B.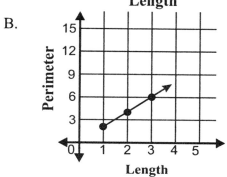

   C.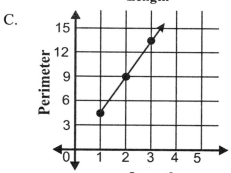

   D.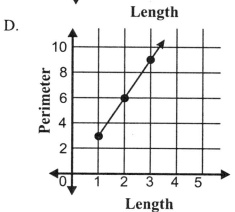

33.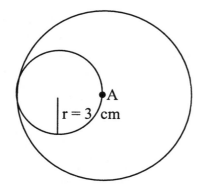

Point A is in the center of the larger circle. The radius of the smaller circle is 3 cm. What is the area of the larger circle? Use 3.14 for π.

A. 9.42 cm²
B. 18.84 cm²
C. 28.26 cm²
D. 113.04 cm²

34. $x^6 \cdot \dfrac{1}{x^3}$

A. x^2
B. x^3
C. x^{-2}
D. x^{-3}

35. Solve: $-6 - x \geq 7$

A. $x \geq -13$
B. $x \leq 13$
C. $x \leq -13$
D. $x \geq 13$

36. Portia drove 50 miles per hour for 5 hours. How fast must she average during the next two hours to average 52 miles per hour for the 7 hours she will have driven?

A. 54 miles per hour
B. 55 miles per hour
C. 56 miles per hour
D. 57 miles per hour

37. Find the <u>circumference</u> of the circle below. Use 3.14 for π.

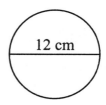

A. 18.84 cm
B. 37.68 cm
C. 75.36 cm
D. 113.04 cm

38. Find the area of the triangle below.

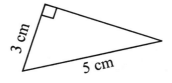

A. 6 cm²
B. 7.5 cm²
C. 12 cm²
D. 15 cm²

39. Miyoki is an executive in an architectural firm. She estimates that drawing the plans for a skyscraper will require approximately 1000 worker·hours. She believes that her architects can finish the drawings in 125 hours. How many architects are working on this project?

A. 7 architects
B. 8 architects
C. 9 architects
D. 10 architects

40. Find the area of the rectangle below.

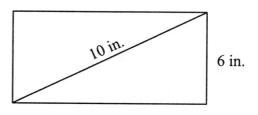

A. 42 sq. in.
B. 48 sq. in.
C. 54 sq. in.
D. 60 sq. in.

41. Which of the following is an accurate graph of the equation $y = 4 - 5x^3$?

A.
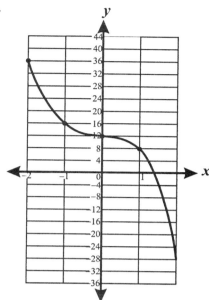

C.
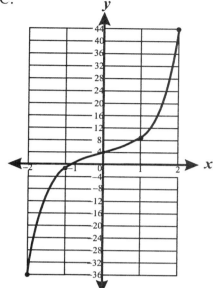

B.
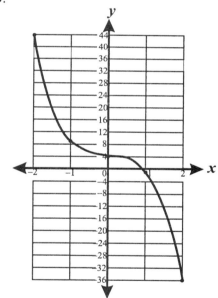

D.
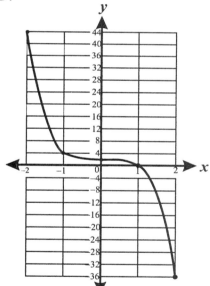

42.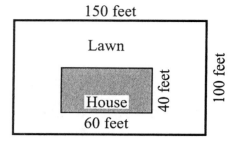

William drew a picture of his property to help figure the area of the lawn. He needed to know how much grass seed and fertilizer to buy. The shaded area represents his house. How many square feet of lawn does he have?

A. 10,000 sq. ft.
B. 12,600 sq. ft.
C. 14,760 sq. ft.
D. 15,000 sq. ft.

43. Consider the following set of data:
{1, 2, 3, 3, 4, 5, 5, 5, 6, 7, 8}
If you constructed a box-and-whisker plot of the data, the box would extend from _____ to _____.

A. 1 to 8   C. 3 to 6
B. 2 to 7   D. 3 to 5

44. There were three brothers. Fernando was two years older than Pedro. Pedro was two years older than Samuel. Together their ages add up to 63 years. How old is Samuel?

A. 17   C. 21
B. 19   D. 23

45. $\dfrac{b + 27}{7} = 11$. Find b.

A. 18   C. 50
B. 31   D. 77

46.

	Fresno	Los Angeles	Sequoia N. P.	San Bernardino	Bakersfield	Santa Barbara
Fresno	0	213	95	263	93	250
Los Angeles	213	0	233	61	105	92
Sequoia N. P.	95	233	0	305	128	285
San Bernardino	263	61	305	0	170	160
Bakersfield	103	105	128	170	0	149
Santa Barbara	250	92	285	160	149	0

Ben left Santa Barbara and arrived in San Bernardino 3 hours later. He later left San Bernardino and arrived back in Santa Barbara 2 hours later. What was his average driving speed for the total round trip?

A. 46 miles per hour
B. 64 miles per hour
C. 53 miles per hour
D. 50 miles per hour

47.

According to the sale ad above, about how much could you save on a board game regularly priced at $12.47?

A. $ 3.00
B. $ 6.00
C. $ 9.00
D. $15.00

48. Which of the following describes two independent events?

    A. Omar reaches into a bag containing 3 red, 3 blue, and 2 green marbles. What is the probability he draws out a red marble, and then a blue marble?
    B. Omar reaches into a bag containing 3 red, and 3 blue marbles and draws out a red marble. Then he reaches into another bag containing 2 green marbles and 3 white marbles and draws out a white marble. What is the probability he draws out a red marble, and then a white marble?
    C. What is the probability Dan will throw two dice, and they will add up to 12?
    D. What is the probability Dan will flip a coin twice and get heads both times?

49. How many centimeters would equal 1 yard?   1 inch = 2.54 cm

    A. 91.44 cm
    B. 23.16 cm
    C. 69.49 cm
    D. 36 cm

50. 1 cubic inch is approximately 16.38 cm^3. About how many cubic centimeters are in a cubic foot?

    A. 28,304.64 cm^3
    B. 244.72 cm^3
    C. 393.12 cm^3
    D. 2,358.72 cm^3

51.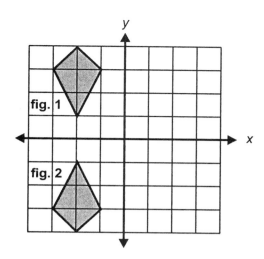

Figure 1 goes through a transformation to form figure 2. Which of the following descriptions correctly describes the transformation shown?

A. reflection across the x-axis
B. reflection across the y-axis
C. $\frac{3}{4}$ clockwise rotation around the origin
D. translation down 2 units

52.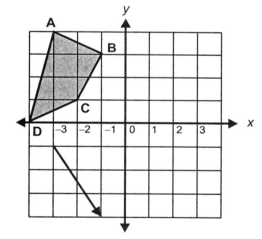

If the figure above is translated in the direction described by the arrow, what will be the new coordinates of point B after the transformation?

A. (2, 1)   C. (1, 0)
B. (0, 1)   D. (1, 1)

53. What is the length of the master bedroom?

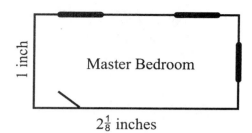

Scale: $\frac{1}{8}$ inch = 1 foot

A. 14 feet C. 16 feet
B. 15 feet D. 17 feet

54. The diagram below shows a box without a top. Each of the sides is a rectangle. What is the total surface area of the five sides?

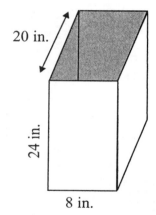

A. 832 in^2 C. 1504 in^2
B. 992 in^2 D. 1664 in^2

55.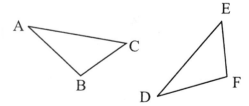

$\triangle ABC$ is congruent to $\triangle DEF$. $\angle A$ is congruent to

A. $\angle D$ C. $\angle F$
B. $\angle E$ D. $\angle C$

56. What is the mode of the following data set?

42, 44, 38, 37, 44, 39, 38, 38

A. 37
B. 38
C. 39
D. 44

57.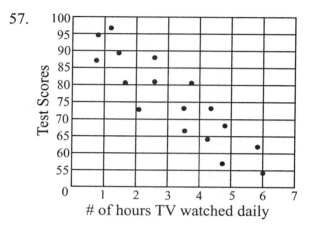

Which of the following best describes the relationship in the data points above?

A. Positive relationship
B. Negative relationship
C. No relationship
D. Cannot be determined

58. Mary made handmade Christmas tree ornaments to sell at a craft fair. She sold 92 ornaments for $3.00 each in 8 hours. She spent $18.00 for materials. For Mary to figure out how much she made per hour, what other information does she need to know?

A. how many days a week she worked at home making ornaments
B. how many ornaments she sold to each person
C. how much she spent buying other people's crafts at the fair
D. the number of hours she worked making the ornaments

59. Which argument is valid?

    If I oversleep, I miss breakfast. If I miss breakfast, I cannot focus well in class. If I do not concentrate in class, I make bad grades.

    A. I made bad grades today, so I missed breakfast.
    B. I got up on time, so I made good grades today.
    C. I could not focus in class today, so I overslept.
    D. I had no breakfast today, so I overslept.

60. What is the mean of the following set of data?

    {82, 84, 84, 81, 82, 79}

    A. 83
    B. 81
    C. 84
    D. 82

61. Solve: $\dfrac{3x + 6}{-2} > -12$

    A. $x < 24$
    B. $x > 0$
    C. $x > 6$
    D. $x < 6$

62. Solve: $2(x + 5) + 4(2x - 1) = -14$

    A. $x = -2$
    B. $x = -1$
    C. $x = -1\frac{4}{5}$
    D. $x = -1\frac{2}{10}$

63. The sum of two numbers is fourteen. The sum of six times the smaller number and two equals four less than the product of three and the larger number. Find the two numbers.

    A. 6 and 8
    B. 5 and 9
    C. 3 and 11
    D. 4 and 10

64. Greg wonders how many fish are in his pond, so he sets a trap and catches 8 fish. He puts a tag on the tail fin of each fish and releases them back into the pond. The next time he sets the trap, he catches 8 fish, but only 2 of them have a tag on their fin. What is a good estimate of the number of fish in the pond?

    A. 16   C. 64
    B. 32   D. 805

65. Consider the following set of data:

    | 28 | 32 | 42 | 37 |
    | 30 | 25 | 57 | 39 |
    | 24 | 32 | 33 | 44 |
    | 38 | 34 | 30 | 44 |
    | 31 | 28 | 31 | 29 |

    If you made a stem-and-leaf plot for the data, what numbers would you use for the stems?

    A. 2, 3, 4
    B. 1, 2, 3, 4
    C. 20, 30, 40
    D. 2, 3, 4, 5

66. Simplify:

    $$\dfrac{\sqrt{20}}{\sqrt{35}}$$

    A. $\dfrac{2\sqrt{7}}{7}$ C. $\dfrac{2\sqrt{5}}{\sqrt{7}}$
    B. $\dfrac{2}{\sqrt{7}}$ D. $\dfrac{4}{7}$

67. **15 Height Comparisons**

    Based on the scatter plot above, what is the relationship between the heights of mothers and their daughters?

    A. No Relationship
    B. Negative
    C. Cannot be Determined
    D. Positive

68. Claire wanted to divide 58.59 by 6.4, but she forgot to enter the decimal points when she put the numbers into the calculator. Using estimation, where should Claire put the decimal point?

    A. 9154.6875
    B. 915.46875
    C. 91.546875
    D. 9.1546875

69. Service Plus charges a fixed amount for a service call plus a charge for each hour worked. The charge for each hour is the same. Use the chart below to determine the cost of a service call plus 8 hours of work.

Number of Hours	Cost
1	$63
2	$81
3	$99
4	$117

    A. $26
    B. $98
    C. $144
    D. $189

70. The graph of which pair of equations below will be parallel?

    A. $x + 3y = 3$
       $3x + y = 3$

    B. $x + 3 = y$
       $x - 3 = y$

    C. $x - 3y = 3$
       $3y - x = -3$

    D. $3x + 3y = 6$
       $9x - 3y = 6$

71. A study has shown that ice-cream sales increase as the temperature increases above 60°. Assuming this generalization to be true, which of the following statements would most likely be true?

    A. As the temperature decreases, more orange juice is sold.
    B. If the home team is winning the football game, more colas will be sold.
    C. The warmer the temperature outside, the more people listen to the radio.
    D. As the temperature increases, the more suntan lotion is sold.

72. Simplify and solve the following inequality:

    $3x - 4 + 7(2 - x) > -6$

    A. $x < 4$
    B. $x > 4$
    C. $x > -4$
    D. $x < -4$

73. Solve the following inequality:
    $40 < -15 + |46x - 83|$

    A. $\frac{10}{13} > x$

    B. $-3 < x < 3$

    C. $3 < x, \ x < \frac{14}{23}$

    D. $-3 > x, \ x > \frac{14}{23}$

74. Which of the following is the graph of $2x - 3y = -3$?

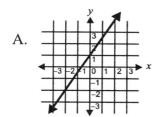

75. What are the x- and y-intercepts for the equation: $5x - y = 35$?

    A. The x-intercept is $(5,0)$ and the y-intercept is $(0,-1)$.
    B. The x-intercept is $(0,-35)$ and the y-intercept is $(7,0)$.
    C. The x-intercept is $(7, 0)$ and the y-intercept is $(0,-35)$.
    D. The x-intercept is $(-35,0)$ and the y-intercept is $(0,7)$.

76. Which ordered pair is a solution for the following system of equations?

    $2x - 8y = 6$
    $-x + 4y = -3$

    A. $(7, 1)$ C. $(15, 3)$
    B. $(-5, -2)$ D. $(11, 2)$

77.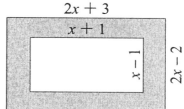

    Which of the following expresses the area of the shaded region?

    A. $5x^2 + 6x - 7$
    B. $3x^2 + 2x - 5$
    C. $5x^2 + 6x - 5$
    D. $5x^2 + 2x - 5$

78. Solve.

    $7 - \left(\frac{3}{4}\right)^2 =$

    A. $6\frac{1}{4}$ C. $4\frac{3}{4}$
    B. $6\frac{7}{16}$ D. $5\frac{1}{2}$

79. Which of the following is the graph of the equation $x + 3 = y$?

A.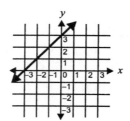

B.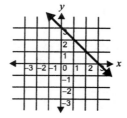

C.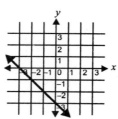

D.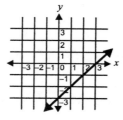

80.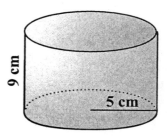

Sandra calculated the volume of the cylinder to be 225cm^3. She knew her answer was wrong because the correct answer should be about

A. $9 \times 10 = 90$
B. $3 \times 10 \times 9 = 270$
C. $3 \times 5 \times 5 \times 9 = 675$
D. $10 \times 10 \times 9 = 900$

California Math Review Progress Test 2

1. **PAXTON FAMILY**

Month	Electric Bill
January	$89.15
February	$99.59
March	$78.99
April	$72.47
May	$99.23
June	$124.69

    Look at the chart above. What is the median electric bill for the Paxton family from January through June?

    A. $ 55.22 C. $ 94.19
    B. $ 94.02 D. $ 99.23

2. Darin is playing a dart game at the county fair. At the booth, there is a spinning board completely filled with different colors of balloons. There are 6 green, 4 burgundy, 5 pink, 3 silver, and 8 white balloons. Darin aims at the board with his dart and pops one balloon. What is the probability that the balloon popped is **not** green?

    A. $\frac{1}{13}$ C. $\frac{3}{13}$
    B. $\frac{10}{13}$ D. $\frac{2}{13}$

3. This year, $\frac{7}{8}$ of all the graduating seniors have signed up to go to the graduation dance. What percent of the seniors will be going to the dance?

    A. 0.78%   C. 8.75%
    B. 0.875% D. 87.5%

4. Which is the equivalent multiplication problem for $\frac{3}{4} \div \frac{2}{3}$?

    A. $\frac{4}{3} \times \frac{3}{2}$ C. $\frac{4}{3} \times \frac{2}{3}$
    B. $\frac{3}{4} \times \frac{2}{3}$ D. $\frac{3}{4} \times \frac{3}{2}$

5. In a family of 4 children, what is the probability that all four will be girls? (Making a tree diagram on your scrap paper can help you determine this.)

    A. $\frac{1}{4}$ C. $\frac{1}{16}$
    B. $\frac{1}{8}$ D. $\frac{1}{24}$

6. Simplify:
    $$\frac{(2^3)^2}{(3)^{-1}}$$

    A. $21\frac{1}{3}$ C. 48
    B. 96   D. 192

7. How would you write 12,000,000,000 in scientific notation?

    A. $12 \times 1,000,000,000$
    B. 12×1^9
    C. 12×10^9
    D. 1.2×10^{10}

8. Choose the statement below that best describes two dependent events.

   A. Mahala brought home two bags of seashells. One bag contains 3 white, two brown, and four blue seashells. The second bag contains four brown and two blue seashells. All these seashells are about the same size. What is the probability that without looking she will draw a brown seashell from each bag?
   B. Melanie's little brother pushed her coin collection out of the second-floor window of their home. She had 3 coins from Mexico, 2 coins from Canada, and 4 coins from Russia. What is the probability she will find the two coins from Canada first?
   C. Dan needs to pay for a lunch in the drive-through at Burger Heaven. Stuffed in his pocket he has four $1 bills, two $5 bills, one $10 bill, and two $20 bills. What is the probability he will pull out a $5 bill and then a $1 bill?
   D. Yar's freezer contains one box of popsicles with 3 different flavors. Of the 3 flavors, 7 are grape, 5 are cherry, and 4 are orange. Yar reaches his hand into the box looking for an orange popsicle and pulls out a grape popsicle instead. Disgusted, Yar places the grape popsicle back into the box and tries again. What is the probability that Yar will pull out an orange popsicle on his second try?

9. Cynda wants to buy a daycare center with the measurements below. How many square feet are in the building?

   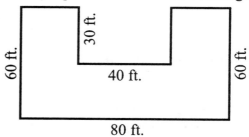

   A. 340 square feet
   B. 480 square feet
   C. 3,600 square feet
   D. 4,800 square feet

10. Diego bought a box of nails of various sizes. In the box were 30 three inch nails, 40 two inch nails, 60 one inch nails, and 20 half inch nails. If he dumps out the nails on the table and one falls off, what is the probability that a one inch nail fell off the table?

    A. $\frac{1}{2}$ C. $\frac{2}{5}$
    B. $\frac{1}{60}$ D. $\frac{3}{5}$

11. Alphonso saw a stereo on sale for $\frac{1}{3}$ off the regular price of $630.00. How much money could he save if he bought the stereo on sale?

    A. $210.00   C. $600.00
    B. $410.00   D. $630.00

12. It is −3° outside right now, and tonight the temperature is expected to drop another 21°. How cold is it expected to get?

    A. −18°   C. −24°
    B. 18°   D. −21°

13. Use correct order of operations to evaluate the following expression.

    $$4(4x-3)^2$$

    A. $64x^2 - 96x + 36$
    B. $400x^2 - 225$
    C. $80x - 45$
    D. $16x^2 - 24x + 9$

14. Marla found a dress she liked for $80. The next week it was on the sale rack for $44. What percent discount is that?

    A. 44%   C. 55%
    B. 36%   D. 45%

15. Michael borrowed $4,500.00 from his Dad to buy a used car. He agreed to pay his Dad back in one year with 6% simple interest. How much interest will Michael pay?

    A. $ 27.00   C. $270.00
    B. $ 24.00   D. $240.00

16. Hanna earns 12% commission on any jewelry sales she makes. About how much is her commission on a $45 sale?

    A. $1.00   C. $5.40
    B. $4.00   D. $12.00

17. $3^{-3} \cdot 3^5 =$

    A. $\frac{1}{9^{15}}$ C. 9
    B. 81   D. $\frac{1}{3^{15}}$

18. Which of the following can be used to compute $\frac{4}{5} + \frac{5}{6}$?

    A. $\frac{4+5}{5+6}$
    B. $\frac{4 \cdot 6}{5 \cdot 6} + \frac{5 \cdot 5}{6 \cdot 5}$
    C. $\frac{4}{5 \cdot 1} + \frac{5}{6 \cdot 2}$
    D. $\frac{4 \cdot 5}{5 \cdot 5} + \frac{5 \cdot 6}{6 \cdot 6}$

19. $\left(\frac{3}{4}\right)^3 =$

    A. $\frac{27}{64}$ C. $\frac{6}{7}$
    B. $\frac{3}{4}$ D. $\frac{27}{4}$

20. $\sqrt{77}$ lies between

    A. 7 and 8
    B. 8 and 9
    C. 76 and 78
    D. 5 and 6

21. $-|-6| =$

    A. 6   C. 36
    B. -6 D. 16

22. Jim had a pack of paper (P) that had 250 sheets. Which expression shows the number of sheets remaining after he used S sheets of paper?

    A. $P + S$ sheets
    B. $P - S$ sheets
    C. $P \div S$ sheets
    D. PS sheets

23. Dane is three years older than twice his brother's age, **b**. Which algebraic expression below represents Dane's age?

    A. $b + 3$
    B. $2b + 3$
    C. $3 - 2b$
    D. $2b - 3$

24. Which expression represents the total area of the 3 rooms shown below?

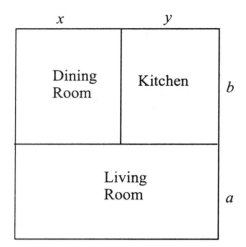

- A. $xy(a + b)$
- B. $ab(x + y)$
- C. $(a + b)(x + y)$
- D. $b(x + y) + ay$

25. According to the chart, how many more feet does it take to stop a car traveling at 70 miles per hour than at 55 miles per hour?

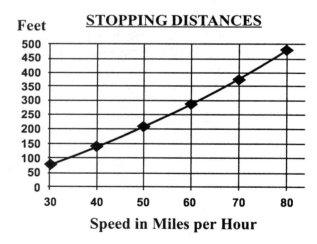

- A. 100 feet
- B. 125 feet
- C. 150 feet
- D. 175 feet

26. According to the graph below, approximately what percent of people in the USA had only a high school diploma in 1994?

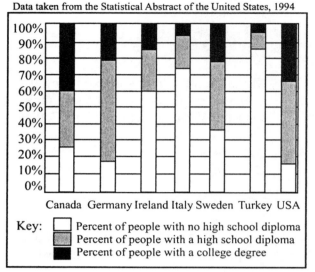

- A. 15%
- B. 50%
- C. 65%
- D. 75%

27. **DUNLAP'S CAR SALES**
(First 8 Months)

For how many months were the actual sales at Dunlap's above the predicted sales?

- A. 2
- B. 3
- C. 4
- D. 5

28. Simplify the expression shown below.

$$3x^{-4}$$

A. $\dfrac{81}{x^4}$

B. $(3x)^{-1}(3x)^{-1}(3x)^{-1}(3x)^{-1}$

C. $\dfrac{3}{x^4}$

D. $\dfrac{1}{3x^4}$

29. Simplify the following monomial.

$$5 \cdot x^4 \cdot y^5 \cdot z^{-3}$$

A. $\dfrac{625x^4y^5}{z^3}$

B. $(5xyz)^6$

C. $\dfrac{5x^4y^5}{z^3}$

D. $x^{20}y^{25}z^{-15}$

30.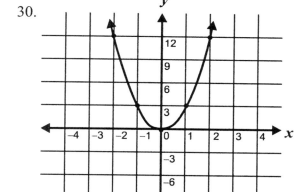

What is the equation graphed above?

A. $y = 3x^2$
B. $y = |2x|$
C. $y = 3x$
D. $y = 6x$

31. What is the slope of the line in the graph below?

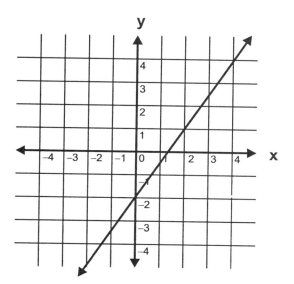

A. 3
B. -3
C. $\dfrac{2}{3}$
D. $\dfrac{3}{2}$

32. What is the slope of the line in the graph below?

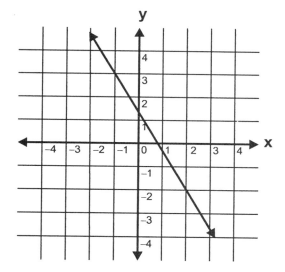

A. 3
B. $-\dfrac{1}{2}$
C. $-\dfrac{5}{3}$
D. $\dfrac{3}{4}$

33. Beth did a stem and leaf plot of her daily math grades for first semester.

Stem	Leaves
3	1
4	5
5	2
6	0,2,3,5,5,5,8,9,9
7	1,3,3,3,4,4,5,5,5,5,5,5,6,6,7,7,7,9,9,9
8	0,1,2,2,4,4,4,4,5,5,6,6,7,7,7,8,9
9	0,2,2,3,6,8,9

What grade was the mode of her data?

A. 31   C. 75
B. 65   D. 77

34. 5 miles per hour is the same as how many feet per second?

A. $\dfrac{5(5280)}{1}$ C. $\dfrac{5(5280)}{(60)}$

B. $\dfrac{5(5280)}{2(60)}$ D. $\dfrac{5(5280)}{(60)^2}$

35. Solve for x in the following equation.

$$\dfrac{6x - 40}{2} = 4$$

A. 6   C. 8
B. $\dfrac{32}{6}$ D. $7\dfrac{1}{3}$

36. $7(2x + 6) - 4(9x + 6) < -26$

A. $x > -2$ C. $x < -2$
B. $x > 2$ D. $x < -1$

37. Which of the following is equal to 300 kilograms?

A. 0.3 grams
B. 0.0003 milligrams
C. 300,000 grams
D. 30,000,000 milligrams

38. A coin bank contains dimes and quarters. The number of dimes is three less than four times the number of quarters. The total amount in the bank is $8.15. How many dimes are in the bank?

A. 13   C. 49
B. 41   D. 65

39.

	Duluth	Minneapolis	Moorhead	Rochester	St. Cloud	St. Paul
Duluth	0	156	253	232	142	150
Minneapolis	156	0	237	90	72	10
Moorhead	253	237	0	328	173	246
Rochester	232	90	238	0	163	82
St. Cloud	142	72	173	163	0	81
St. Paul	150	10	246	82	81	0

Steve drove from Rochester to Duluth in 4 hours and 15 minutes. What was his average speed?

A. 50 miles per hour
B. 54.59 miles per hour
C. 55.9 miles per hour
D. 58 miles per hour

40. The speed of sound in dry air at 32°F is 331.6 m/sec. How far would sound travel in 2 minutes?

A. 663.2 meters
B. 10,611.2 meters
C. 19,896 meters
D. 39,792 meters

41. Madison is reading the floor plans of her new house. What is the perimeter of the room shown below?

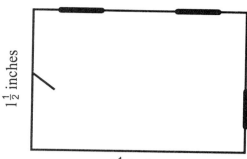

Scale: $\frac{1}{8}$ inch = 1 foot

A.   20 feet
B.   60 feet
C.   40 feet
D.   30 feet

42. If you have a 6 inch cube and decide when you make the next cube, you are going to double the length of each side, how will the volume be affected?

A.   The volume will be 3 times larger.
B.   The volume will be twice as large.
C.   The volume will be 8 times larger.
D.   The volume will be 9 times larger.

43. Jack is going to paint the ceiling and four walls of a room that is 10 feet wide, 12 feet long, and 10 feet from floor to ceiling. How many square feet will he paint?

A.   120 square feet
B.   560 square feet
C.   680 square feet
D.   1,200 square feet

44. What is the volume of the following oil tank? Round your answer to the nearest hundredth.

Use the formula $V = \pi r^2 h$ $\pi = 3.14$

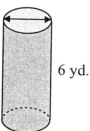

A.   18.84 yd³
B.   37.68 yd³
C.   44.48 yd³
D.   75.36 yd³

45.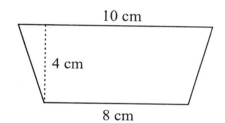

Find the area of the trapezoid above.

A.   22 square centimeters
B.   36 square centimeters
C.   72 square centimeters
D.   320 square centimeters

46. A back yard is 160 feet wide. What is the width of the field in yards?

A.   50 yards
B.   53 yards
C. $53\frac{1}{3}$ yards
D. $53\frac{2}{3}$ yards

47. Which of the following Cartesian planes is an accurate graph of the point values below?

Cups	Ounces
1	8
2	16
3	24

A.
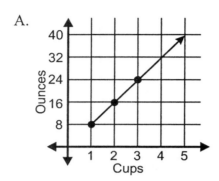

B.
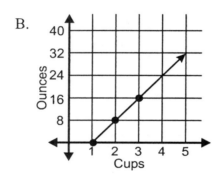

C.
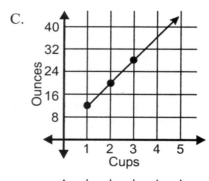

D.
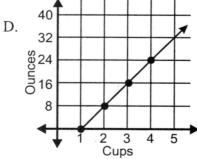

48.
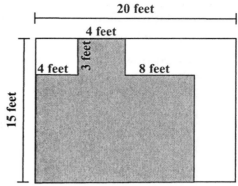

Find the area of the shaded area above. (Diagram not drawn to scale.)

A. 204 ft.^{2} C. 276 ft.^{2}
B. 264 ft.^{2} D. 300 ft.^{2}

49.
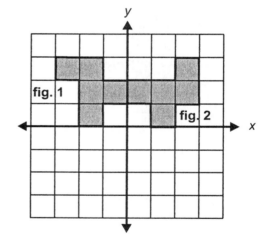

Figure 1 goes through a transformation to form figure 2. Which of the following descriptions fits the transformation shown?

A. reflection across the x-axis
B. reflection across the y-axis
C. $\frac{1}{4}$ clockwise rotation around the origin
D. translation right 3 units

50. Which of the following is the correct area of the shape drawn from the points (−1, −2)(−1, 1)(4,−2)?

    A. 15
    B. 5
    C. 12.5
    D. 7.5

51. In the right triangle below, what is the value of *x*?

    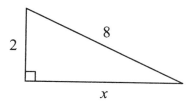

    A. $2\sqrt{15}$
    B. 68
    C. $2\sqrt{17}$
    D. $\sqrt{10}$

52. Find the length of the missing side of the triangle below.

    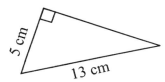

    A. 10 cm
    B. 11 cm
    C. 12 cm
    D. 15 cm

53. Which of the following best describes the relationship in the data points above?

    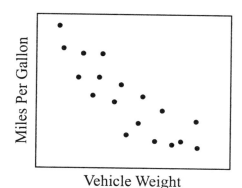

    A. Positive
    B. Negative
    C. No relationship
    D. Cannot be determined

54.

    Which of the following best describes the relationship in the data points above?

    A. Positive
    B. Negative
    C. No relationship
    D. Cannot be determined

55. Consider the set of data below. What is the median of the set of numbers?

    $\{1, 2, 4, 5, 6, 7, 7\}$

    A. 5
    B. 4.6
    C. 6
    D. 7

56. If you made a box-and-whisker plot for the following sets of data, which set would have only one whisker?

    A. $\{1, 2, 4, 5, 6, 7, 7\}$
    B. $\{2, 3, 5, 7, 10, 10, 12\}$
    C. $\{0, 1, 3, 4, 7, 8, 9\}$
    D. $\{1, 3, 3, 4, 5, 6, 9\}$

57. Consider the following set of data. $\{1, 2, 3, 4, 6, 7, 8, 9, 9, 11, 11, 11, 12\}$ What is the mode of this set of data?

    A. 7.3
    B. 10
    C. 11
    D. 12

58. Julie has a part-time business selling cosmetics out of her home. This week she spent 5 hours on her business, and cosmetics sales totaled $147.50. To find out how much Julie made per hour, you also need to know

   A. how many samples she gave away.
   B. how much she spent at the grocery store.
   C. how many customers she had.
   D. how much she paid for the cosmetics and supplies.

59. Water hyacinths were introduced into the swamps of Louisiana to put oxygen back in the water. The hyacinths reproduced rapidly and soon became a nuisance. Read the growth table below, and then answer the question that follows.

Number of Hyacinths	Number of Days
2	1
4	21
8	41
16	61

   Assuming the pattern continues, how many hyacinths will there be at 121 days?

   A. 32 C. 128
   B. 64 D. 256

60. Simplify: $\sqrt{45} \times \sqrt{27}$

   A. $3\sqrt{15}$
   B. $\sqrt{72}$
   C. $9\sqrt{15}$
   D. $\sqrt{9} \times \sqrt{15}$

61. Examine the chart below.

   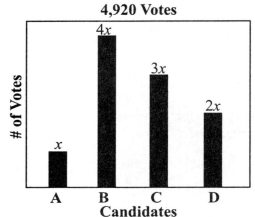

   Based on this graph, about how many votes did candidate C receive?

   A. 2,300 C. 800
   B. 1,500 D. 500

62. Jason wanted to know how many honey bees he had in his hive. He lured 40 bees into a trap and colored their wings with a harmless green dye and released them back into the hive. The next day, he lured 10 bees into the trap and found 2 bees had dyed wings. About how many bees must have been in the hive?

   A. 80   C. 200
   B. 100 D. 400

63. Pat wanted to divide 7.86 by 3.9, but he forgot to enter the decimal points when he put the numbers into the calculator. Using estimation, where should Pat put the decimal point?

   A. 0.2015386
   B. 2.015386
   C. 20.15386
   D. 201.5386

64. If two triangles have all corresponding sides and all corresponding angles congruent, then they are congruent triangles. If two triangles are congruent, then they are similar triangles. Given these facts, which of the following statements is valid?

    A. Similar triangles have all sides and all angles congruent.
    B. If two triangles are similar, then they are congruent.
    C. If two triangles are not congruent, then they are not similar.
    D. If two triangles have all corresponding sides and angles congruent, then they are similar triangles.

65. Which of the following is the correct solution set for the problem below?

    $21 + 7|a| \geq -14$

    A. $\{-5, -4, -3, -2, -1, 0, 1, 2, 3, 4, 5\}$
    B. $\{-4, -3, -2, -1, 1, 2, 3, 4\}$
    C. $\{1, 2, 3, 4\}$
    D. $\{-4, -3, -2, -1, 0, 1, 2, 3, 4\}$

66. Solve: $3(5x + 3) + 5(4x - 9) = 34$

    A. $x = 1$ C. $x = -1$
    B. $x = 2$ D. $x = -2$

67. Solve: $-4(2x + 7) > 3(4x + 5) + 27$

    A. $x > \frac{7}{2}$ C. $x < \frac{7}{2}$
    B. $x > -\frac{7}{2}$ D. $x > \frac{1}{4}$

68. If the equation below were graphed, which of the following points would lie on the line?

    $4x + 7y = 56$

    A. (8, 0)   C. (7, 4)
    B. (0, 14)   D. (4, 7)

69. Find the x- and y- intercept for the following equation: $2x + 5y = 30$

    A. x-intercept = 15
       y-intercept = 6
    B. x-intercept = 5
       y-intercept = 4
    C. x-intercept = 6
       y-intercept = 15
    D. x-intercept = 4
       y-intercept = 5

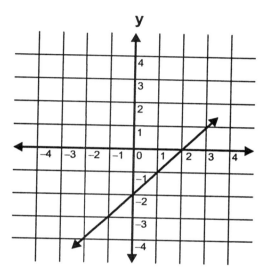

70. What are the x and y intercepts of the equation graphed above?

    A. x-intercept = 1
       y-intercept = -1
    B. x-intercept = -2
       y-intercept = 2
    C. x-intercept = -1
       y-intercept = 1
    D. x-intercept = 2
       y-intercept = -2

71.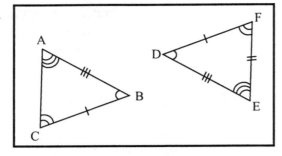

Are the two figures shown above congruent? Why or Why not?

A. The two figures are congruent because the corresponding sides are congruent.
B. The two figures are congruent because both the corresponding sides and the corresponding angles are congruent.
C. The two figures are incongruent because the corresponding angles are different measures.
D. The two figures are incongruent because the corresponding sides have different lengths.

72. The perimeter of a rectangle is 160 feet. The length of the rectangle is 20 feet less than three times the width. What is the width and length of the rectangle?

A. $w = 15, l = 25$
B. $w = 25, l = 15$
C. $w = 30, l = 15$
D. $w = 25, l = 55$

73. The graph of which pair of equations below will be parallel?

A. $x + 4y = 3$
$3x + 4y = 3$

B. $x - 4y = 3$
$4y - x = -3$

C. $2x - 8 = 2y$
$2x + 8 = 2y$

D. $6x + 6 = 6y$
$11x - 12 = 7y$

74. Which ordered pair is a solution for the following system of equations?

$$-3x + 7y = 25$$
$$3x + 3y = -15$$

A. $(-13, -2)$
B. $(-6, 1)$
C. $(-3, -2)$
D. $(-20, -5)$

75. To prepare a soup, one cook added 60 grams of salt to a three liter container of water. What is the density of salt in this solution?

A. $60 \frac{grams}{liter}$
B. $.05 \frac{grams}{liter}$
C. $20 \frac{grams}{liter}$
D. $50 \frac{grams}{liter}$

76.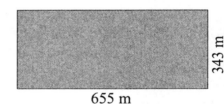

The coach calculated that the perimeter of the playing field was 224,665 meters. Lee knew this was wrong because

A. $340 + 660 = 1000$
B. $2 \times (340 + 660) = 2000$
C. $2 \times 340 + 660 = 1340$
D. $34 \times 66 = 2244$

77. A chef needs to prepare a meal for a large banquet. The preparation of this meal requires 12 worker·hours. How many additional workers does he need to prepare the meal in three hours? (The chef is also working on the meal.)

   A. 3 workers
   B. 4 workers
   C. 9 workers
   D. 36 workers

78. Jack and David work as window washers for high rise buildings. Jack can clean one side of a 9 floor square building in 5 hours. David can clean one side of the same building in 7 hours. How long would it take Jack, and David to clean the windows of this entire building working together? (Each side of the building has the same number of windows.)

   A. 11 hours
   B. 12 hours
   C. 24 hours
   D. $11\frac{2}{3}$ hours

79. Natural gas companies have announced that if there are more than 20 nights below 45 degrees this winter, the price of natural gas will increase by 50%. Based on this information, which of the following statements is true?

   A. The average monthly gas bill will rise from $270 to $440.
   B. This winter will be a colder than normal winter.
   C. A monthly gas bill of $184 will increase to $276 when there are 21 winter nights below 44 degrees.
   D. The price of natural gas will continue to go up until other sources of energy can be found.

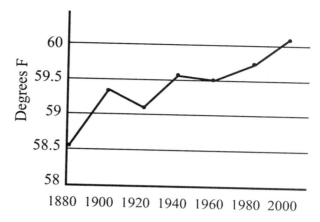

CURRENT WARMING TREND
Global average temperature in Fahrenheit degrees

80. Why is the graph shown above misleading?

   A. The bar graph does not take other factors such as industrialization into account.
   B. The graph begins at 58°, not 0°, on the y-axis.
   C. The lines on the graph do not connect the values.
   D. The graph values are measured in °F instead of the more scientific measurement °C.